SPRINGER HANDBOOK OF
AUDITORY RESEARCH

Series Editors: Richard R. Fay and Arthur N. Popper

Springer

New York
Berlin
Heidelberg
Hong Kong
London
Milan
Paris
Tokyo

Fan-Gang Zeng
Arthur N. Popper
Richard R. Fay
Editors

Cochlear Implants: Auditory Prostheses and Electric Hearing

With 100 Illustrations

 Springer

Fan-Gang Zeng
University of California-Irvine
Departments of Otolaryngology, Anatomy
  and Neurobiology, Biomedical Engineering
  and Cognitive Sciences
Irvine, CA 92697
USA
fzeng@uci.edu

Arthur N. Popper
Department of Biology and
  Neuroscience and
  Cogntive Science Program
University of Maryland
College Park, MD 20742-4415, USA
apopper@umd.edu

Richard R. Fay
Department of Psychology and
  Parmly Hearing Institute
Loyola University of Chicago
Chicago, IL 60626, USA
rfay@wpo.it.luc.edu

Series Editors: Richard R. Fay and Arthur N. Popper

Cover illustration: The inset figure has been taken from chapter 4 (in text, Figure 4.1); authors for this chapter are Patricia A. Leake and Stephen J. Rebscher.

Library of Congress Cataloging-in-Publication Data
Zeng, Fan-Gang.
  Cochlear implants : auditory prostheses and electric hearing / Fan-Gang Zeng,
  Arthur N. Popper, Richard Fay.
     p. cm.—(Springer handbook of auditory research; v. 20)
  Includes bibliographical references and index.
  ISBN 0-387-40646-8 (alk. paper)
    1. Cochlear implants. 2. Hearing impaired–Rehabilitation. 3. Electrocochleography.
  I. Popper, Arthur N. II. Fay, Richard R. III. Title. IV. Series.
RF305.Z46 2004
617.8'82—dc22   2003058950

ISBN 0-387-40646-8   Printed on acid-free paper.

Printed in the United States of America.   (MVY)

9 8 7 6 5 4 3 2 1   SPIN 10940358

www.springer-ny.com

Springer-Verlag is a part of *Springer Science+Business Media*

springeronline.com

Contents

Contributors

PAUL J. ABBAS
Audiology and Speech Pathology, University of Iowa, Iowa City IA 52242, USA

MIRANDA CLEARY
Speech Research Laboratory, Department of Psychology, Indiana University, Bloomington, IN 47405, USA

LENDRA FRIESEN
Auditory Implants and Perception, House Ear Institute, Los Angeles, CA 90057, USA

QIAN-JIE FU
Auditory Implants and Perception, House Ear Institute, Los Angeles, CA 90057, USA

JOHN GALVIN
Auditory Implants and Perception, House Ear Institute, Los Angeles, CA 90057, USA

RAINER HARTMANN
Institute of Sensory Physiology and Neurophysiology, J.W. Goethe University, D-60590 Frankfurt am Main, Germany

ANDREJ KRAL
Institute of Sensory Physiology and Neurophysiology, J.W. Goethe University, D-60590 Frankfurt am Main, Germany

PATRICIA A. LEAKE
Epstein Laboratory, Department of Otolaryngology, University of California–San Francisco, San Francisco, CA 94143-0526, USA

COLETTE M. McKAY
Department of Otolaryngology, University of Melbourne, East Melbourne 3002, Australia

CHARLES A. MILLER
Department of Otololaryngology, Head and Neck Surgery, University of Iowa, Iowa City, IA 52242, USA

JOHN K. NIPARKO
Division of Otology, Neurotology, and Skull Base Surgery, Johns Hopkins Hospital, Baltimore, MD 21287, USA

DAVID B. PISONI
Speech Research Laboratory, Department of Psychology, Indiana University, Bloomington, IN 47405, USA

STEPHEN J. REBSCHER
Epstein Laboratory, Department of Otolaryngology, University of California–San Francisco, San Francisco, CA 94143-0526, USA

ROBERT V. SHANNON
Auditory Implants and Perception, House Ear Institute, Los Angeles, CA 90057, USA

BLAKE S. WILSON
Center for Auditory Prosthesis Research, Research Triangle Institute/Duke University, North Corolina 27709-2194, USA

FAN-GANG ZENG
Departments of Otolaryngology, Anatomy and Neurobiology, Biomedical, Engineering, and Cognitive Sciences, University of California–Irvine, Irvine, CA 92697, USA

Series Preface

The *Springer Handbook of Auditory Research* presents a series of comprehensive and synthetic reviews of the fundamental topics in modern auditory research. The volumes are aimed at all individuals with interests in hearing research including advanced graduate students, post-doctoral researchers, and clinical investigators. The volumes are intended to introduce new investigators to important aspects of hearing science and to help established investigators to better understand the fundamental theories and data in fields of hearing that they may not normally follow closely.

Each volume is intended to present a particular topic comprehensively, and each chapter serves as a synthetic overview and guide to the literature. As such, the chapters present neither exhaustive data reviews nor original research that has not yet appeared in peer-reviewed journals. The volumes focus on topics that have developed a solid data and conceptual foundation rather than on those for which a literature is only beginning to develop. New research areas will be covered on a timely basis in the series as they begin to mature.

Each volume in the series consists of approximately five to nine substantial chapters on a particular topic. In some cases, the topics will be ones of traditional interest for which there is a substantial body of data and theory, such as auditory neuroanatomy (Vol. 1) and neurophysiology (Vol. 2). Other volumes in the series will deal with topics that have begun to mature more recently, such as development, plasticity, and computational models of neural processing. In many cases, the series editors will be joined by a co-editor having special expertise in the topic of the volume.

Arthur N. Popper, College Park, MD
Richard R. Fay, Chicago, IL

Volume Preface

The auditory prosthesis is perhaps the most successful neural prosthesis that uses electric stimulation to enhance or restore human neural, sensory, and motor function. Auditory prostheses have been used successfully to partially restore hearing in more than 60,000 hearing-impaired people worldwide. The auditory prostheses range from early single-electrode to modern multiple-electrode cochlear implants that bypass the damaged cochlea and stimulate the auditory nerve with electric currents. They also include specifically designed short electrodes and signal processing schemes that optimally combine the residual low-frequency acoustic hearing and the high-frequency electric hearing. Furthermore, the auditory prostheses may stimulate the auditory brainstem or cortex in patients whose auditory nerve is not accessible because of, for example, acoustic tumors. This volume focuses on cochlear implants, but also discusses the design principles and performance data for other types of auditory prostheses.

In Chapter 1, Zeng gives a fascinating and highly useful historical overview of the auditory prostheses and sets the content for the rest of this volume. Wilson (Chapter 2) then details the engineering design of the auditory prostheses from microphones to speech processors and from transmission links to electrodes. Niparko (Chapter 3) provides a detailed account of the evolving criteria in patient selection and additionally the surgical, cost-utility, educational, pre- and postoperative issues in cochlear implants. In Chapter 4, Leake and Rebscher address the anatomical and histological changes as a consequence of auditory deprivation as well as electric stimulation. Abbas and Miller (Chapter 5) then focus on the biophysical and physiological aspects of electrical stimulation of the auditory nerve, and Hartmann and Kral (Chapter 6) emphasize the central auditory system's responses to electric stimulation. McKay (see Chapter 7) compares the psychophysical measures between acoustic and electric hearing in humans and provides critical information regarding the stimulus coding in electric hearing. Shannon, Fu, Galvin, and Friesen (Chapter 8) discuss speech processing and performance with cochlear implants, auditory brainstem implants, bilateral cochlear implants, and combined acoustic and electric hearing. Finally, Pisoni and Cleary (Chapter 9) discuss the view that the individual differences observed in the present cochlear implant users may be to a large extent due to differences in learning, memory, and cognitive processes in these individuals.

Chapters in a number of previous volumes in the *Springer Handbook of Auditory Research* series have direct relevance to chapters in this volume. For example, the material in Chapter 6 of this volume can be expanded upon by reading chapters by Oertel and by Covey and Cassedy on the auditory CNS in volume 15 of the series (*Integrative Functions in the Mammalian Auditory Pathway*). Chapter 7 in this volume is complemented by chapters on psychophysics by Oxenham and Bacon in volume 17 (*Compression: From Cochlea to Cochlear Implants*) and in volume 3 (*Human Psychophysics*). Other perspectives on auditory prostheses can also be found in chapters by Levitt and Zeng in volume 17 and as related to speech processing in chapters by Edwards and Clarke in volume 18 (*Speech Processing in the Auditory System*).

Fan-Gang Zeng, Irvine, CA
Arthur N. Popper, College Park, MD
Richard R. Fay, Chicago, IL

1
Auditory Prostheses: Past, Present, and Future

1. Introduction

The auditory prosthesis is perhaps the most successful neural prosthesis that uses electric stimulation to enhance or restore human neural, sensory, and motor function. Although a visual prosthesis is still in an early experimental stage, auditory prostheses have been used successfully to partially restore hearing in more than 60,000 hearing-impaired people worldwide. The auditory prostheses range from early single-electrode to modern multiple-electrode cochlear implants that bypass the damaged cochlea and stimulate the auditory nerve with electric currents. They also include specifically designed short electrodes and signal processing schemes that can be used to stimulate the auditory nerve while preserving the residual low-frequency hearing for optimally combined acoustic and electric hearing. Furthermore, the auditory prostheses may stimulate the auditory brainstem or cortex in patients whose auditory nerve is not accessible because of, for example, acoustic tumors. This volume focuses on cochlear implants, but also discusses the design principles and performance data for other types of auditory prostheses.

A multidisciplinary approach is needed to ensure the successful application of an auditory prosthesis. Hearing and speech scientists, physiologists, and psychologists have to derive the working principles underlying signal conversion, processing, transmission, and delivery to the neural tissue. Biomedical engineers and material scientists have to design and fabricate the actual device so that it is safe, biocompatible, functional, power efficient, and cosmetically appealing. Clinicians including audiologists, speech pathologists, and physicians have to assess the patient's candidacy, perform the implantation, and conduct postsurgical rehabilitation. This volume emphasizes the basic principles underlying this multidisciplinary approach.

With the exponential growth in both patient population and scientific output, the auditory prosthesis has become an active and important area in auditory research. In addition to addressing questions specifically related to the safety and efficiency in electric stimulation of the auditory nerve, the

1

auditory prosthesis has been used extensively as a unique research tool to probe the basic mechanisms in auditory processing. For example, auditory prostheses have been used to infer coding mechanisms of loudness and pitch (Zeng and Shannon 1994; Carlyon et al. 2002; Zeng 2002), relative contribution of temporal and spectral cues to speech and auditory processing (Shannon et al. 1995; Smith et al. 2002), language development and critical periods in children (Svirsky et al. 2000; Sharma et al. 2002), as well as brain plasticity and across-modality processing (Ponton et al. 1996; Giraud et al. 2001; Lee et al. 2001). This volume reviews the latest development in these areas of exciting research and also forecasts future directions in an attempt to stimulate new research.

2. History and Present Status

The history of auditory prostheses is relatively short yet inspirational, and filled with pioneering spirits. The Italian scientist Alessandro Volta (1800) is credited with being the first to demonstrate that electric stimulation could directly evoke auditory, visual, olfactory, and touch sensations in humans. When he invented the battery about 200 years ago, one thing he did was to place one of the two ends of a 50-volt battery in each ear. He then observed, "At the moment when the circuit was completed, I received a shock in the head, and some moments after I began to hear a sound, or rather noise in the ears, which I cannot well define: it was a kind of crackling with shocks, as if some paste or tenacious matter had been boiling . . . The disagreeable sensation, which I believe might be dangerous because of the shock in the brain, prevented me from repeating this experiment."

Perhaps because of Volta's warning, electric stimulation of the auditory system was seldom reported for another 150 years until modern electronic technology emerged. Equipped with vacuum-tube–based oscillators and an amplifier, as well as a 10-cm copper wire with a roughened end serving as an electrode placed in the ear canal and absorbent cotton as insulation, S.S. Stevens and his colleagues conducted a series of studies to reexamine the electric stimulation of hearing (Stevens 1937; Stevens and Jones 1939; Jones et al. 1940). They identified at least three mechanisms that were responsible for the "electrophonic perception." The first mechanism is related to what Kiang and Moxon (1972) call an "electromechanical effect," in which electric stimulation causes the hair cells in the cochlea to vibrate, resulting in a perceived tonal pitch at the signal frequency as it was acoustically stimulated. The second mechanism is related to the tympanic membrane's conversion of the electric signal into an acoustic signal, resulting in a tonal pitch perception but at the doubled signal frequency. Stevens and his colleagues were able to isolate the second mechanism from the first because they found that only the original signal pitch was perceived with electric stimulation in patients lacking the tympanic membrane. The third mechanism is related to direct electric activation of the auditory nerve, because some patients reported a noise-like sensation in response to sinusoidal electrical stimulation, much steeper loudness growth with electric currents, and occasionally activation of facial nerves. However, the first direct evidence of electric stimulation of the auditory nerve was offered by a group of Russian scientists who reported hearing sensations with electric stimulation in a deaf patient whose middle and inner ears were damaged (Andreev et al. 1935).

The modern era of cochlear implants started with a French physician, Djourno, and his colleagues, who reported in 1957 successful electric stimulation of hearing in two totally deafened patients (Djourno and Eyries 1957). Their success spurred an intensive level of activity in attempts to restore hearing to deaf people on the U.S. west coast in the 1960s and 1970s (Doyle et al. 1964; Simmons 1966; Michelson 1971; House and Urban 1973). Animal models of electric hearing were also developed to compare the neural activation patterns between acoustic and electric stimulation of the inner ear (Kiang and Moxon 1972; Simmons and Glattke 1972; Merzenich et al. 1973). Although the methods were crude compared with today's technology, these early studies identified critical problems and limitations that needed to be considered and overcome for successful implementation of electric hearing. For example, they observed that compared with acoustic hearing, electric hearing of the auditory nerve produced a much narrower dynamic range, much steeper loudness growth, temporal pitch limited to several hundred hertz, and much broader or no tuning. Bilger et al. (1977) provided a detailed account of these earlier activities.

On the commercial side, the House-3M single-electrode implant became the first Food and Drug Administration (FDA)-approved device in 1984 and had several hundred users. The University of Utah developed a six-electrode implant with a percutaneous plug interface and also had several hundred users. The Utah device was called either the Ineraid or Symbion device in the literature and was well suited for research purposes (Eddington et al. 1978; Wilson et al. 1991; Zeng and Shannon 1994). The University of Antwerp in Belgium developed the Laura device that could deliver either 8-channel bipolar or 15-channel monopolar stimulation. The MXM laboratories in France also developed a 15-channel monopolar device, the Digisonic MX20. These devices were later phased out and are no longer available commercially. At present, there are three major cochlear implant manufacturers including Advanced Bionics Corporation, U.S. (Clarion); Med-El Corporation, Austria; and Cochlear Corporation, Australia (Nucleus).

The number of cochlear implant users has grown exponentially to a total of 60,000 worldwide, including 20,000 children. Functionally, the cochlear implant has evolved from the single-electrode device that was used mostly as an aid for lip-reading and sound awareness to modern multielectrode devices that can allow an average user to talk on the telephone. Figure 1.1

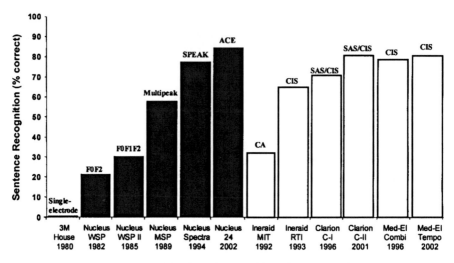

FIGURE 1.1. Sentence recognition in quiet conditions in cochlear implant users. The x-axis labels show the type of device (3 M House, Nucleus, Ineraid, Clarion, C-II, and Med-El), the processor model (WSP, WSP II, MSP, Spectra, Combi, and Tempo), the place where the studies were conducted [Massachusett Institute of Technology (MIT), and Research Triangle Institute (RTI)], and the year of the studies published. The y-axis is the percent correct scores for sentence recognition in quiet. The insert text on top of each bar indicates the type of electrode (House/3 M) or speech processing strategies used in these devices (ACE: advanced combination encoders; CA: compressed analog; CIS: continuous interleaved sampling; SAS: simultaneous analog stimulation, SPEAK, spectral peak). The scores in earlier cochlear implants (House/3 M, Nucleus WSP, WSP II, MSP, Ineraid MIT, and RTI) were averaged from investigative studies published in peer-reviewed journals. The scores in later devices were obtained from relatively large-scale company-sponsored clinical trials that had also been published in peer-reviewed journals.

documents the advances made in cochlear implants in terms of speech performance over the past 20 years. The early single-electrode device provided essentially no open-set speech recognition except in a few subjects. The steady improvement at a rate of about 20 percentage points in speech recognition was particularly apparent with the Nucleus device. Despite the differences in speech processing and electrode design, there appears to be no significant difference in performance among present cochlear implant users.

Cochlear implant research has also improved and matured as a scientific field. Figure 1.2 shows the annual number of publications on cochlear implants retrieved from the MEDLINE database. As of December 31, 2002, there were a total of 2489 publications in the database that were related to cochlear implants. For comparison, the entry "hearing AND aid" yielded a total of 2460 publications, while that for "auditory" yielded 54,194. The

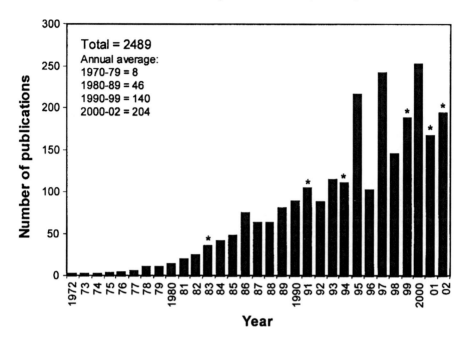

FIGURE 1.2. Annual number of publications containing "cochlear AND implant" in MEDLINE (http://www.pubmed.gov) search; the data were retrieved on March 18, 2003. The total number of publications from January 1, 1972 through December 31, 2002 was 2489. The accumulative number of publications in the 1970s, 1980s, 1990s, and 2000s was 32, 455, 1391, and 611, respectively. The resulting annual average number of publications is displayed in the inserted text. The asterisk on top of the bar indicates a year in which a cochlear implant–related article was published in either *Science* or *Nature*.

annual number of publications clearly shows an exponential growth pattern, mirroring the growth pattern in the number of cochlear implant users and also, most likely, in the amount of funding. The growth and maturity of the implant field are also reflected by the number and density of articles published in broad interest–based journals like *Science* and *Nature* (indicated by the asterisk on top of the bar in Fig. 1.2). The entire 1960s, 1970s, and 1980s produced two articles published in *Science* (Simmons et al. 1965; Tong et al. 1983). Both were case reports, with Simmons et al. characterizing basic perceptual consequences in electric hearing and Tong et al. showing fused auditory perception by two-electrode stimulation. The 1990s produced three *Science/Nature* articles including the Wilson et al. (1991) study on improved speech performance with the continuous interleaved sampling strategy, the Zeng and Shannon (1994) study on loudness-coding mechanisms inferred from electric stimulation of the auditory nerve and brainstem, and the Klinke et al. (1999) study on cortical plasticity in deaf cats induced by long-term electric stimulation. While it is still early in the new millennium, there is already one *Nature* article, by Lee et al. (2001), describing the cross-modality plasticity in cochlear implant users, and another *Science* article, by Rauschecker and Shannon (2002), reviewing the cochlear and brainstem implants.

3. Critical Issues in Auditory Prostheses

More than three decades ago, Simmons (1969) forecasted that we have to resolve the following six major problems "before artificial hearing can become a reality": electronics, patient selection, tissue tolerance, long-term effects of electric stimulation, selective stimulation of the discrete neurons, and stimulus coding and auditory learning. Today, most of these problems have been successfully addressed, but several additional issues are also identified. This volume deals with both old and new issues that are briefly discussed below, and then elaborated upon in the following chapters.

3.1 Engineering and Clinical Issues

Taking advantage of the rapid development and innovation in microelectronics in the last several decades, auditory prostheses have become a safe, efficient, and effective medical device that not only restores partial hearing to totally deafened persons but also enhances residual hearing in hearing-impaired individuals. Wilson (Chapter 2) details the engineering design of the auditory prostheses from microphones to speech processors and from transmission links to electrodes. Of particular interest to beginners is his summary of critical components in the design of cochlear implants (Table 2.1) and of processing strategies with implant devices in present use (Table 2.2).

Niparko (Chapter 3) provides a detailed account of the evolving criteria in patient selection as well as the surgical, cost-utility, educational, pre- and postoperative issues in cochlear implants. Most notably, the audiological criteria for cochlear implantation has relaxed from bilateral total deafness (>110 dB HL) in the early 1980s, to severe hearing loss (>70 dB HL) in the 1990s, and then to the current suprathreshold speech-based criteria (<50% open-set sentence recognition with properly fitted hearing aids). His treatment on the topics of presurgical evaluations, surgical approach and procedure, and electrode insertion are both valuable and informative (Figs. 3.3–3.8).

3.2 Anatomical and Physiological Issues

Leake and Rebscher (Chapter 4) address issues in tissue tolerance and long-term effects of electric stimulation. In addition, their chapter provides an excellent summary of the anatomical and histological changes as a consequence of auditory deprivation as well as electric stimulation. They convincingly demonstrate the protective effect of electric stimulation on the auditory nerve (Figs. 4.3–4.5), but more importantly they identify several critical problems that directly challenge the present design of auditory prostheses, including surgical trauma due to electrode insertion and the significant influence of electric stimulation on the central auditory system, particularly the developing brain.

Abbas and Miller (Chapter 5) focus on the biophysical and physiological aspects of electrical stimulation of the auditory nerve. Their treatment of the general approach in physiological studies of electric hearing is systematic and informative (sections 2 and 3). These sections are particularly useful for those who do not have an engineering and quantitative background. Their thorough description of the response properties of auditory-nerve stimulation (section 4) and spatial patterns of neural excitation (section 5) provides the current viewpoint on one of Simmons' problems in stimulus coding and electrode interaction (i.e., "selective stimulation of the discrete neurons"). Their chapter also provides a timely treatment on changes in response properties with neural degeneration (section 6) and with viable hair cells (section 7).

On the other hand, Hartmann and Kral (Chapter 6) emphasize the central auditory system's responses to electric stimulation. They have addressed and also extended Simmons' "selective stimulation of the discrete neurons" problem at the central level (Figs. 6.11 and 6.14). Their description of the central responses to electric stimulation in animal models of acquired and congenital hearing loss (section 4) is illuminating and highly relevant to human implantation in pre- and postlingually deafened adults as well as in congenitally deafened children. Their discussion on brainstem implants (section 3.6) should be of great value to those who are interested in directly stimulating the central auditory structure.

3.3 Perceptual and Cognitive Issues

McKay (Chapter 7) compares the psychophysical measures between acoustic and electric hearing in humans. She highlights the differences in intensive, spectral, and temporal processing between acoustic and electric stimulation, providing critical information regarding the stimulus coding in electric hearing. She also systematically documents the psychophysical responses to simple stimuli on a single electrode, more complex multielectrode stimulation, and finally bilateral electric stimulation. She has effectively addressed Simmons' "selective stimulation of the discrete neurons" problem at the psychophysical level by relating it to the spread of the excitation pattern, pitch perception, and stimulus space (Figs. 7.1–7.3).

Shannon et al. (Chapter 8) discuss speech processing and performance with cochlear implants, auditory brainstem implants, bilateral cochlear implants, and combined acoustic and electric hearing. It is unlikely that Simmons could have anticipated that 35 years later the most pressing issue facing auditory prostheses is not how to make artificial hearing a reality but rather how to perfect it. Shannon et al. assess the relative contribution of amplitude and spectral and temporal cues to speech perception, and conclude that the improvement of speech performance in the last 20 years is due to faster processors and that more electrodes may have reached the "point of diminishing returns." They believe that customized fitting procedures, particularly the frequency-to-electrode mapping, need to be developed to meet the different capacities of each individual user.

However, Pisoni and Cleary (Chapter 9) believe that the individual differences observed in the present cochlear implant users may be to a large extent due to differences in learning, memory, and cognitive processes in these individuals. Their perspective on the large individual variability issue is fresh and unique, providing a "top-down" approach complementing the traditional "bottom-up" approach. Moreover, their discussion on the basic principles underlying the information processing approach (section 3) and some measure of the information processing capacity (section 5) should stimulate those readers who come to auditory research from the traditional biology and engineering fields. Finally, we are beginning to address Simmons' last problem in auditory learning.

4. Summary and Future Directions

Thanks to the visionary and inspirational pioneers in electric stimulation of hearing, the cochlear implant has proven to be a safe and effective medical device. The implant research as a scientific field is also maturing, evidenced by the increasing number and quality of publications and scientists in the field. As a typical scientific endeavor, we have solved many of the initial problems to make artificial hearing into a reality, but more problems have emerged to challenge us. There are cultural, educational, ethical, social, and economical issues that restrict the widespread application of the cochlear implant and demand our attention not only as scientists but also as human beings (Niparko, Chapter 3), for example, whether the cochlear implant will wipe out the deaf culture and why it cannot be made affordable to the majority of deaf people in the world. In addition, there are tremendously challenging engineering problems that need to be resolved to make the auditory prostheses smaller, more power efficient, more biomimic, and even usable as a drug delivery device (Wilson, Chapter 2). However, due to the limited scope and nature of this volume, only scientific issues are considered in the future directions section.

Now that artificial hearing has become a reality, the new problems that need to be addressed arise from trying to perfect the cochlear implant.

Some are long-standing problems in the literature (1, 3, and 4 on the list below) while others are emerging (2 and 5):

1. Restore normal pitch sensation in cochlear implant users. We know from Stevens et al.'s pioneering experiments in the 1930s that direct stimulation of the auditory nerve does not give rise to tonal pitch even with sinusoidal stimuli. At present, cochlear implant users can perceive only very coarse pitch by either stimulating different electrodes or stimulating the same electrode at different rates. While the place pitch resolution is severely limited by the available number of electrodes and their ability to stimulate a discrete neural population, it is puzzling that the rate pitch resolution, even at low frequencies, is still an order of magnitude poorer in electric stimulation than acoustic stimulation. For example, a normal-hearing listener can discriminate reliably a pitch difference of 1 to 2 Hz at 100 Hz, but a cochlear-implant listener requires a 10- to 20-Hz difference (Zeng 2002). In other words, we have not been able to reproduce a simple pure tone sensation in the present cochlear implant users, let alone harmonics and music perception. The problem here first at all is to restore pure tone perception as measured by the objective frequency discrimination task in cochlear implant listeners, and then to restore harmonic pitch perception as measured by both the psychophysical measures, such as resolved versus unresolved harmonics, and the functional measures, such as melody recognition.

2. Introduce the fine structure to the cochlear implant. Acoustic signals can be divided into a slow-varying temporal envelope component and a fast-varying fine structure component (related to the phase of the stimulus). Recent perceptual experiments have highlighted the importance of the fine structure cues in realistic listening situations (e.g., Smith et al. 2002). However, the current cochlear implants either discard entirely the fine structure information [e.g., continuous interleaved sampling (CIS), SPEAK, and advanced combination encoder (ACE) strategies] or deliver the unprocessed, convolved temporal and fine-structure cues [e.g., compressed analog (CA) and simultaneous analog stimulation (SAS)]. As a result, the present cochlear implant users perform poorly in speech recognition in noise, cannot accurately recognize speakers, melodies, and tonal languages, and are severely limited in achieving their full potential with the bilateral implants. One way to alleviate the requirement for the encoding of the fine structure is to increase the number of functional electrodes (see problem 3, below), and the other way is to utilize innovative signal processing to slow down the fast fine structure to frequency modulations that may be perceived by the cochlear implant users via the existing hardware (Nie et al. 2003).

3. Increase the number of functional channels. It is clear that the multi-electrode cochlear implant provides better performance than the single-electrode implant. However, the numbers game is pretty tricky. Recent

studies have revealed that essentially all current implant users achieve an asymptotic level of performance functionally equivalent to four to eight channels despite the fact that up to 22 electrodes are being stimulated physically (Dorman and Loizou 1997; Fishman et al. 1997; Friesen et al. 2001; Garnham et al. 2002). Mysteriously, the cause of the four to eight functional channel equivalent in the present implant listeners is not their inability to discriminate more than that number of electrodes. Clearly the number of electrodes does not equal the number of channels. To address this problem, we need to understand the relationships among the number of physical electrodes, the number of psychophysically discriminable electrodes, and the number of functional channels. We also need to understand the electrode-neuron interface at the peripheral level (Leake and Rebscher, Chapter 4; Abbas and Miller, Chapter 5) and the central level (Hartmann and Kral, Chapter 6). To demonstrate how little we understand these relationships, we can go back to the Simmons' original problem on "selective stimulation of the discrete neurons." The traditional wisdom has been that more discrete neural stimulation will result in less electrode interaction and thus better performance. However, research results directly challenge this assumption, as they show that the broader electrode configurations actually produce performance equal to or better than the narrower electrode configurations (Zwolan et al. 1996; Pfingst et al. 1997).

4. Improve the implant fitting procedure. The current fitting system needs to be significantly modified to achieve improved efficiency and effectiveness. At present, audiologists spend most of their time adjusting the threshold and maximal comfortable loudness on a single electrode stimulation basis. This labor-intensive and time-consuming fitting procedure is not necessary because accurate estimation of the threshold and loudness on a single electrode is neither required for high levels of speech performance nor indicative of the realistic wide-band listening situations. The loudness measures can be achieved with a combined electrophysiological and modeling approach, in which the electrically evoked potentials can be used to estimate the dynamic range on individual electrodes (Seyle and Brown 2002), and a computational model can predict the overall loudness of multielectrode stimulation (McKay, Chapter 7). It is hoped that together they can achieve speech performance equivalent to the subjectively measured maps. This approach is particularly useful for the growing pediatric cochlear implant population. However, to achieve improved performance, particularly in speech recognition in noise, we need to solve the customized fitting problem that allows efficient and accurate frequency-to-electrode mapping in individual cochlear implant listeners (Shannon et al., Chapter 8).

5. Improve presurgical evaluation and postsurgical rehabilitation. Despite numerous attempts, we are still far from predicting presurgically with reasonable confidence and accuracy the level of postsurgical performance (Niparko, Chapter 3). This has become an increasingly important

problem as people with residual hearing and various etiology (e.g., auditory neuropathy; see Starr et al. 1996) become candidates for cochlear implants. If we cannot predict the outcome, how can we advise a hearing-impaired person with 50% or more intelligibility to either get or not get a cochlear implant. The solution to this problem may require a combination of the old-fashioned promontory stimulation (Abbas and Miller, Chapter 5) and the newly developed brain imaging (Hartmann and Kral, Chapter 6) and cognitive measures (Pisoni and Cleary, Chapter 9). Postsurgical rehabilitation presents an equally challenging problem in terms of explaining large individual variability and improving performance particularly in poor users. At present there are essentially no postsurgical rehabilitation protocols, but limited data have shown improvement with auditory training (Fu et al. 2001). Formal, structured, and systematic learning protocols ought to be developed to help cochlear implant users adapt to the new modality with electric hearing.

Acknowledgments. I thank Monita Chatterjee, Ackland Jones, Ying-Yee Kong, Art Popper and Dick Fay for helpful comments. Preparation of this manuscript was partially supported by a grant from the National Institutes of Health (RO1-DC-02267).

References

Andreev AM, Gersuni GV, Volokhov AA (1935) On the electrical excitability of the human ear: on the effect of alternating currents on the affected auditory apparatus. J Physiol USSR 18:250–265.

Bilger RC (1977) Psychoacoustic evaluation of present prostheses. Arch Otorhinolaryngol 86:92–104.

Carlyon RP, van Wieringen A, Long CJ, Deeks JM, Wouters J (2002) Temporal pitch mechanisms in acoustic and electric hearing. J Acoust Soc Am 112:621–633.

Djourno A, Eyries C (1957) Prosthese auditive par excitation electique a distance du nerf sensorial a l'aide d'un bobinage inclus a demeure. Presse Med 35:14–17.

Dorman M, Loizou P (1997) Speech intelligibility as a function of the number of channels of stimulation for signal processors using sine-wave and noise-band outputs. J Acoust Soc Am 102:2403–2411.

Doyle JH, Doyal JB, Turnbull FM (1964) Electrical stimulation of the eighth cranial nerve. Arch Otolaryngol 80:388–391.

Eddington DK, Dobelle WH, Brackmann DE, Mladejovsky MG, Parkin JL (1978) Auditory prostheses research with multiple channel intracochlear stimulation in man. Arch Otorhinolaryngol 87:1–39.

Fishman KE, Shannon RV, Slattery WH (1997) Speech recognition as a function of the number of electrodes used in the SPEAK cochlear implant speech processor. J Speech Lang Hear Res 40:1201–1215.

Friesen LM, Shannon RV, Baskent D, Wang X (2001) Speech recognition in noise as a function of the number of spectral channels: comparisons of acoustic hearing and cochlear implants. J Acoust Soc Am 110:1150–1163.

Fu QJ, Shannon RV, Galvin III JJ (2001) Perceptual learning following changes in the frequency-to-electrode assignment with the Nucleus-22 cochlear implant. J Acoust Soc Am 112:1664–1674.

Garnham C, O'Driscoll M, Ramsden R, Saeed S (2002) Speech understanding in noise with a Med-El COMBI 40+ cochlear implant using reduced channel sets. Ear Hear 23:540–552.

Giraud AL, Price CJ, Graham JM, Truy E, Frackowiak RS (2001) Cross-modal plasticity underpins language recovery after cochlear implantation. Neuron 30:657–663.

House WF, Urban J (1973) Long term results of electrode implantation and electronic stimulation of the cochlea in man. Ann Otol Rhinol Laryngol 82:504–517.

Jones RC, Stevens SS, Lurie MH (1940) Three mechanisms of hearing by electrical stimulation. J Acoust Soc Am 12:281–290.

Kiang NY, Moxon EC (1972) Physiological considerations in artificial stimulation of the inner ear. Ann Otol Rhinol Laryngol 81:714–730.

Klinke R, Kral A, Heid S, Tillein J, Hartmann R (1999) Recruitment of the auditory cortex in congenitally deaf cats by long-term cochlear electrostimulation. Science 285:1729–1733.

Lee DS, Lee JS, Oh SH, Kim SK, et al. (2001) Cross-modal plasticity and cochlear implants. Nature 409:149–150.

Merzenich MM, Michelson RP, Pettit CR, Schindler RA, Reid M (1973) Neural encoding of sound sensation evoked by electrical stimulation of the acoustic nerve. Ann Otol Rhinol Laryngol 82:486–503.

Michelson RP (1971) Electrical stimulation of the human cochlea. A preliminary report. Arch Otolaryngol 93:317–323.

Nie KB, Stickney GS, Zeng FG (2003) Independent Contributions of amplitude Modulation and frequency modulation to auditory perception: I. Consonant, vowel and sentence recognition. Abstract of the 26th annual midwinter research meeting 213.

Pfingst BE, Zwolan TA, Holloway LA (1997) Effects of stimulus configuration on psychophysical operating levels and on speech recognition with cochlear implants. Hear Res 112:247–260.

Ponton CW, Don M, Eggermont JJ, Waring MD, Kwong B, Masuda A (1996) Auditory system plasticity in children after long periods of complete deafness. Neuroreport 8:61–65.

Rauschecker JP, Shannon RV (2002) Sending sound to the brain. Science 295: 1025–1029.

Seyle K, Brown CJ (2002) Speech perception using maps based on neurla response telemetry measures. Ear Hear 23(1 suppl):72S–79S.

Shannon RV, Zeng FG, Kamath V, Wygonski J, Ekelid M (1995) Speech recognition with primarily temporal cues. Science 270:303–304.

Sharma A, Dorman MF, Spahr AJ (2002) A sensitive period for the development of the central auditory system in children with cochlear implants: implications for age of implantation. Ear Hear 23:532–539.

Simmons FB (1966) Electrical stimulation of the auditory nerve in man. Arch Otolaryngol 84:2–54.

Simmons FB (1969) Cochlear implants. Arch Otolaryngol 89:61–69.

Simmons FB, Glattke TJ (1972) Comparison of electrical and acoustical stimulation of the cat ear. Ann Otol Rhinol Laryngol 81:731–737.

Simmons FB, Epley JM, Lummis RC, Guttman N, et al. (1965) Auditory nerve: electrical stimulation in man. Science 148:104–106.

Smith ZM, Delgutte B, Oxenham AJ (2002) Chimaeric sounds reveal dichotomies in auditory perception. Nature 416:87–90.

Starr A, Picton TW, Sininger Y, Hood LJ, Berlin CI (1996) Auditory neuropathy. Brain 119:741–753.

Stevens SS (1937) On hearing by electrical stimulation. J Acoust Soc Am 8:191–195.

Stevens SS, Jones RC (1939) The mechanism of hearing by electrical stimulation. J Acoust Soc Am 10:261–269.

Svirsky MA, Robbins AM, Kirk KI, Pisoni DB, Miyamoto RT (2000) Language development in profoundly deaf children with cochlear implants. Psychol Sci 11:153–158.

Tong YC, Dowell RC, Blamey PJ, Clark GM (1983) Two-component hearing sensations produced by two-electrode stimulation in the cochlea of a deaf patient. Science 219:993–994.

Volta A (1800) On the electricity excited by mere contact of conducting substances of different kinds. R Soc Philos Trans 90:403–431.

Wilson BS, Finley CC, Lawson DT, Wolford RD, Eddington DK, Rabinowitz WM (1991) Better speech recognition with cochlear implants. Nature 352:236–238.

Zeng FG (2002) Temporal pitch in electric hearing. Hear Res 174:101–106.

Zeng FG, Shannon RV (1994) Loudness-coding mechanisms inferred from electric stimulation of the human auditory system. Science 264:564–566.

Zwolan TA, Kileny PR, Ashbaugh C, Telian SA (1996) Patient performance with the Cochlear Corporation "20 + 2" implant: bipolar versus monopolar activation. Am J Otol 17:717–723.

2
Engineering Design of Cochlear Implants

Blake S. Wilson

1. Introduction

The cochlear implant is the most successful neural prosthesis developed to date. Approximately 60,000 people have received cochlear implants as of this writing. This number exceeds by orders of magnitude the numbers for any other type of neural prosthesis. According to the recent National Institutes of Health (NIH) Consensus Statement on Cochlear Implants (1995), "A majority of those individuals with the latest speech processors for their implants will score above 80 percent correct on high-context sentences, even without visual cues." This level of success is truly remarkable, given the relatively crude representations of speech and other sounds provided by present implant systems.

Although much progress has been made in the engineering design of implant systems, much remains to be done. Patients with the best results still do not hear as well as listeners with normal hearing, especially in challenging situations such as speech presented in competition with noise or other talkers. In addition, some patients still do not enjoy much benefit from implants, even with the current speech processing strategies and electrode arrays.

This chapter provides an overview of prior and current designs for cochlear implants, and presents possibilities for improvements in design.

2. Components of Implant Systems

In the great majority of cases, deafness is caused by the absence or degeneration of sensory hair cells in the inner ear. Such loss may be produced by gene defects, viral or bacterial labyrinthitis, various autoimmune diseases, Meniere's disease, ototoxic drugs, overexposure to loud sounds, trauma, and other causes (see Niparko, Chapter 3). The function of a cochlear implant is to bypass the hair cells via direct electrical stimulation of surviving neurons in the auditory nerve. In general, at least some neurons survive

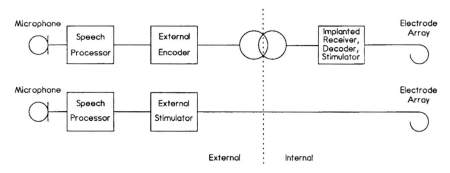

FIGURE 2.1. Components of cochlear implant systems: a system with a transcutaneous transmission link (top), and a system with a percutaneous connector (bottom). (From Wilson 1993, with permission from the Singular Publishing Group.)

even in cases of prolonged deafness and even for virulent etiologies such as meningitis (see Leake and Rebscher, Chapter 4; Hinojosa and Marion 1983).

The essential components of implant systems are shown in Figure 2.1. A microphone converts sound into an electrical signal for input to the speech processor. The processor transforms that input into a set of stimuli for an implanted electrode or array of electrodes. The stimuli are sent to the electrodes through a transcutaneous link (top) or through a percutaneous connector (bottom). A typical transcutaneous link includes encoding of the stimulus information for efficient radiofrequency transmission from an external transmitting coil to an internal (implanted) receiving coil. The signal received by the internal coil is decoded to specify stimuli for the electrodes. A cable connects the internal receiver/stimulator package to the implanted electrodes. In the case of a percutaneous connector, a cable connects pins in the connector to the electrodes.

These components are shown in a different way in Figure 2.2, which is a diagram of an implant system that uses a transcutaneous link. In this particular system a speech processor is worn on the belt or in a pocket. The processor is relatively light and small. A cable connects the output of the speech processor (a radiofrequency signal with encoded stimulus information) to a head-level unit that is worn behind the ear. A standard behind-the-ear (BTE) housing is used. A microphone is included within the BTE housing, and its output signal is amplified and then sent to the speech processor through one of the wires in the cable connecting the processor and the head-level unit. The external transmitting coil is connected to the base of the BTE housing with a separate cable. The external coil is held in place over the internal receiver/stimulator package (which includes the internal coil) with a pair of external and internal magnets. The receiver/stimulator package is implanted in a flattened or recessed portion of the skull, posterior to and slightly above the pinna (see Niparko, Chapter

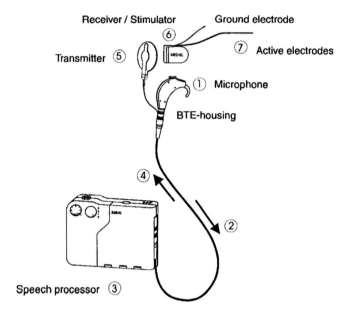

FIGURE 2.2. Schematic drawing of the Med-El Combi 40 implant system, with the major components indicated in Figure 2.1. The microphone is mounted in a behind-the-ear (BTE) headset. (Courtesy of Med-El GmbH, Innsbruck, Austria.)

3). The ground electrode is implanted at a location remote from the cochlea, usually in the temporalis muscle. For some implant systems, a metallic band around the outside of the receiver/stimulator package serves as the ground (or reference) electrode. An array of active electrodes is inserted into the scala tympani (ST) through the round window membrane or through a larger drilled opening at or near the round window.

An expanded view of the implanted cochlea is presented in Figure 4.1 in Chapter 4. This figure shows a cutaway drawing of an electrode array inserted into the first turn and part of the second turn of the ST. Different electrodes, or closely spaced pairs of "bipolar" electrodes (illustrated), ideally stimulate different subpopulations of cochlear neurons. Neurons near the base of the cochlea (first turn and lower part of the drawing) respond to high-frequency sounds in normal hearing, and neurons near the apex of the cochlea respond to low-frequency sounds. Most implant systems attempt to mimic this tonotopic encoding by stimulating basal electrodes to indicate the presence of high-frequency sounds and by stimulating apical electrodes to indicate the presence of low-frequency sounds.

Figure 4.1 in Chapter 4 indicates a partial insertion of the electrode array. This is a characteristic of all available ST implants—no array has been inserted farther than about 30 mm from the round window membrane, and typical insertion depths are much less than that, e.g., 20 to 26 mm. The figure also shows a complete presence of hair cells (in the organ of Corti) and a pristine survival of cochlear neurons. However, the number of hair cells is either zero or close to it in the deafened cochlea. In addition, survival of neural processes peripheral to the ganglion cells (the dendrites, projecting from the ganglion cells to the organ of Corti) is rare in the deafened cochlea. Survival of the ganglion cells and central processes (the axons) ranges from sparse to substantial. The pattern of survival generally is not uniform, with reduced or sharply reduced counts of spiral ganglion cells in certain regions of the cochlea. In all, the neural "target" for a cochlear implant can be quite different from one patient to the next (see Leake and Rebscher, Chapter 4).

Figure 2.2 in the present chapter shows components of the Med-El Combi 40 implant system. Other systems share the same basic components but are different in detail. For example, systems recently introduced by Advanced Bionics Corp., Cochlear Ltd., and Med-El GmbH incorporate the speech processor within a BTE housing, eliminating the separate and much larger speech processor of prior systems and the cable connecting the processor to the BTE housing. The details of processing and of techniques for the transmission stimuli or stimulus information across the skin differ widely among implant systems. The details of the electrode design also vary widely across systems.

Some of the choices and unknowns faced by designers of implant systems are summarized in Table 2.1. Each choice may affect performance, and each

TABLE 2.1. Principal options and considerations in the design of cochlear implant systems.

Processing strategy	Transmission link
Number of channels	Percutaneous
Number of electrodes and channel-to-electrode assignments	Transcutaneous
Stimulus waveform	Maximum stimulus update rate, within and across channels
Pulsatile	Back telemetry of implant status, electrode impedances, and/or intracochlear evoked potentials
Analog	
Approach to speech analysis	
Filter-bank representation	
Feature extraction	

Electrodes	Patient
Placement	Survival of neurons in the cochlea and auditory nerve
Extracochlear	Proximity of electrodes to target neurons
Intracochlear	Function of central auditory pathways
Within the modiolus	Cognitive and language skills
Cochlear nucleus	
Bilateral	
Number and spacing of contacts	
Orientation with respect to excitable tissue	

Adapted from Wilson et al. 1995, with permission from Mosby–Year Book, Inc.

choice may interact with other choices. The variability imposed by the patient may be even more important than the particulars of implant design (Wilson et al. 1993). Even though any of several implant devices can support high levels of speech reception performance for some patients, other patients have poor outcomes with each of those same devices. Factors contributing to this variability may include differences among patients in the survival of neural elements in the implanted cochlea, proximity of the electrodes to the target neurons, integrity of the central auditory pathways, and cognitive skills.

The hardware of implant systems is described in greater detail in the remainder of this section. Further discussion of strategies for representing speech information with implants is presented in section 3, and further discussion about the patient variable is presented in section 4.

2.1 Microphone

The microphone typically is placed in the BTE housing or the speech processor enclosure. A separate "tie-tack" or "clip-on" microphone also can be placed remotely and connected to the speech processor with a thin cable. (Connection of a remote microphone typically disables the microphone in the BTE housing or speech processor enclosure.) A good microphone for an implant system has a broad frequency response but not extending to very low frequencies, so as to minimize responses to low-frequency vibrations that can be produced by head movements and walking.

A directional microphone can help in listening to speech under adverse conditions, such as attending to one speaker in competition with other speakers or in competition with background noise (e.g., as in a cafeteria). The directional pattern of sensitivity for a single microphone is determined largely by its housing and placement on the body. For example, the head can act as a baffle for high-frequency sounds from the contralateral side when the microphone is mounted at the side of the head (in a BTE unit). The length and orientation of the tube in front of the microphone can affect its frequency response and directional pattern.

The selectivity of the directional pattern can be increased substantially with the use of multiple microphones. With two microphones, for example, sounds originating between and in front of the microphones produce microphone outputs that are in phase with each other, producing a large summed output. In contrast, sounds originating from other locations produce microphone outputs that are not in phase with each other and lower summed outputs. The summed signals emphasize sounds in front of the microphones and suppress sounds from other locations.

Use of dual microphones in conjunction with adaptive filtering techniques (to enhance the above summation and cancellation effects) has been evaluated by Hamacher et al. (1997), and by Wouters and Vanden Berghe (2001). Although the processing algorithms and test procedures differed between these two studies, results from each demonstrated highly significant gains in recognition of speech presented in competition with various types of noise and at various speech-to-noise ratios (S/Ns). (Another dual-microphone system, the Audallion Beamformer, has been made available for use with devices manufactured by Cochlear Ltd. Results from evaluation of this system for speech reception by implant patients have not been published; however, a description of the system is presented in Figueiredo et al. 2001.)

One of the future directions for cochlear implants described later in this chapter is the use of bilateral implants, for which placement of the microphones in each of the ear canals may be helpful. Sensing of sound pressure within the canal would include the spatially dependent frequency filtering provided by the pinna on both sides. (The principal component of the filtering is a deep spectral notch at high frequencies, produced by summation of sound directly entering the canal and sound reflected off the pinna/concha surfaces. The reflection path depends on the location of the sound source, and different locations produce different path lengths and different frequency positions of the notch.) Such cues might augment other cues to sound source location, that would be represented in any case with microphones at the standard location or within the canal.*

2.2 Speech Processor

The function of the speech processor is to convert a microphone or other input (e.g., direct inputs from a telephone, TV, CD player, or FM system) into patterns of electrical stimulation. Ideally, the outputs of the speech processor represent the information-bearing elements of speech in a way that they can be perceived by implant patients. Strategies for achieving this objective are described in section 3, below.

The processor is powered with batteries. Hearing aid batteries are used for the head-level processors (processors incorporated into the BTE housing), and larger batteries (e.g., two AA batteries) are used for the body-worn processors. Battery life typically exceeds 12 to 16 hours, allowing patients to use their devices during the waking hours without the need for recharging or replacing the batteries.

Adequate battery life for the head-level processors is made possible through the use of low-power integrated circuit technology, particularly low-power digital signal processing (DSP) chips that have become available in the past 10 years or so. Present head-level processors have all or most of

---

*The other principal cues are the interaural differences in amplitude and timing produced with sounds at various locations in the lateral plane. Pinna cues vary with position in the three-dimensional space, including the vertical plane and front–back positions.

the capabilities of the body-worn processors for each of the systems. Use of head-level processors is rapidly replacing use of body-worn processors, as the former are more cosmetic and more convenient than the latter.

Advances in battery, integrated circuit, and DSP chip technologies have been driven by huge commercial markets for mobile phones, portable computers, and other hand-held or portable instruments. The economic incentives to develop better batteries and power-efficient chips are enormous.

Recipients of cochlear implants have benefited from such developments, in that the developments have made possible progressively smaller and more capable speech processors and implanted receiver/stimulators. Even greater reductions in size and increases in capabilities may be available in the near future. Fully implantable systems, with the speech processor placed in the middle ear cavity, may well be available within the next several years.[†]

2.3. Transmission Link

A percutaneous connector or transcutaneous link is used to convey stimuli or stimulus information from the external speech processor to the implanted electrodes. A principal advantage of a percutaneous connector is signal transparency, i.e., the specification of stimuli is in no way constrained by the limitations imposed with any practical design of a transcutaneous transmission link. Also, the percutaneous connector allows high-fidelity recordings of intracochlear evoked potentials, which may prove to be useful in assessing the physiological condition of the auditory nerve on a sector-by-sector basis (Brown et al. 1990, 1998; Wilson et al. 1997a; Abbas et al. 1999) and for programming the speech processor (Brown et al. 1998, 2000; Shallop et al. 1999; Abbas et al. 2000; Mason et al. 2001; Franck 2002; Gordon et al. 2002; Seyle and Brown 2002).[‡]

An important advantage of transcutaneous links is that the skin is closed over the implanted components, which may reduce the risk of infection compared with systems using a percutaneous connector. A disadvantage is that only a limited amount of information can be transmitted across the

---

[†] Development of fully implantable systems is well under way or nearing completion at two of the major implant companies, at a new company in Canada, and at the University of Michigan. Two of the principal problems facing designers of fully implantable systems are (1) specification or design of batteries that do not need to be replaced any sooner than every 5 to 10 years, and (2) specification or design of microphones that can be implanted under the skin or elsewhere and yet still have an adequate signal-to-noise ratio.

[‡] Recordings of intracochlear evoked potentials may be helpful in setting currents for threshold and comfortably loud percepts for processors using relatively low rates of stimulation. However, the recordings are not good predictors of those currents for processors using high rates of stimulation (Zimmerling and Hochmair 2002). Although high-fidelity recordings of intracochlear evoked potentials are desirable, reduced-fidelity recordings may suffice for certain applications.

skin with a transcutaneous link. This usually means that the rates at which stimuli can be updated are limited and that the repertoire of stimulus waveforms is limited (e.g., restricted to biphasic pulses only for some systems).

All commercially available implant systems use a transcutaneous link. In some cases, the link is bidirectional, allowing transmission of data from the implanted components out to the external coil and speech processor or speech processor interface, as well as transmission of data from the speech processor to the implanted receiver/stimulator and electrode array. The data sent from the implanted components to the external components can include:

- information about the status of the receiver/stimulator, such as measures of critical voltages
- impedances of the implanted electrodes
- voltages at unstimulated electrodes
- neural evoked potentials, as recorded using unstimulated electrodes

Rates of transmission required for the first three measures are relatively low. However, high-fidelity recordings of intracochlear evoked potentials require high sampling rates (e.g., 50,000 samples/s), high resolution (e.g., a resolution of 12 bits or higher for the analog-to-digital converter), and rapid recovery of the recording amplifiers from the saturation produced by the presentation of stimulus pulses (van den Honert et al. 1997; Wilson 1997). The CI24M implant system, manufactured by Cochlear Ltd., has a capability to record intracochlear evoked potentials and to send the results from the internal receiver/stimulator to the external coil and speech processor interface (Brown et al. 1998). The arrangement for recording intracochlear evoked potentials is shown in Figure 2.3. A separate computer is used in conjunction with the speech processor interface and transcutaneous link to specify and transmit the stimuli to the selected stimulus electrode (or electrode pair) via the forward path of the link. Following delivery of the stimulus pulses, voltages recorded at the selected (unstimulated) electrode are encoded for transmission from the implanted receiver/stimulator back out to the external coil and speech processor interface. The computer then is used to reconstruct and plot the data received from the internal components.

The "neural response telemetry" feature of the CI24M implant does not fulfill the requirements for high-fidelity recordings noted above, and the types of stimuli that can be specified are limited. (For example, the feature has a maximum sampling rate of 20 kHz in an "interlaced" mode and a sampling resolution of approximately 9 bits.) However, the data obtained with its use may well be helpful in assessing the status of the nerve and for the fitting of speech processors (Abbas et al. 1999; Shallop et al. 1999; Brown et al. 2000; Mason et al. 2001; Franck 2002; Gordon et al. 2002; Seyle and Brown 2002).

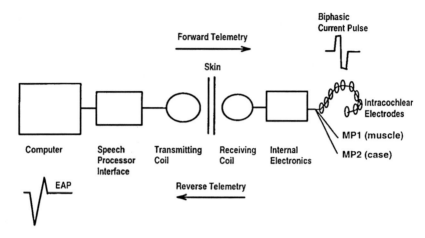

FIGURE 2.3. Schematic diagram of the system used for recording intracochlear evoked potentials in the CI24M device. EAP: Electrically evoked whole nerve Action Potential. (Courtesy of Cochlear Corporation, Englewood, CO, the U.S. subsidiary of Cochlear Ltd., Sydney, Australia.)

A capability for recording intracochlear evoked potentials also has been included in the new Clarion implant system (Frijns et al. 2002), and is called "neural response imaging" for that system. The recordings are made with a 9-bit analog-to-digital converter, and with sampling rates as high as 60 kHz. The recording amplifier also has an exceptionally fast recovery from prior saturation (less than 20 microseconds). These specifications allow better recordings than those obtained with the back-telemetry feature of the CI24M implant system. Such recordings may prove to be especially useful for assessing the physiological status of the nerve on a sector-by-sector basis.

2.4 Electrodes

The electrodes for the great majority for current implant systems are placed in the ST. The ST offers an accessible site that is relatively close to the spiral ganglion, which is not readily accessible with present surgical techniques (see Niparko, Chapter 3). The electrodes and electrode carrier (together called the electrode array) must be biocompatible and remain so over the life span of the patient. The array must also be mechanically stable and facilitate atraumatic insertion. Surgical handling of the array is determined by its stiffness and cross-sectional area. In general, flexible arrays and narrow cross-sectional areas facilitate insertion. Also, use of biocompatible lubricants, such as hyaluronic acid, can facilitate insertion (Laszig et al. 2002).

Intracochlear electrodes can be stimulated in either a monopolar or bipolar configuration. In the monopolar configuration, each intracochlear electrode is stimulated with reference to a remote electrode, usually in the temporalis muscle or outside of the case of the implanted receiver/stimulator. In the bipolar configuration, one intracochlear electrode is stimulated with reference to another (nearby) intracochlear electrode. Different pairs of electrodes are used to stimulate different sites along the electrode array.

The spatial specificity of stimulation, for selective activation of different populations of cochlear neurons, depends on many factors, including:

- whether neural processes peripheral to the ganglion cells are present or absent
- the number and distribution of surviving ganglion cells
- the proximity of the electrodes to the target neurons
- the electrode coupling configuration

These factors can interact in ways that produce selective excitation fields for monopolar or bipolar stimulation and in ways that produce broad excitation fields for either type of stimulation. For example, highly selective fields can be produced with bipolar electrodes oriented along the length of surviving neural processes peripheral to the ganglion cells (e.g., van den Honert and Stypulkowski 1987). Highly selective fields also can be produced with close apposition of monopolar electrodes to the target neurons (e.g., Ranck 1975) or through use of "field steering" arrangements in which the field produced by a central electrode is sharpened with the simultaneous application of opposite-phase fields at neighboring electrodes (e.g., Jolly et al. 1996).

In general, broad fields become more likely with increasing distance between electrodes and target neurons for either coupling configuration. Broad fields (produced by high stimulus levels) may be required for adequate stimulation of cochleas with sparse nerve survival.

An important goal of implant design is to maximize the number of largely non-overlapping populations of neurons that can be addressed with the electrode array. This might be accomplished through the use of a bipolar coupling configuration for some situations or through positioning of electrode contacts immediately adjacent to the inner wall of the ST. Such positioning would minimize the distance between the contacts and the ganglion cells.

Several new designs of electrode arrays produce close positioning of the array next to the inner wall of the ST (e.g., Gstoettner et al. 2001; Balkany et al. 2002). Such placements can not only increase the spatial specificity of stimulation, as noted above, but also produce reductions in threshold and increases in the dynamic range of stimulation (Ranck 1975; Shepherd et al. 1993; Cohen et al. 2001).

Although placements next to the inner wall appear to produce beneficial effects in the basal turn of the cochlea, the relative efficacy of the placements may be limited in higher turns. In particular, perimodiolar placements may not be any more effective in the higher turns than standard

FIGURE 2.4. Depiction of the anatomic courses of the basilar membrane (outer spiral) and the spiral ganglion (inner spiral). (Diagram courtesy of Darlene Ketten and adapted from Ketten et al. 1997.)

placements (electrode array out against the lateral wall of the ST, see Shepherd et al. 1985, 1993; Gstoettner et al. 2001), due to (1) different anatomic courses of the ST and spiral ganglion, and (2) a relatively non-differentiated "clustering" of spiral ganglion cells at about the level of the second turn of the ST. Figure 2.4 illustrates these different anatomic courses (Ketten et al. 1997; also see Ariyasu et al. 1989). The course of the basilar membrane is depicted by the outer spiral and the course of Rosenthal's canal (and, within it, the spiral ganglion) is depicted by the inner spiral. Rosenthal's canal has $1^{3}/_{4}$ turns, whereas the ST has $2^{3}/_{4}$ turns. Closer apposition of electrodes next to the medial wall of the ST may well reduce the distance between the electrodes and target ganglion cells in the basal turn, but the distance may not be substantially reduced for higher turns. In addition, stimulation by electrodes at the second turn and higher is likely to excite the cluster of cells at the apex of the spiral ganglion (not illustrated in the figure or in Figure 4.1 in Chapter 4). Thus, different stimulus sites at and beyond the second turn may not address significantly different populations of neurons. (However, some survival of peripheral processes in the apical region of the cochlea might allow selective activation of those neural elements with deeply inserted electrodes.)

A further possible limitation of perimodiolar electrode arrays has been suggested by Frijns et al. (2001). Results from their modeling studies have indicated that perimodiolar electrodes beyond the first turn may stimulate axons in the modiolus at lower current levels than the nearest ganglion cells. The axons are "fibers of passage," from ganglion cells and peripheral processes that innervate higher turns of the cochlea. Exclusive stimulation of such fibers at relatively low current levels would be expected to produce unintended (and tonotopically misplaced) percepts for the patient. Stimulation of both the fibers and nearby ganglion cells at higher levels would be expected to produce complex percepts, that would correspond to excitation in multiple turns of the cochlea.

Although perimodiolar placements may not be a panacea, such placements can increase the spatial specificity and dynamic range of stimulation at least in the basal turn. The placements also can reduce thresholds and increase dynamic range for most or all electrodes in the array. These changes may in turn produce improvements in the speech reception performance of implant systems.

An alternative to perimodiolar placements is to implant electrodes directly within the auditory nerve (Simmons 1966). This relatively old concept has been resurrected by a team at the University of Utah (Maynard et al. 2001). The development includes refinement of surgical approaches and placements, and fabrication of an 8×10 array of pin electrodes suitable for a lifetime of use in humans. The dimensions of the array (1.4 mm $\times$ 1.8 mm, with 200 μm spacing between adjacent pins) and graded lengths of the pins (with the longest pins at 1.5 mm) approximate the cross-sectional dimensions of the auditory nerve at the level of the basal turn, where the array is to be implanted.

An intramodiolar implant offers the likely advantages of lower currents required for threshold stimulation, greater spatial selectivity of stimulation, and a greater number of stimulus sites, compared with ST implants, including ST implants with perimodiolar placements of electrodes. On the other hand, mapping of processor channel outputs onto stimulus electrodes is likely to be far more complex with intramodiolar implants. The "roping" structure of the auditory nerve presents a complex anatomy compared with the cochleotopic organization of the spiral ganglion in Rosenthal's canal, and that complexity without doubt will complicate the fitting of speech processors used in conjunction with intramodiolar electrodes. (This problem might be addressed by placing a temporary ST implant at surgery, following placement of the intramodiolar implant. Each electrode of the ST implant would then be stimulated in sequence while recording the pattern of neural responses across all electrodes in the intramodiolar implant. Maps of the intramodiolar pin positions that correspond to the different sites of ST stimulation could be constructed from the recordings. Upon completion of the recordings, the temporary ST implant would be withdrawn and the remainder of the surgery completed. The maps of the pin positions could greatly facilitate the fitting of the speech processor at a later time.)

3. Strategies for Representing Speech Information with Implants

The performance of cochlear implants has improved dramatically since the introduction of single-channel devices in the mid-1970s. A large increment in performance was obtained with the use of multiple channels of processing and stimulation in the early 1980s (Gantz et al. 1988; Cohen et al. 1993). Steady and large improvements since that time have been produced with developments in speech processor design (e.g., Loizou 1998, 1999; Wouters et al. 1998; Clark 2000; Wilson, 2000b; David et al. 2003). Differences in the design of the multiple-electrode arrays have not produced obvious differences in performance to date, although in some studies a monopolar coupling configuration provided better results than a bipolar coupling configuration, using the Nucleus 22 electrode array (Zwolan et al. 1996; Franck et al. 2003).

Present strategies for representing speech information with cochlear implants are listed in Table 2.2. They include the continuous interleaved sampling (CIS; see Wilson et al. 1991), spectral peak (SPEAK; see Skinner et al. 1994; Seligman and McDermott 1995; Patrick et al. 1997), advanced combination encoder (ACE; see Arndt et al. 1999; Vandali et al. 2000; Kiefer et al. 2001), "n-of-m" (corresponding to selection of n among m channels for stimulation in each "sweep" across channels; see Wilson et al. 1988; McDermott et al. 1992; Lawson et al. 1996; Ziese et al. 2000), and simultaneous analog stimulation (SAS; see Battmer et al. 1997; Osberger and Fisher 2000) strategies. As described in the references above, each of these strategies can support relatively high levels of speech reception for a sub-

TABLE 2.2. Processing strategies used with implant systems in present use.

System	CIS	n-of-m	ACE	SPEAK	SAS
Combi 40, 40+	•	•			
Laura	•				
CI22				•	
CI24M	•		•	•	
Clarion	•				•

Note: The Combi 40 and Combi 40+ systems are manufactured by Med-El GmbH of Innsbruck, Austria; the Laura system was recently manufactured by Philips Hearing Instruments and before that by Antwerp Bionic Systems, in Antwerp, Belgium (manufacture of this system has been discontinued); the CI22 was manufactured by Cochlear Ltd. of Sydney, Australia, up until 1997; the CI24M system is manufactured by Cochlear Ltd. now; and several versions of the Clarion system have been manufactured by Advanced Bionics Corp. of Sylmar, CA. As described in Wilson, 2000b, the ACE strategy can be regarded as a close variation of the n-of-m strategy. Systems are listed in the leftmost column and the strategies used with each are indicated in the remaining columns. See text for full names of the strategies. Note that the strategies indicate broad categories and that the details of implementation for a given strategy can vary (sometimes widely) across systems.

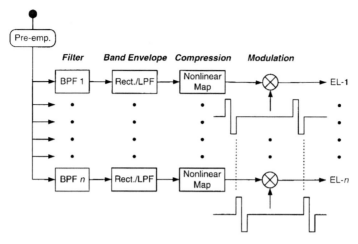

FIGURE 2.5. Block diagram of a continuous interleaved sampling (CIS) processor. Pre-emp.: preemphasis; BPF: band pass filter; Rect.: rectifier; LPF: low-pass filter; EL: electrode. (Adapted from Wilson et al. 1991, with permission from Macmillan Publishers Ltd.)

stantial subpopulation of users. However, a wide range of outcomes persists with any of the strategies, with some users obtaining only modest if any benefit from their implants.

3.1 CIS Strategy

As indicated in Table 2.2, the CIS strategy is available as a processing option for each of the three major implant systems in current use. A block diagram of the strategy is presented in Figure 2.5. Inputs from a microphone and optional automatic gain control (AGC) are directed to a preemphasis filter, which attenuates frequency components below 1.2 kHz at 6 dB/octave. This preemphasis helps relatively weak consonants (with a predominant frequency content above 1.2 kHz) compete with vowels, which are intense compared with most consonants and have strong components below 1.2 kHz.

The output of the preemphasis filter is directed to a bank of bandpass channels. Each channel includes stages of bandpass filtering, envelope detection, and compression. Envelope detection typically is accomplished with a rectifier, followed by a low-pass filter (a Hilbert transform also has been used for envelope detection, see, e.g., Helms et al. 2001). A logarithmic transformation is used to map the relatively wide dynamic range of derived envelope signals onto the narrow dynamic range of electrically evoked hearing. The channel outputs are used to modulate trains of biphasic pulses. This transformation produces a normal or nearly normal growth

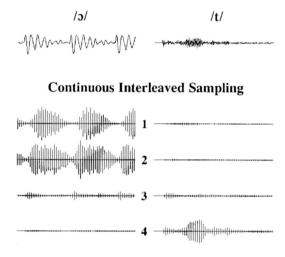

FIGURE 2.6. Stimuli produced by a simplified implementation of a continuous interleaved sampling (CIS) processor. The top panels show preemphasized (6 dB/octave attenuation below 1.2 kHz) speech inputs. An input corresponding to a voiced speech sound ("aw") is shown in the left panel and an input corresponding to an unvoiced speech sound ("t") is shown in the right panel. The bottom panels show stimulus pulses produced by a CIS processor for these inputs. The numbers indicate the electrodes to which the stimuli are delivered. The lowest number corresponds to the apical-most electrode and the highest number to the basal-most electrode. The pulse amplitudes in the figure reflect the amplitudes of the envelope signals for each channel. In actual implementations, the range of pulse amplitudes is compressed using a logarithmic or power-law transformation of the envelope signal for each channel. The duration of each of the trances is 25.4 ms. (Adapted from Wilson et al. 1991, with permission from Macmillan Publishers Ltd.)

of loudness with increases in sound level (Eddington et al. 1978; Zeng and Shannon 1992; Dorman et al. 1993). The modulated pulses for each channel are applied through a percutaneous or transcutaneous link to a corresponding electrode in the cochlea. Stimuli derived from channels with low center frequencies for the bandpass filter are directed to apical electrodes in the implant, and stimuli derived from channels with high center frequencies are directed to basal electrodes in the implant.

Patterns of stimulation for a simplified implementation of a four-channel CIS processor are illustrated in Figure 2.6. Speech inputs are shown in the top panels, and stimulus pulses are shown for each of four electrodes (and channels) in the bottom panels. The four electrodes are arranged in an apex-to-base order, with electrode 4 being the most basal. The amplitudes of the pulses for each of the electrodes are derived from the envelope signals in the corresponding bandpass channels. The envelope signal in the bandpass channel with the lowest center frequency controls the amplitudes of pulses

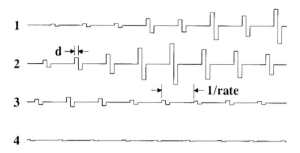

FIGURE 2.7. Expanded display of continuous interleaved sampling (CIS) stimuli. Pulse duration per phase (d) and the period between pulses on each electrode (1/rate) are indicated. The sequence of stimulated electrodes is 4-3-2-1 (a base-to-apex order). The duration of each trace is 3.3 ms. (Adapted from Wilson et al. 1991, with permission from Macmillan Publishers Ltd.)

delivered to the apical-most electrode, and the envelope signal in the band-pass channel with the highest center frequency controls the amplitude of pulses delivered to the basal-most electrode. This arrangement mimics the tonotopic organization of the cochlea in normal hearing, with high-frequency sounds exciting neurons at basal locations and low-frequency sounds exciting neurons at apical locations.

Continuous interleaved sampling processors use relatively high rates of stimulation to represent rapid temporal variations within channels. The pulse rate must be higher than twice the cutoff frequency of the low-pass filters (in the envelope detectors) to avoid aliasing effects (Rabiner and Shafer 1987; Wilson 1997). Results from recent recordings of auditory nerve responses to sinusoidally amplitude modulated pulse trains indicate that the rate should be even higher—four to five times the cutoff frequency—to avoid other distortions in the neural representations of the modulation waveforms (Wilson et al. 1994; Wilson 1997, 2000c). A typical CIS processor might use a pulse rate of 1000 pulses/s/electrode or higher, in conjunction with a 200-Hz cutoff for the low-pass filters.

An expanded display of stimuli during a short segment of the vowel input is presented in Figure 2.7. This display shows the interlacing and order of stimulus pulses across electrodes. In this particular implementation of a CIS processor, stimulus pulses are delivered in a non-overlapping sequence from the basal-most electrode (electrode 4) to the apical-most electrode (electrode 1). The rate of pulses on each electrode may be varied through manipulation in the duration of the pulses and the time between sequential pulses. Any ordering of electrodes may be used in the stimulation sequence, such as an apex-to-base order or a staggered order (i.e., an order designed to produce on average the maximum spatial separation between sequentially stimulated electrodes).

Unlike some prior processing strategies for implants, no specific features of speech are extracted or represented with CIS processors. Instead, envelope variations in each of multiple bands are presented to the electrodes through modulated trains of interleaved pulses. The rate of stimulation for each channel and electrode does not vary between voiced and unvoiced sounds (see illustration of this by comparing the left and right panels of Fig. 2.6). This "waveform" or "filter-bank" representation does not make any assumptions about how speech is produced or perceived.

A key feature of the CIS, n-of-m, ACE, and SPEAK strategies is the interlacing of stimulus pulses across electrodes. This eliminates a principal component of interaction among electrodes that otherwise would be produced through vector summation of the electric fields from different (simultaneously stimulated) electrodes (e.g., Favre and Pelizzone 1993). Such interaction, if allowed to stand, would be expected to reduce the salience of channel-related cues.

An additional aspect of the CIS, n-of-m, and ACE strategies is relatively high cutoff frequencies for the low-pass filters in the envelope detectors, along with rates of stimulation that are sufficiently high to represent the highest frequencies without aliasing or other distortions. The cutoff frequencies generally are in the range of 200 to 400 Hz. This range encompasses the fundamental frequency of voiced speech sounds (see periodicity of envelope variations in the lower left panels of Fig. 2.6) and rapid transitions in speech such as those produced by stop consonants. The range also does not exceed the perceptual space of typical patients. In particular, most patients perceive changes in frequency or rate of stimulation as changes in pitch up to about 300 Hz or 300 pulses/s, respectively (e.g., Shannon 1983; Tong et al. 1983; Townshend et al. 1987; Zeng 2002). Further increases in frequency or rate do not produce further increases in pitch for a majority of patients, for loudness-balanced stimuli. Some subjects have higher pitch saturation limits, as high as about 1000 Hz, but these subjects are the exception rather than the rule (Hochmair-Desoyer et al. 1983; Townshend et al. 1987; Wilson et al. 1997b). Thus, representation of frequencies much beyond 300 Hz probably would not convey any additional information that could be perceived or utilized by typical patients.

3.2. n-of-m *and ACE Strategies*

The CIS, n-of-m, and ACE strategies share (1) nonsimultaneous stimulation and (2) high cutoff frequencies for the envelope detectors. The principal difference between the CIS and the other strategies is that the channel outputs are "scanned" in the latter strategies to select the n channels with the highest envelope signals prior to each frame of stimulation across electrodes. Stimulus pulses are delivered only to the subset of m electrodes that correspond to the n selected channels. This "peak picking" scheme is designed to reduce the density of stimulation while still representing the most important aspects of the acoustic environment. It also allows a lower overall stimulus rate, that might be accommodated with a limited-bandwidth transcutaneous transmission system. The deletion of low-amplitude channels for each frame of stimulation may reduce the overall level of masking across electrode and stimulus regions in the implanted cochlea. To the extent that the omitted channels do not contain significant information, such "unmasking" may improve the perception of the input signal by the patient.

The n-of-m and ACE strategies are quite similar in design. The n-of-m approach was first described in 1988 (Wilson et al. 1988) and has been refined in several lines of subsequent development (e.g., McDermott et al. 1992; Lawson et al. 1996; McDermott and Vandali 1997). Current implementations include those listed in Table 2.2, along with laboratory implementations. Comparisons between n-of-m (or ACE) and CIS have indicated roughly equivalent performances for the two (e.g., Lawson et al. 1996; Ziese et al. 2000) or somewhat better performance for the n-of-m approach (e.g., Kiefer et al. 2001). The results have varied from subject to subject and also among different implementations of the strategies, using the Combi 40+, CI24M, or laboratory hardware and software.

Work to evaluate and refine n-of-m processors is still in progress. The choices of n and m probably are important, as the m electrodes most likely should be perceptually distinct for the best performance (limiting the number for practical electrode arrays), and n should be high enough to include all essential information but lower than m to provide any reduction in the density or overall rate of stimulation. In addition, there may be better ways to choose the n channels and electrodes for each frame of stimulation. Flanagan (1972), for example, describes various alternative procedures for identifying the channels for analogous "peak picking" vocoders (see also Loizou 1998). Those procedures were more effective for vocoders than a simple selection of maxima, and they might be more effective for implants as well.

3.3 SPEAK Strategy

The SPEAK strategy uses an adaptive n-of-m approach, in which n may vary from one stimulus frame to the next. The input is filtered into as many as 20 bands. Envelope signals are derived as in the CIS, n-of-m, and ACE strategies above, with an envelope cutoff frequency of 200 Hz. The number of bandpass channels selected in each scan (the adaptive n) depends on the number of envelope signals exceeding a preset "noise threshold" and on details of the input such as the distribution of energy across frequencies. In many cases, six channels are selected. However, the number can range from one to a maximum that can be set as high as 10. Cycles of stimulation, which include the selected channels and associated electrodes, are presented at rates between 180 and 300/s. The amount of time required to complete each cycle depends on the number of electrodes and channels included in the cycle (*n*) and the pulse amplitudes and durations for each of the electrodes. In general, inclusion of relatively few electrodes in a cycle allows relatively high rates, whereas inclusion of many electrodes reduces the rate.

A diagram illustrating the operation of the SPEAK strategy is presented in Figure 2.8. The speech input is directed to a bank of bandpass filters and envelope detectors, whose outputs are scanned for each cycle of stimulation. In this diagram, six channels are selected in each of two scans, and the corresponding electrodes are stimulated sequentially in a base-to-apex order. The diagram does not illustrate the compressive mapping of envelope signals onto pulse amplitudes.

The SPEAK strategy was designed for use with the N22 implant system. The transcutaneous link in that system is relatively slow, e.g., the maximum cycle rate for six channels of stimulation is about 400/s under ideal conditions (Crosby et al. 1985; Shannon et al. 1990). This limitation constrained the design of the strategy such that the average rate was set at about 250 cycles/s, and the range of rates is between 180 and 300/s, as noted above.

The rates used in the SPEAK strategy, in combination with the 200-Hz cutoff for the envelope detectors, are substantially lower than the minimum required to prevent aliasing and other distortions. For the average rate of 250/s, the representation of frequencies in the modulation waveforms above 125 Hz is subject to aliasing effects, and the representation of frequencies in the range of one-fourth to one-half the pulse rate probably is distorted to a lesser extent (Busby et al. 1993; McKay et al. 1994; Wilson 1997).

Results from recent comparisons of the SPEAK and ACE strategies generally have indicated superiority of the latter (e.g., Arndt et al. 1999; Kiefer et al. 2001; Pasanisi et al. 2002). The performance of the SPEAK strategy may be affected by (1) relatively low rates of stimulation, (2) aliasing and other distortions arising from the use of low rates in conjunction with a high cutoff frequency for the envelope detectors, or (3) some combination of these factors. At present, the ACE strategy is regarded by many clinicians as the "default" or "first-choice" strategy for the CI24M implant system.

3.4 SAS Strategy

The SAS strategy was derived from a compressed analog (CA) strategy (Eddington 1980; Merzenich et al. 1984), originally used in somewhat different implementations with the now-discontinued University of California at San Francisco (UCSF)/Storz and Ineraid implant systems. In contrast to the other strategies described above, the CA and SAS strategies use "analog" or continuous waveforms for stimuli, instead of biphasic pulses. A block diagram of the CA strategy is presented in Figure 2.9. A microphone or other input is compressed with a fast-acting AGC. The AGC output is filtered into contiguous bands (usually four in the UCSF/Storz and Ineraid implementations), that span the range of speech frequencies. The signal

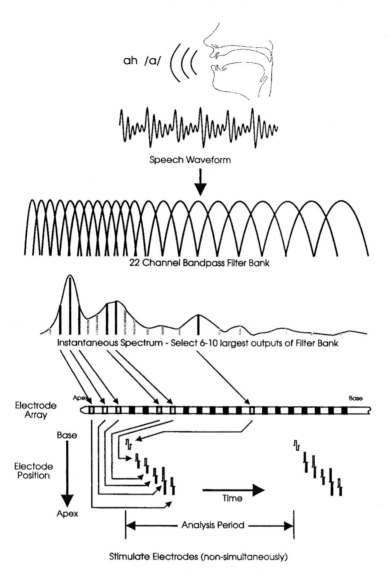

FIGURE 2.8. Key steps in the spectral peak (SPEAK) processing strategy. Speech inputs are directed to a bank of up to 20 bandpass filters and envelope detectors. The envelope signals are scanned just prior to each cycle of stimulation across electrodes. Between 1 and 10 of the highest-amplitude signals are selected in each scan, depending on characteristics of the input (i.e., overall level and spectral composition). Electrodes associated with the selected envelope signals (and bandpass channels) are stimulated in a base-to-apex order. (From Patrick et al. 1997, with permission from the Singular Publishing Group, Inc.)

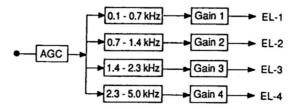

FIGURE 2.9. Block diagram of a compressed analog (CA) processor. The CA strategy uses a broadband automatic gain control (AGC), followed by a bank of bandpass filters. The outputs of the filters are adjusted with independent gain controls. Compression is achieved through rapid action of the AGC, and high-frequency emphasis and limited mapping to individual electrodes is accomplished through adjustments of the channel gains. (Adapted from Wilson et al. 1991, with permission from Macmillan Publishers Ltd.)

from each bandpass filter is amplified and then directed to a corresponding electrode in the implant. The gains of the amplifiers for the different bandpass filters can be adjusted to produce stimuli that do not exceed the upper end of the dynamic range of percepts for each electrode and that provide a high-frequency emphasis (e.g., percepts for high-frequency channel 4 can be made as loud as percepts for low-frequency channel 1, even though high-frequency sounds in speech generally are much less intense than low-frequency sounds).

The compression provided by the AGC can reduce the dynamic range of the input to approximate the dynamic range of electrically evoked hearing. The dynamic range of stimulation also can be restricted with a clipping circuit before or after the amplifiers, for some or all of the bandpass channels (Merzenich et al. 1984; Merzenich 1985). Of course, such "front-end" compression introduces spectral components that are not present in the input. The severity of this distortion depends on the time constants of the AGC (attack and release) and the compression ratio of the AGC (e.g., White 1986). Thus, a balance must be met between sufficient compression for mapping the wide dynamic range of the input onto the narrow dynamic range of electrically evoked hearing, and introduction of spurious frequency components, principally in the high-frequency channels.

Stimuli produced by a simplified implementation of a CA processor are shown in Figure 2.10. This figure has the same format as Figure 2.6 (which shows stimuli for a CIS processor). The CA stimuli represent a large portion of the information in the unprocessed speech input. Spectral and temporal patterns of speech are represented in the relative amplitudes of the stimuli across electrodes and in the temporal variations of the stimuli for each of the electrodes.

Although a large amount of information is presented with CA stimuli, much of it may not be available to implant patients. As mentioned before,

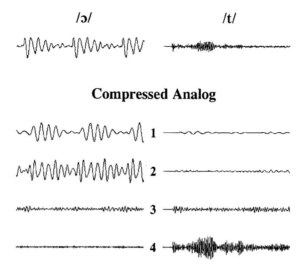

FIGURE 2.10. Stimuli produced by a simplified implementation of a compressed analog (CA) processor. The organization and speech inputs for this figure are the same as those in Figure 2.6. This simplified implementation does not include effects of a front-end AGC or adjustment of gains for the different channels. (Adapted from Wilson et al. 1991, with permission from Macmillan Publishers Ltd.)

within-channel changes in frequency above about 300 Hz are not perceived as changes in pitch by many patients. In addition, the simultaneous stimulation of multiple electrodes can produce large and uncontrolled interactions through vector summation (at sites of neural excitation) of the electric fields from each of the electrodes (White et al. 1984; Favre and Pelizzone 1993). The resulting degradation of independence among electrodes would be expected to reduce the salience of channel-related cues. In particular, the neural response to stimuli presented at one electrode may be significantly distorted or even counteracted by coincident stimuli presented at other electrodes. The pattern of interaction also may vary according to the instantaneous phase relationships among the stimuli for each of the electrodes. Phase is not controlled in CA processors, and this may degrade further the representation of channel-by-channel amplitudes.

Certain likely limitations of the CA approach have been addressed in the design of the SAS strategy. The SAS strategy includes a logarithmic mapping function at the output of each bandpass channel, to provide "back-end" rather than "front-end" compression. As with the CIS and other pulsatile strategies above, such back-end compression allows the mapping of stimuli on a channel-by-channel basis and also eliminates the spurious, across-channel spectral components introduced by front-end compression. As with the other strategies, the SAS strategy can include a front-end AGC, but with long time constants and a low compression ratio, to minimize any distortions.

The SAS strategy also is used in conjunction with electrode arrays designed to produce spatially selective patterns of stimulation, either through a particular orientation of electrodes for each channel of stimulation (the "offset radial" orientation, see Loeb et al. 1983, or the "enhanced bipolar" configuration, see Battmer et al. 1999) or through the use of techniques to place electrodes close to the target neurons in the spiral ganglion (e.g., Lenarz et al. 2000; Zwolan et al. 2001; Balkany et al. 2002; Frijns et al. 2002). To the extent that the design goals of these electrode arrays are fulfilled, interactions among electrodes may be relatively low compared with other designs (c.g., an array of monopolar electrodes not deliberately positioned immediately adjacent to the inner wall of the ST).

Present implementations of the SAS strategy include those in the Clarion CI and CII implant systems. The CI implant uses a "precurved" electrode array and has eight independent current sources. The CII implant uses an electrode positioning device (inserted at surgery behind and after a flexible electrode array with 16 contacts) and has 16 current sources. Up to eight channels of bipolar stimulation (between closely spaced pairs of electrodes in the array) can be supported with the CI implant, and up to 16 can be supported with the CII implant. (The number of channels is limited to seven when the enhanced bipolar configuration is used in conjunction with the CI implant.) The electrode array with the positioning device is called the "HiFocus" electrode.

Prior comparisons between the CA and CIS strategies showed a marked superiority of the latter (e.g., Wilson et al. 1991; Boëx et al. 1996; Pelizzone et al. 1999). Recent comparisons of the SAS and CIS strategies, as implemented in the Clarion hardware and software, have produced varied results across studies, but generally have demonstrated high levels of performance for some subjects using SAS. The studies conducted to date have included those listed in Table 2.3. The studies of Battmer et al. (1999), Osberger and Fisher (2000), and Stollwerck et al. (2001) used the CI implant and associated electrode array, and the study of Frijns et al. (2002) used the CII implant with the HiFocus electrode. The study of Zwolan et al. (2001) used both systems, in separate groups of subjects. In these latter studies (Frijns et al. and Zwolan et al.), new capabilities of the speech processor and transcutaneous link in the CII implant were not utilized as such use had not yet been approved by the U.S. Food and Drug Administration at the time of the studies. Thus, for example, CIS implementations used the relatively low pulse rates of the CI implant and a maximum of eight channels of processing and stimulation. (The maximum number of channels for SAS also was set at eight.)

An additional strategy, paired pulsatile stimulation (PPS), was included in some of the studies. It is a variation of CIS in which pairs of distant electrodes are stimulated simultaneously, with stimulation of the pairs in a

TABLE 2.3. Comparisons among continuous interleaved sampling (CIS), paired pulsatile stimulation (PPS), and simultaneous analog stimulation (SAS) strategies, as implemented in either the Clarion CI or CII implant systems.

| Study | Device | Subjects | Percent preferring | | |
			CIS	PPS	SAS
Battmer et al. 1999	CI	22[a]	50	N/A	50
Osberger and Fisher 2000	CI	58	72	N/A	28
Zwolan et al. 2001	CI	56	50	11	39
	CII	56	15	33	52
Stollwerck et al. 2001[b]	CI	55	75	N/A	25
Frijns et al. 2002	CII	10	90	10	0

Note: Subject preferences are indicated. For studies that included multiple test intervals, results for the final interval are given. The PPS strategy was not included in some of the comparison studies, as indicated by N/A in the PPS column.
[a] Two of the subjects could not use SAS, due to inadequate loudness. The data in the "percent preferring" columns reflect findings for the 20 subjects who could use both the SAS and CIS strategies.
[b] Within-subject comparisons demonstrated that the subjects performed better with their preferred strategy.

nonsimultaneous sequence. This approach doubles the rate of stimulation across all electrodes, while possibly minimizing interactions between the simultaneously stimulated electrodes by choosing electrodes in each pair that are far apart from each other. [The PPS strategy also has been called the "multiple pulsatile sampler" (MPS) strategy.]

In each of the studies the subjects have been asked to indicate their preference between the SAS and CIS strategies, or among the SAS, CIS, and PPS strategies. The results are listed in Table 2.3. They vary widely across studies. In the study of Battmer et al. (1999), approximately half of the subjects preferred SAS to CIS and also achieved scores with SAS that were comparable to the scores achieved by the other subjects with CIS. In the study of Osberger and Fisher (2000), less than 30% of the subjects preferred SAS, but some of those subjects had exceptionally high scores (on average, scores for the SAS and CIS groups were not different at the final, 6-month test interval). The subjects preferring SAS had short durations of deafness compared to the CIS group. In the study of Zwolan et al. (2001), SAS was compared with both CIS and PPS. At the final (again, 6-month) test interval, approximately 50% of the 56 subjects using the CI implant and precurved electrode array preferred CIS, and about 40% and 10% of those subjects preferred SAS and PPS, respectively. For the additional 56 subjects using the CII implant and HiFocus electrode array, approximately 15% preferred CIS, and about 52% and 33% preferred SAS and PPS, respectively. Zwolan et al. attributed this reversal in preference, for the two types of electrode array, to a closer positioning of electrodes next to the inner wall of the ST with the new electrode array. Presumably such positioning, if present, would reduce interactions among electrodes and thereby perhaps allow simultaneous stimulation strategies to become more useful. In another group of 55 subjects using the CI implant system, Stollwerck et al. (2001) found that 75% of the subjects preferred CIS over SAS. Among the 10 subjects studied by Frijns et al. (2002), nine preferred CIS, one PPS, and none SAS. These subjects used the CII implant and HiFocus electrode array (but with the positioner inserted only along the basal turn, see Frijns et al. 2002).

Speech reception data also have been collected for at least the preferred strategy for each subject in each of the studies. Controlled comparisons between or among strategies, balancing experience and other variables across strategies, also have been made in the study of Stollwerck et al. (2001). In broad terms, results from all studies, except the study of Frijns et al. (2002) (in which none of subjects stated a preference for SAS), indicated levels of performance for the group preferring SAS that were comparable to those of the group preferring CIS. In addition, some of the subjects in each of the groups had especially high scores. Results from the study of Stollwerck et al. also showed that subjects performed better with their preferred strategy.

The SAS strategy may convey more temporal information than the CIS strategy, especially in the apical (low-frequency) channels and especially for patients who can perceive changes in frequency as changes in pitch over relatively wide ranges. If so, that advantage may outweigh the likely disadvantage of higher electrode interactions, which would be expected even with a spatially selective electrode array (Loizou et al. 2003).

3.5 "HiRes" Strategy

The CIS implementations in the above comparisons used eight channels and a carrier rate of about 800 pulses/s/electrode. This may have limited the performance of the strategy in that more channels and/or higher rates may have been helpful. Indeed, an increase in the number of perceptually separable channels beyond eight would be expected to improve speech reception, especially for speech presented in competition with noise (Friesen et al. 2001; Dorman et al. 2002). In addition, results from studies using other implant systems indicate significant benefits of higher rates of stimulation for at least some patients (e.g., Brill et al. 1997; Kiefer et al. 2000; Loizou et al. 2000; Wilson et al. 2000).

A variation of CIS, called the HiRes strategy, has been implemented for use with the Clarion CII implant system (Frijns et al. 2002). It utilizes the full capabilities of the device. The HiRes strategy can present pulses at rates up to 90,000/s across the addressed electrodes, and the strategy can support up to 16 channels of processing and stimulation. A typical fitting would include 16 channels and a carrier rate of 2800 to 5600 pulses/s/electrode. The fidelity of the temporal representation provided with HiRes might match or exceed that provided with SAS. At the same time, electrode inter-actions are minimized in HiRes through the use of nonsimultaneous stimuli. This might increase the number of perceptually separable channels well beyond the maximum available with SAS.

Initial studies to evaluate HiRes have included a clinical trial sponsored by the Advanced Bionics Corp. In that study, HiRes was compared with the preferred strategy among the prior CIS, PPS, or SAS strategies for each of 80 subjects, implanted with the HiFocus electrode and positioner across 19 centers in North America. Phase I of the study included use by each subject of their preferred prior strategy for 3 months. Phase II included use of HiRes for the subsequent 3 months. Speech reception data were collected at the conclusions of the two phases, and a preference questionnaire was given at the end of phase II. The speech reception tests included recognition of monosyllabic words and recognition of key words in the Central Institute for the Deaf (CID) and Hearing in Noise Test (HINT) sentences. The speech items were presented in quiet for all tests, and also at the S/N of +10 dB for the HINT sentence test. As of September 2002, 51 of the 80 subjects had completed phase II (Osberger et al. 2002). The mean scores for the 51 were significantly higher with HiRes compared to the control strategies for each of the administered tests. The greatest gains were observed for speech reception in noise, with an increase from 47% to 61% correct in the mean scores for the HINT sentences at the S/N +10 dB. Subjects with relatively low scores using the control strategy showed a larger gain overall than subjects with relatively high scores. (Ceiling effects may have limited the sensitivity of the test for some of the latter subjects.) Ninety percent of the subjects indicated a preference for HiRes at the conclusion of phase II.**

The initial studies also have included comparisons of CIS processors using 833 pulses/s/electrode versus processors using 1400 pulses/s/electrode (Frijns et al. 2003). Each of the nine subjects had used a processor with the lower rate and eight channels (and electrodes) for a period of 3 to 11 months. They then were fitted in randomized orders with 8-, 12-, or 16-channel processors using the higher rate. Each of the processors was used for 1 month prior to testing, and then the next processor was fitted. Tests with all processors included recognition of monosyllabic words in quiet and in competition with noise at the S/Ns of +10, +5, 0, and −5 dB. The results demonstrated significant differences in scores between the rates and across the numbers of channels. Some patients achieved their best scores with the higher rate and eight channels, whereas others achieved their best scores with the higher rate and 12 or 16 channels. These best scores showed a substantial and significant advantage of the higher rate for speech reception in noise, consistent with the results of the prior studies cited above.

** Note that the additional experience gained with HiRes, following the initial experience with the control strategies, may have favored HiRes in these comparisons.

HiRes has become the default strategy for the CII system.[‡] More information about its performance should be available in the near future.

3.6 Summary

All present processing strategies for cochlear implants use a "filter-bank" or "waveform" approach. Explicit extraction and representation of specific features of speech was abandoned in the early 1990s. All but one of the present strategies use biphasic pulses as stimuli, presented in a nonsimultaneous sequence across electrodes. The SAS strategy presents "analog" or continuous waveforms as stimuli simultaneously to all (utilized) electrodes. This strategy produces good results for some subjects that are comparable to results produced with CIS using the Clarion CI implant system.

4. Design Considerations

Many factors can affect the performance of implant systems. Some of those factors are described above, e.g., choices among processing strategies and various electrode designs. Additional important factors include the patient variable, fitting procedures, and the hardware and software implementations of processing strategies.

4.1 Patient Variable

One of the most striking findings from research on implants is that the range of performance across patients is large for any of a variety of implant systems. Some patients score at or near 100% correct on standard audiological tests of sentence and word recognition in quiet, whereas other patients obtain zero scores using an identical speech processor and electrode array. Although average scores across patients have increased substantially with the introductions of new processing strategies, a wide range of performance still remains.

Data indicating the importance of the patient variable are presented in Figure 2.11, which shows a scatter plot of scores for recognition of Northwestern University Auditory Test 6 (NU-6) monosyllabic words, from within-subject comparisons of CA and CIS processors (Wilson et al. 1993). These scores were obtained with careful fittings of the two processing strategies and with high-quality implementations of both (see below). Note that relatively low scores for one strategy are associated with relatively low scores for the other strategy and vice versa. The data points in the figure

---

[‡]The electrode positioner device is no longer used with the CII system, due to a possible association between such use and meningitis (e.g., Arnold et al. 2002).

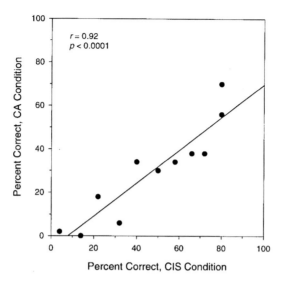

FIGURE 2.11. Scatter plot of percent-correct scores for the recognition of Northwestern University Auditory Test 6 (NU-6) monosyllabic words. Each point represents the scores obtained with the compressed analog (CA) and continuous interleaved sampling (CIS) processing strategies for one subject. Different lists of words were used for the tests with the different strategies. The Pearson correlation coefficient, level of significance, and regression line are also shown in the figure. (Data from Wilson et al. 1993.)

are highly correlated ($r = 0.92$), indicating that 85% of the variance in the results is explained by the subject (or patient) variable ($r^2 = 0.85$). Correlations for other tests not limited by ceiling effects also are quite high for these subjects and processors, in a range between 0.87 and 0.92 (Wilson et al. 1993).

Identification of the factors that underlie the effects of the patient variable may lead to the development of reliable prognostic tests for prospective patients (e.g., Rubinstein et al. 1999a). The factors may be related to the condition of the auditory nerve (see Leake and Rebscher, Chapter 4), the integrity of the central auditory system (see Hartmann and Kral, Chapter 6), the cognitive abilities of a patient (see Pisoni and Cleary, Chapter 9), or some combination of these possibilities. Knowledge of the factors should help in the design of better implant systems that take these factors into account and minimize or eliminate their deleterious effects for patients who otherwise would have relatively poor outcomes.

4.2 Fitting Procedures

Large improvements in the performance of a cochlear implant system can be obtained for individual patients through informed choices of parameter values. In studies with CIS processors, for example, quite large gains have been produced through choices of pulse rate, pulse duration, electrode update order, the range of frequencies spanned by the bandpass filters, and other parameter values (e.g., Wilson et al. 1995; Loizou et al. 2000). Although predetermined values for some parameters may be appropriate for virtually all patients, the values of other parameters should be varied over certain ranges to optimize performance for individuals.

4.3 Strategy Implementations

The performance of a processing strategy also can be affected by the quality of its implementation. Seemingly subtle changes in hardware and the programming of that hardware can produce large changes in performance. Such effects have been demonstrated in comparisons between versions of commercial implant systems that were designed to implement the same processing strategy. Dowell et al. (1991) compared two versions of the Nucleus device (the WSP III and MSP versions), each of which was designed to implement the F0/F1/F2 processing strategy, and found significant increases in speech reception scores with the newer implementation (the MSP version). Similarly, Battmer et al. (1997) found significant improvements in speech test scores when the Clarion version 1.2 implementation of a CIS strategy was substituted for the prior version 1.1 implementation of that strategy.

Examples of ways in which an implementation can go awry include use of microphones with poor frequency response or high levels of noise, use of amplifier and AGC circuits with low dynamic ranges or high levels of noise, use of digital filters with a reduced number of elements compared with conventional and well-functioning digital filters (a reduced number of elements has been used in devices with small memories or slow DSP chips), use of current sources that are especially noisy, use of current sources that saturate or begin to saturate in the dynamic ranges of the electrodes for some or all patients (current sources saturate when the commanded current requires a voltage at the electrodes that approaches or exceeds the "compliance" or voltage limit of the device), and an excessive amount of digital or switching noise that appears at the electrodes. Any one of these factors can degrade or destroy the performance of an otherwise good strategy.

5. Future Directions

A number of efforts are under way to improve the design and performance of present implant systems. They include:

- Use of bilateral implants to (1) increase the number of perceptually independent stimulus sites and thereby improve speech reception in noise or (2) restore at least to some extent sound localization abilities and the signal-to-noise advantages that accompany such abilities (e.g., Lawson et al. 2000, 2001a; Gantz et al. 2002; Müller et al. 2002; Tyler et al. 2002a; van Hoesel et al. 2002).
- Combined electric and acoustic stimulation of the auditory system for patients with some remaining (low-frequency) hearing, with the two modes of stimulation used for opposite ears (e.g., Armstrong et al. 1997; Lawson et al. 2001b; Tyler et al. 2002b) or with both acoustic and electric stimuli delivered to the same ear (von Ilberg 1999; Lawson et al. 2001b; Kiefer et al. 2002; Wilson et al. 2003b).
- Drug delivery through the implant (e.g., Clark 2001), to preserve neurons or even promote the growth of neurites (toward the electrode array) from existing neurons or regeneration of neurons and associated structures (also see Qun et al. 1999, for descriptions of, and studies with, various neurotrophic factors).
- Development of new processing strategies that emphasize the representation of important transients in speech (Geurts and Wouters 1999; Vandali 2001).
- Use of high carrier rates, or high-rate "conditioner" pulses, in conjunction with CIS and other strategies, to (1) increase the correspondence between modulation waveforms and temporal patterns of neural responses, (2) increase the dynamic range of responses, and (3) reinstate at least to some degree a normal pattern of stochastic or "spontaneous" activity in auditory neurons (Wilson et al. 1997a; Rubinstein et al. 1999b; Wilson 2000c; Litvak et al. 2001).
- Development of new processing strategies designed to replicate closely the signal transformations in normal cochlea, including (1) the nonlinear filtering that occurs at the basilar membrane and outer hair cell complex, and (2) the stages of instantaneous and noninstantaneous compression that occur at the inner hair cells and at the synapses between the cells and adjacent fibers of the auditory nerve (Wilson et al., 2003a).
- Applications of new and emerging knowledge about factors that are highly correlated with outcomes for implants, in the design of approaches or training procedures to help patients presently at the low end of the performance spectrum (e.g., Fu 2002).

Each of these efforts, except for drug delivery through the implant, is described in detail in a recent article by Wilson et al. (2003a). Drug delivery and additional possibilities for the further development of implant systems are presented in two recent reviews by Clark (2000, 2001). Selected developments also are described in somewhat older but still relevant articles by Klinke and Hartmann (1997), Lenarz (1997), and Wilson (1997, 2000c).

6. Summary

Cochlear implant systems include a microphone, speech processor, transcutaneous or percutaneous link, and an electrode array. In general, design choices for any one of these components may affect or interact with choices for other components. The design problem should be viewed at the system level.

In addition to choices in the design of the hardware (and associated software), factors presented by the patient can exert large effects on performance. Indeed, the patient variable probably is the most important of all variables in implant design.

Current processing strategies for implants all use a "waveform" or "filterbank" approach. In addition, all but one of the strategies use biphasic pulses as stimuli, presented in a sequential and non-overlapping sequence across electrodes.

The design of cochlear implants is changing rapidly, with the advent of combined electric and acoustic stimulation of the auditory system, perimodiolar electrode arrays, and high-rate and conditioner-pulses stimuli. New processing strategies, designed to provide a closer mimicking of signal transformations in the normal auditory periphery, are being developed. Additional developments that are under way include those for intramodiolar implants and fully implantable devices. Implants of the near future may be quite unlike today's implants, with better performance overall and, it is hoped, with a much smaller range of outcomes across patients.

Acknowledgments. Preparation of this chapter was supported by National Institutes of Health (NIH) projects N01-DC-8-2105 and N01-DC-2-1002. Portions of text in the chapter were updated or adapted from prior publications (Wilson, 2000a,b).

Many people have contributed to the ideas and findings presented here. I am most grateful to have had the grand opportunity to work with such talented and dedicated colleagues.

References

Abbas PJ, Brown CJ, Shallop JK, Firszt JB, et al. (1999) Summary of results using the nucleus CI24M implant to record the electrically evoked compound action potential. Ear Hear 20:45–59.

Abbas PJ, Brown CJ, Hughes ML, Gantz BJ, et al. (2000) Electrically evoked compound action potentials recorded from subjects who use the nucleus CI24M device. Ann Otol Rhinol Laryngol 185(suppl):6–9.

Ariyasu L, Galey FR, Hilsinger R Jr, Byl FM (1989) Computer-generated three-dimensional reconstruction of the cochlea. Otolaryngol Head Neck Surg 100:87–91.

Armstrong M, Pegg P, James C, Blamey P (1997) Speech perception in noise with implant and hearing aid. Am J Otol 18(suppl):S140–S141.

Arndt P, Staller S, Arcaroli J, Hines A, Ebinger K (1999) Within-subjects comparison of advanced coding strategies in the Nucleus 24 cochlear implant. Technical Report, Cochlear Corporation, Englewood, CO.

Arnold W, Bredberg G, Gstöttner W, Helms J, et al. (2002) Meningitis following cochlear implantation: pathomechanisms, clinical symptoms, conservative and surgical treatments. ORL J Otorhinolaryngol Relat Spec 64:382–389.

Balkany TJ, Eshraghi AA, Yang N (2002) Modiolar proximity of three perimodiolar cochlear implant electrodes. Acta Otolaryngol 122:363–369.

Battmer RD, Feldmeier I, Kohlenberg A, Lenarz T (1997) Performance of the new Clarion speech processor 1.2 in quiet and in noise. Am J Otol 18:S144–S146.

Battmer RD, Zilberman Y, Haake P, Lenarz T (1999) Simultaneous analog stimulation (SAS)–continuous interleaved sampler (CIS) pilot comparison study in Europe. Ann Otol Rhinol Laryngol 177(suppl):69–73.

Boëx C, Pelizzone M, Montandon P (1996) Speech recognition with a CIS strategy for the Ineraid multichannel cochlear implant. Am J Otol 17:61–68.

Brill S, Gstöttner W, Helms J, von Ilberg C, et al. (1997) Optimization of channel number and stimulation rate for the fast continuous interleaved sampling strategy in the COMBI 40+. Am J Otol 18:S104–S106.

Brown CJ, Abbas PJ, Gantz B (1990) Electrically evoked whole-nerve action potentials: data from human cochlear implant users. J Acoust Soc Am 88:1385–1391.

Brown CJ, Abbas PJ, Gantz BJ (1998) Preliminary experience with neural response telemetry in the nucleus CI24M cochlear implant. Am J Otol 19:320–327.

Brown CJ, Hughes ML, Luk B, Abbas PJ, Wolaver A, Gervais J (2000) The relationship between EAP and EABR thresholds and levels used to program the nucleus 24 speech processor: data from adults. Ear Hear 21:151–163.

Busby PA, Tong YC, Clark GM (1993) The perception of temporal modulations by cochlear implant patients. J Acoust Soc Am 94:124–131.

Clark GM (2000) The cochlear implant: a search for answers. Cochlear Implants Int 1:1–15.

Clark GM (2001) Cochlear implants: climbing new mountains. Cochlear Implants Int 2:75–97.

Cohen LT, Saunders E, Clark GM (2001) Psychophysics of a prototype perimodiolar cochlear implant electrode array. Hear Res 155:63–81.

Cohen NL, Waltzman SB, Fisher SG (1993) A prospective, randomized study of cochlear implants. N Engl J Med 328:233–237.

Crosby PA, Daly CN, Money DK, Patrick JF, Seligman PM, Kuzma JA (1985) Cochlear implant system for an auditory prosthesis. U.S. patent 4532930.

David EE, Ostroff JM, Shipp D, Nedzelski JM, et al. (2003) Speech coding strategies and revised cochlear implant candidacy: an analysis of post-implant performance. Otol Neurotol 24:228–233.

Dorman MF, Smith L, Parkin JL (1993) Loudness balance between acoustic and electric stimulation by a patient with a multichannel cochlear implant. Ear Hear 14:290–292.

Dorman MF, Loizou PC, Spahr AJ, Maloff E (2002) A comparison of the speech understanding provided by acoustic models of fixed-channel and channel-picking signal processors for cochlear implants. J Speech Lang Hear Res 45:783–788.

Dowell RC, Dawson PW, Dettman SJ, Shepherd RK, et al. (1991) Multichannel cochlear implantation in children: a summary of current work at the University of Melbourne. Am J Otol 12(suppl 1):137–143.

Eddington DK (1980) Speech discrimination in deaf subjects with cochlear implants. J Acoust Soc Am 68:885–891.

Eddington DK, Dobelle WH, Brackman DE, Mladejovsky MG, Parkin JL (1978) Auditory prosthesis research with multiple channel intracochlear stimulation in man. Ann Otol Rhinol Laryngol 87(suppl 53):1–39.

Favre E, Pelizzone M (1993) Channel interactions in patients using the Ineraid multichannel cochlear implant. Hear Res 66:150–156.

Figueiredo JC, Abel SM, Papsin BC (2001) The effect of the audallion BEAMformer noise reduction preprocessor on sound localization for cochlear implant users. Ear Hear 22:539–547.

Flanagan JL (1972) Speech Analysis, Synthesis and Perception, 2nd ed. Berlin, Heidelberg and New York: Springer-Verlag, pp. 330–331.

Franck KH (2002) A model of a nucleus 24 cochlear implant fitting protocol based on the electrically evoked whole nerve action potential. Ear Hear 23(suppl):67S–71S.

Franck KH, Xu L, Pfingst BE (2003) Effects of stimulus level on speech perception with cochlear prostheses. J Assoc Res Otolaryngol 4:49–59.

Friesen LM, Shannon RV, Baskent D, Wang X (2001) Speech recognition in noise as a function of the number of spectral channels: comparison of acoustic hearing and cochlear implants. J Acoust Soc Am 110:1150–1163.

Frijns JH, Briaire JJ, Grote JJ (2001) The importance of human cochlear anatomy for the results with modiolus-hugging multichannel cochlear implants. Otol Neurotol 22:340–349.

Frijns JH, Briaire JJ, de Laat JA, Grote JJ (2002) Initial evaluation of the Clarion CII cochlear implant: speech perception and neural response imaging. Ear Hear 23:184–197.

Frijns JH, Klop WM, Bonnet RM, Briaire JJ (2003) Optimizing the number of electrodes with high-rate stimulation of the Clarion CII cochlear implant. Acta Otolaryngol 123:138–142.

Fu Q-J (2002) Temporal processing and speech recognition in cochlear implant users. Neuroreport 13:1–5.

Gantz BJ, Tyler RS, Knutson JF, Woodworth G, et al. (1988) Evaluation of five different cochlear implant designs: audiologic assessment and predictors of performance. Laryngoscope 98:1100–1106.

Gantz BJ, Tyler RS, Rubinstein JT, Wolaver A, et al. (2002) Binaural cochlear implants placed during the same operation. Otol Neurotol 23:169–180.

Geurts L, Wouters J (1999) Enhancing the speech envelope of continuous interleaved sampling processors for cochlear implants. J Acoust Soc Am 105:2476–2484.

Gordon KA, Ebinger KA, Gilden JE, Shapiro WH (2002) Neural response telemetry in 12- to 24-month-old children. Ann Otol Rhinol Laryngol 189(suppl):42–48.

Gstoettner WK, Adunka O, Franz P, Hamzavi J Jr, et al. (2001) Perimodiolar electrodes in cochlear implant surgery. Acta Otolaryngol 121:216–219.

Hamacher V, Doering WH, Mauer G, Fleischman H, Hennecke J (1997) Evaluation of noise reduction systems for cochlear implant users in different acoustic environments. Am J Otol 18:S46–S49.

Helms J, Müller J, Schön F, Winkler F, et al. (2001) Comparison of the TEMPO+ ear-level speech processor and the CIS pro+ body-worn processor in adult MED-EL cochlear implant users. ORL J Otorhinolaryngol Relat Spec 63:31–40.

Hinojosa R, Marion M (1983) Histopathology of profound sensorineural deafness. Ann N Y Acad Sci 405:459–484.

Hochmair-Desoyer IJ, Hochmair ES, Burian K, Stiglbrunner HK (1983) Percepts from the Vienna cochlear prosthesis. Ann N Y Acad Sci 405:295–306.

Jolly CN, Spelman FA, Clopton BM (1996) Quadrupolar stimulation for cochlear prostheses: modeling and experimental data. IEEE Trans Biomed Eng 43:857–865.

Ketten DR, Skinner MW, Gates GA, Nadol JB Jr, Neely JG (1997) In vivo measures of intracochlear electrode position and Greenwood frequency approximations. Abstracts for the 1997 Conference on Implantable Auditory Prostheses, Pacific Grove, CA, p. 33. (The book of abstracts is available from the House Ear Institute, Los Angeles, CA.)

Kiefer J, von Ilberg C, Hubner-Egener J, Rupprecht V, Knecht R (2000) Optimized speech understanding with the continuous interleaved sampling speech coding strategy in cochlear implants: effect of variations in stimulation rate and number of channels. Ann Otol Rhinol Laryngol 109:1009–1020.

Kiefer J, Hohl S, Sturzebecher E, Pfennigdorff T, Gstoettner W (2001) Comparison of speech recognition with different speech coding strategies (SPEAK, CIS, and ACE) and their relationship to telemetric measures of compound action potentials in the Nucleus CI 24M cochlear implant system. Audiology 40:32–42.

Kiefer J, Tillein J, von Ilberg C, Pfennigdorff T, et al. (2002) Fundamental aspects and first results of the clinical application of combined electric and acoustic stimulation of the auditory system. In: Kubo T, Takahashi Y, Iwaki T, eds. Cochlear Implants—An Update. The Hague: Kugler Publications, pp. 569–576.

Klinke R, Hartmann R (1997) Basic neurophysiology of cochlear implants. Am J Otol 18:S7–S10.

Laszig R, Ridder GJ, Fradis M (2002) Intracochlear insertion of electrodes using hyaluronic acid in cochlear implant surgery. J Laryngol Otol 116:371–372.

Lawson DT, Wilson BS, Zerbi M, Finley CC (1996) Speech processors for auditory prostheses: 22 electrode percutaneous study—results for the first five subjects. Third Quarterly Progress Report, NIH project N01-DC-5-2103, Neural Prosthesis Program, National Institutes of Health, Bethesda, MD. (This report is available online at http://scientificprograms.nidcd.nih.gov/npp/index.html.)

Lawson DT, Brill S, Wolford RD, Wilson BS, Schatzer R (2000) Speech processors for auditory prostheses: binaural cochlear implant findings—summary of initial studies with eleven subjects. Ninth Quarterly Progress Report, NIH project N01-DC-8-2105, Neural Prosthesis Program, National Institutes of Health, Bethesda, MD. (This report is available online at http://scientificprograms.nidcd.nih.gov/npp/index.html.)

Lawson DT, Wolford RD, Brill SM, Schatzer R, Wilson BS (2001a) Speech processors for auditory prostheses: further studies regarding benefits of bilateral cochlear implants. Twelfth Quarterly Progress Report, NIH project N01-DC-8-2105, Neural Prosthesis Program, National Institutes of Health, Bethesda, MD. (This report is available online at http://scientificprograms.nidcd.nih.gov/npp/index.html.)

Lawson DT, Wolford RD, Brill SM, Wilson BS, Schatzer R (2001b) Speech processors for auditory prostheses: cooperative electric and acoustic stimulation of the auditory periphery—comparison of ipsilateral and contralateral implementations. Thirteenth Quarterly Progress Report, NIH project N01-DC-8-2105, Neural Prosthesis Program, National Institutes of Health, Bethesda, MD. (This report is available online at http://scientificprograms.nidcd.nih.gov/npp/index.html.)

Lenarz T (1997) Cochlear implants—What can be achieved? Am J Otol 18:S2–S3.

Lenarz T, Kuzma J, Weber BP, Reuter G, et al. (2000) New Clarion electrode with positioner: insertion studies. Ann Otol Rhinol Laryngol 185(suppl):16–18.

Litvak L, Delgutte B, Eddington D (2001) Auditory nerve fiber responses to electric stimulation: modulated and unmodulated pulse trains. J Acoust Soc Am 110:368–379.

Loeb GE, Byers CL, Rebscher SJ, Casey DE, et al. (1983) Design and fabrication of an experimental cochlear prosthesis. Med Biol Eng Comput 21:241–254.

Loizou PC (1998) Mimicking the human ear: an overview of signal processing strategies for cochlear prostheses. IEEE Sig Proc Mag 15:101–130.

Loizou PC (1999) Signal-processing techniques for cochlear implants. IEEE Eng Med Biol Mag 18:34–46.

Loizou PC, Poroy O, Dorman M (2000) The effect of parametric variations of cochlear implant processors on speech understanding. J Acoust Soc Am 108:790–802.

Loizou PC, Stickney G, Mishra L, Assmann P (2003) Comparison of speech processing strategies used in the Clarion implant processor. Ear Hear 24:12–19.

Mason SM, Cope Y, Garnham J, O'Donoghue GM, Gibbin KP (2001) Intraoperative recordings of electrically evoked auditory nerve action potentials in young children by use of neural response telemetry with the nucleus CI24M cochlear implant. Br J Audiol 35:225–235.

Maynard E, Norman R, Shelton C, Najaragan S (2001) The feasibility of an intraneural auditory prosthesis stimulating array. First Quarterly Progress Report, NIH project N01-DC-1-2108, Neural Prosthesis Program, National Institutes of Health, Bethesda, MD. (This report is available online at http://scientificprograms. nidcd.nih.gov/npp/index.html.)

McDermott HJ, Vandali AE (1997) Spectral maxima sound processor. U.S. patent 5597380.

McDermott HJ, McKay CM, Vandali AE (1992) A new portable sound processor for the University of Melbourne/Nucleus Limited multielectrode cochlear implant. J Acoust Soc Am 91:3367–3391.

McKay CM, McDermott HJ, Clark GM (1994) Pitch percepts associated with amplitude-modulated current pulse trains in cochlear implantees. J Acoust Soc Am 96:2664–2673.

Merzenich MM (1985) UCSF cochlear implant device. In: Schindler RA, Merzenich MM, eds. Cochlear Implants. New York: Raven Press, pp. 121–130.

Merzenich MM, Rebscher SJ, Loeb GE, Byers CL, Schindler RA (1984) The UCSF cochlear implant project. Adv Audiol 2:119–144.

Müller, Schön F, Helms J (2002) Speech understanding in quiet and noise in bilateral users of the MED-EL COMBI 40/40+ cochlear implant system. Ear Hear 23:198–206.

NIH Consensus Statement (1995) Cochlear implants in adults and children. Office of Medical Applications of Research, National Institutes of Health, Bethesda, MD. JAMA 274:1955–1961.

Osberger MJ, Fisher L (2000) New directions in speech processing: patient performance with simultaneous analog stimulation. Am Otol Rhinol Laryngol 185(suppl):70–73.

Osberger MJ, Koch DB, Zimmerman-Phillips S, Segel P, Kruger T, Kessler DK (2002) Clarion CII Bionic Ear: HiRes clinical results. Abstracts for the Seventh International Cochlear Implant Conference, Manchester, U.K., Abstract 1.13.1.

Pasanisi E, Bacciu A, Vincenti V, Guida M, et al. (2002) Comparison of speech perception benefits with SPEAK and ACE coding strategies in pediatric Nucleus CI24M cochlear implant recipients. Int J Pediatr Otorhinolaryngol 64:159–163.

Patrick JF, Seligman PM, Clark GM (1997) Engineering. In: Clark GM, Cowan RSC, Dowell RC, eds. Cochlear Implantation for Infants and Children: Advances. San Diego, CA: Singular Publishing Group, pp. 125–145.

Pelizzone M, Cosendai G, Tinembart J (1999) Within-patient longitudinal speech reception measures with continuous interleaved sampling processors for Ineraid implanted subjects. Ear Hear 20:228–237.

Qun LX, Pirvola, U, Saarma M, Ylikoski J (1999) Neurotrophic factors in the auditory periphery. Ann N Y Acad Sci 884:292–304.

Rabiner LR, Shafer RW (1978) Digital Processing of Speech Signals. Englewood Cliffs, NJ: Prentice-Hall.

Ranck JB Jr (1975) Which elements are excited in electrical stimulation of the mammalian central nervous system: a review. Brain Res 98:417–440.

Rubinstein JT, Parkinson WS, Tyler RS, Gantz BJ (1999a) Residual speech recognition and cochlear implant performance: effects of implantation criteria. Am J Otol 20:445–452.

Rubinstein JT, Wilson BS, Finley CC, Abbas PJ (1999b) Pseudospontaneous activity: stochastic independence of auditory nerve fibers with electrical stimulation. Hear Res 127:108–118.

Seligman P, McDermott H (1995) Architecture of the Spectra 22 speech processor. Ann Otol Rhinol Laryngol 166(suppl):139–141.

Seyle K, Brown CJ (2002) Speech perception using maps based on neural response telemetry measures. Ear Hear 23(suppl):72S–79S.

Shallop JK, Facer GW, Peterson A (1999) Neural response telemetry with the nucleus CI24M cochlear implant. Laryngoscope 109:1755–1759.

Shannon RV (1983) Multichannel electrical stimulation of the auditory nerve in man. I. Basic psychophysics. Hear Res 11:157–189.

Shannon RV, Adams DD, Ferrel RL, Palumbo RL, Grandgenett M (1990) A computer interface for psychophysical and speech research with the Nucleus cochlear implant. J Acoust Soc Am 87:905–907.

Shepherd RK, Clark GM, Pyman BC, Webb RL (1985) Banded intracochlear electrode array: evaluation of insertion trauma in human temporal bones. Acta Otolaryngol 399(suppl):19–31.

Shepherd RK, Hatsushika S, Clark GM (1993) Electrical stimulation of the auditory nerve: the effect of electrode position on neural excitation. Hear Res 66:108–120.

Simmons FB (1966) Electrical stimulation of the auditory nerve in man. Arch Otolaryngol 84:2–54.

Skinner MW, Clark GM, Whitford LA, Seligman PM, et al. (1994) Evaluation of a new spectral peak (SPEAK) coding strategy for the Nucleus 22 channel cochlear implant system. Am J Otol 15(suppl 2):15–27.

Stollwerck LE, Goodrum-Clarke K, Lynch C, Armstrong-Bednall G, et al. (2001) Speech processing strategy preferences among 55 European CLARION cochlear implant users. Scand Audiol Suppl 2001:36–38.

Tong YC, Blamey PJ, Dowell RC, Clark GM (1983) Psychophysical studies evaluating the feasibility of a speech processing strategy for a multiple-channel cochlear implant. J Acoust Soc Am 74:73–80.

Townshend B, Cotter N, Van Compernolle D, White RL (1987) Pitch perception by cochlear implant subjects. J Acoust Soc Am 82:106–115.

Tyler RS, Gantz BJ, Rubinstein JT, Wilson BS, et al. (2002a) Three-month results with bilateral cochlear implants. Ear Hear 23(suppl):80S–89S.

Tyler RS, Parkinson AJ, Wilson BS, Witt S, Preece JP, Noble W (2002b) Patients utilizing a hearing aid and a cochlear implant: speech perception and localization. Ear Hear 23:98–105.

Vandali AE (2001) Emphasis of short-duration acoustic speech cues for cochlear implant users. J Acoust Soc Am 109:2049–2061.

Vandali AE, Whitford LA, Plant KL, Clark GM (2000) Speech perception as a function of electrical stimulation rate: using the Nucleus 24 cochlear implant system. Ear Hear 21:608–624.

van den Honert C, Stypulkowski PH (1987) Single fiber mapping of spatial excitation patterns in the electrically stimulated auditory nerve. Hear Res 29:195–205.

van den Honert C, Finley CC, Wilson BS (1997) Speech processors for auditory prostheses: development of the evoked potentials laboratory. Ninth Quarterly Progress Report, NIH project N01-DC-5-2103, Neural Prosthesis Program, National Institutes of Health, Bethesda, MD. (This report is available online at http://scientificprograms.nidcd.nih.gov/npp/index.html.)

van Hoesel R, Ramsden R, O'Driscoll M (2002) Sound-direction identification, interaural time delay discrimination, and speech intelligibility advantages in noise for a bilateral cochlear implant user. Ear Hear 23:137–149.

von Ilberg C, Kiefer J, Tillein J, Pfennigdorff T, et al. (1999) Electric-acoustic stimulation of the auditory system. ORL J Otorhinolaryngol Relat Spec 61:334–340.

White MW (1986) Compression systems for hearing aids and cochlear prostheses. J Rehabil Res Dev 23:25–39.

White MW, Merzenich MM, Gardi JN (1984) Multichannel cochlear implants: channel interactions and processor design. Arch Otolaryngol 110:493–501.

Wilson BS (1993) Signal processing. In: Tyler RS, ed. Cochlear Implants: Audiological Foundations. San Diego, CA; Singular Publishing Group, pp. 35–85.

Wilson BS (1997) The future of cochlear implants. Br J Audiol 31:205–225.

Wilson BS (2000a) Cochlear implant technology. In: Niparko JK, Kirk KI, Mellon NK, Robbins AM, Tucci DL, Wilson BS, eds. Cochlear Implants: Principles & Practices. Philadelphia: Lippincott Williams & Wilkins, pp. 109–119.

Wilson BS (2000b) Strategies for representing speech information with cochlear implants. In: Niparko JK, Kirk KI, Mellon NK, Robbins AM, Tucci DL, Wilson BS, eds. Cochlear Implants: Principles & Practices. Philadelphia: Lippincott Williams & Wilkins, pp. 129–170.

Wilson BS (2000c) New directions in implant design. In: Waltzman SB, Cohen N, eds. Cochlear Implants. New York: Thieme Medical and Scientific Publishers, pp. 43–56.

Wilson BS, Finley CC, Farmer JC Jr, Lawson DT, et al. (1988) Comparative studies of speech processing strategies for cochlear implants. Laryngoscope 98:1069–1077.

Wilson BS, Finley CC, Lawson DT, Wolford RD, Eddington DK, Rabinowitz WM (1991) Better speech recognition with cochlear implants. Nature 352:236–238.

Wilson BS, Lawson DT. Finley CC, Wolford RD (1993) Importance of patient and processor variables in determining outcomes with cochlear implants. J Speech Hear Res 36:373–379.

Wilson BS, Finley CC, Zerbi M, Lawson DT (1994) Speech processors for auditory prostheses: temporal representations with cochlear implants—modeling, psychophysical, and electrophysiological studies. Seventh Quarterly Progress Report, NIH project N01-DC-2-2401, Neural Prosthesis Program, National Institutes of Health, Bethesda, MD. (This report is available online at http://scientificprograms.nidcd.nih.gov/npp/index.html.)

Wilson BS, Lawson DT, Zerbi M (1995) Advances in coding strategies for cochlear implants. Adv Otolaryngol Head Neck Surg 9:105–129.

Wilson BS, Finley CC, Lawson DT, Zerbi M (1997a) Temporal representations with cochlear implants. Am J Otol 18:S30–S34.

Wilson BS, Zerbi M, Finley CC, Lawson DT, van den Honert C (1997b) Speech processors for auditory prostheses: relationships between temporal patterns of nerve activity and pitch judgments for cochlear implant patients. Eighth Quarterly Progress Report, NIH project N01-DC-5-2103, Neural Prosthesis Program, National Institutes of Health, Bethesda, MD. (This report is available online at http://scientificprograms.nidcd.nih.gov/npp/index.html.)

Wilson BS, Wolford RD, Lawson DT (2000) Speech processors for auditory prostheses: effects of changes in stimulus rate and envelope cutoff frequency for CIS processors. Sixth Quarterly Progress Report, NIH project N01-DC-8-2105, Neural Prosthesis Program, National Institutes of Health, Bethesda, MD. (This report is available online at http://scientificprograms.nidcd.nih.gov/npp/index.html.)

Wilson BS, Lawson DT, Müller JM, Tyler RS, Kiefer J (2003a) Cochlear implants: some likely next steps. Annu Rev Biomed Eng 5:207–249.

Wilson BS, Wolford RD, Lawson DT, Schatzer R (2003b) Speech processors for auditory prostheses: additional perspectives on speech reception with combined electric and acoustic stimulation. Third Quarterly Progress Report, NIH project N01-DC-2-1002, Neural Prosthesis Program, National Institutes of Health, Bethesda, MD. (This report is available online at http://scientificprograms.nidcd.nih.gov/npp/index.html.)

Wouters J, Vanden Berghe J (2001) Speech recognition in noise for cochlear implantees with a two-microphone monaural adaptive noise reduction system. Ear Hear 22:420–430.

Wouters J, Geurts L, Peeters S, Vanden Berghe J, van Wieringen A (1998) Developments in speech processing for cochlear implants. Acta Otorhinolaryngol Belg 52:129–132.

Zeng F-G (2002) Temporal pitch in electric hearing. Hear Res 174:101–106.

Zeng F-G, Shannon RV (1992) Loudness balance between acoustic and electric stimulation. Hear Res 60:231–235.

Ziese M, Stutzel A, von Specht H, Begall K, et al. (2000) Speech understanding with the CIS and the n-of-m strategy in the MED-EL COMBI 40+ system. ORL J Otorhinolaryngol Relat Spec 62:321–329.

Zimmerling MJ, Hochmair ES (2002) EAP recordings in Ineraid patients—correlations with psychophysical measures and possible implications for patient fitting. Ear Hear 23:81–91.

Zwolan TA, Kileny PR, Ashbaugh C, Telian SA (1996) Patient performance with the Cochlear Corporation "20 + 2" implant: bipolar versus monopolar activation. Am J Otol 17:717–723.

Zwolan T, Kileny PR, Smith S, Mills D, Koch D, Osberger MJ (2001) Adult cochlear implant patient performance with evolving electrode technology. Otol Neurotol 22:844–849.

3
Cochlear Implants: Clinical Applications

John K. Niparko

1. Introduction

Over the past 30 years, hearing care clinicians have increasingly relied on cochlear implants to restore auditory functions in selected patients with advanced sensorineural hearing loss (SNHL). Inventory data from manufacturers for the last 10 years indicate consistent yearly increases of more than 20% in numbers of devices placed. In this same period criteria for implantation in candidates of all ages have been broadened based on refined diagnosis and increased benefits with implantation. Thus the evolution of the state of the art of cochlear implantation has been driven by an interchange between technological advances and clinical trends.

Though the lay media have focused attention on the intrigue in placing and activating a cochlear implant, it is increasingly clear that achieving success entails considerations that extend well beyond the device per se. Indeed, success in this setting is highly individualized and should be considered in the context of preimplant history and baseline function, lifestyle, and the candidate's preferred social milieu. Clinicians can facilitate success by identifying candidates with a high probability of improved speech recognition ability, performing safe and effective implantation, and helping them use the implant to improve their quality of life. Clinical challenges stem not so much from the differences between acoustic and electrical hearing as from the deprivation effects caused by time spent without meaningful auditory input prior to implantation.

This chapter examines the many facets of clinical cochlear implantation. It describes practices related to cochlear implantation, including selection of candidates, techniques of device implantation and activation, and the use of cochlear implants as a communication tool. It provides an orientation to basic clinical constructs, an update on technologies for the clinical assessment of candidacy, and interventions that can facilitate successful communication change with a cochlear implant.

2. Candidacy

A host of factors determine whether a cochlear implant is likely to benefit an individual with SNHL. Candidacy is necessarily considered as a composite of criteria because successful use of the cochlear implant depends on a combination of variables in development, cognition, and psychosocial functioning, as well as auditory experience. Moreover, guidelines for implant candidacy are not categorical because of the variety of communication choices available to hearing-impaired populations. Even criteria related to baseline hearing are difficult to specify because of differences in responses to amplification for a given level of SNHL. There is also disparity in the ability to process sensory inputs, and variability in behavioral and attitudinal attributes affecting the use of residual hearing. Beyond this, candidacy criteria are dynamic as they have evolved as advances in microcircuitry and information processing have translated into new designs of cochlear implants and processing strategies.

As advances have improved implant performance, candidacy criteria have, consequently, reflected changes in the expected levels of benefit. Consensus conferences sponsored by the National Institutes of Health in 1988 and 1995 (NIH Consensus Development Statements 1988, 1995) concluded that for properly selected candidates the benefits of cochlear implantation were clear and that, in fact, given the success that had been observed to those dates, relaxed criteria for implant candidacy should be considered. With clinical experience has come recognition of the factors that help to reduce morbidity and to predict the approximate level of communication benefits. Clinical guidelines for implant candidacy thus represent a composite assessment of an individual's age, hearing history, and unaided and aided audition; the circumstances of support surrounding the candidate (particularly patterns of use of spoken language) likely to influence use of the device; and an awareness of potential benefits and constraints of current implantable technologies.

2.1 Principles of Cochlear Implant Candidacy

Cochlear implants convert acoustic signals into codes that preserve those features critical to speech understanding in normal listeners. Parkins (1985) described four principal requirements for restoring speech comprehension with prosthetic stimulation of the auditory periphery.

2.1.1 A Substrate for Peripheral Auditory Processing

The first of the principal requirements is that there be sufficient initial processing of the acoustic signal. That is, appropriately designed patterns of stimulation are required to effect the transfer of information that is physiologically useful to the auditory system (see Wilson, Chapter 2; Leake and

Rebscher, Chapter 4). Other principal requirements, described below, relate to factors that are patient-specific, and often vary widely. Retained neural populations, central processing capabilities, and the interface between stimulating electrodes and neural populations. These three tenets form the foundation of clinical approaches to implant candidacy.

2.1.2 Preserved Auditory Nerve Fibers with Responsivity

The majority of temporal bone studies of cases of profound SNHL reveal substantial retention of populations of spiral ganglion cells within the canal of Rosenthal (Fig. 3.1). Although neuronal survival varies somewhat with etiology, histologic studies reveal surviving spiral ganglion counts that typically range from 10% to 70% of the normal complement of 35,000 to 40,000 cells, and such studies also show that retained neural elements are well disbursed throughout the cochlea outside of the proximal basal turn (Hinojosa and Marion 1983; Nadol 1984). For many etiologies neuronal survival

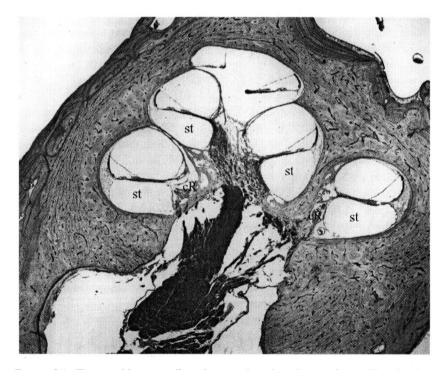

FIGURE 3.1. Temporal bone studies of cases of profound sensorineural hearing loss (SNHL) reveal substantial retention of populations of spiral ganglion cells within the canal of Rosenthal. Current designs of implanted electrode arrays are placed within the scala tympani (st). The scala tympani offers a robust, shielded site to ensure electrode stability. Electrode arrays placed within the scale tympani lie proximate to target cell bodies (and occasionally retained peripheral dendrites) of auditory nerve afferent fibers that lie within the canal of Rosenthal (cR).

varies widely, but it is predictably high in cases of deafness induced by oto-toxicity and low in deafness due to bacterial meningitis. In subjects without clinical SNHL, neuronal loss usually occurs at a rate of 2000 per decade as a result of senescent changes alone (Schuknecht 1989). As expected, then, in profound deafness, older age and longer duration of profound hearing loss are associated with greater reductions in spiral ganglion cell populations (Nadol et al. 1989).

Beyond the requirement for retained populations of auditory neurons, the effect of the size of retained spiral ganglion populations on implant performance is unclear. For instance, clinical surveys have shown no clear relationship between the etiology of deafness and success in speech recognition with a cochlear implant with early designs of multichannel cochlear implants (Gantz et al. 1988a). This suggests that although light microscopy may confirm spiral ganglion cell presence, it provides no evidence of preserved function of auditory afferents. Moreover other neuronal attributes—peripheral processes, myelinization, and terminal axons—are not addressed by temporal bone light microscopy. Although neuronal elements may be present on light microscopy, their electrical responsivity is unknown.

The requisite number of residual, responsive neurons required to effectively encode speech has been estimated in studies that correlate speech audiometry with neuronal reserves. In moderate-to-severe SNHL, histologic correlative studies indicate that approximately one sixth to one third of the normal neuronal complement is required to subserve socially useful speech recognition (Ylikoski and Savolainen 1984). Kerr and Schuknecht (1968) suggested that neurons in the region of the upper basal and second turn (15 to 22 mm from the round window membrane) were most critical in predicting preserved capabilities for speech recognition. Otte et al. (1978) concluded that at least 10,000 spiral ganglion cells—with 3000 or more in the apical 10 mm of the cochlea—were required for discrimination of speech to be preserved in cases of severe SNHL (with residual hair cell populations).

The minimal number of auditory neurons needed to facilitate speech recognition with a cochlear implant is less certain. However, the number is likely to be quite small given observations of speech understanding in cases with only a modest number of residual neurons, with less than 10% of the normal complement of auditory neurons (Linthicum and Galey 1983; Fayad et al. 1991).

2.1.3 Estabished Patterns of Central Neuronal Pathways that Enable Processing Transmitted Signals

Studies of employing morphometry of neurons in animal models (e.g., Ryugo et al. 1997; Niparko 1999) and in humans (e.g., Moore et al. 1997) reveal that neurons within the auditory brainstem undergo measurable change as a consequence of auditory deafferentation (Fig. 3.2). The greatest degree of degenerative changes is associated with deafness that is profound, of early onset, and of long duration wherein changes extend across synapses to affect neuron size. Nonetheless, there is a high degree of nerve survival within the auditory brainstem even under these conditions. The central auditory plasticity may account for observed trends in results with selected implant populations, particularly when a cochlear implant is introduced after substantially prolonged periods of deafness.

Tyler and Summerfield (1996) found evidence suggesting that central auditory factors may affect implant performance even in postlingual adults. They found significant negative correlations between speech perception ability after implantation and the duration of profound/total deafness

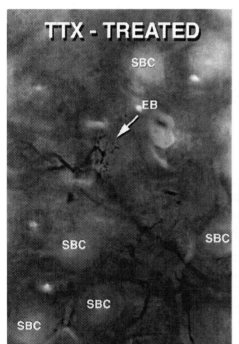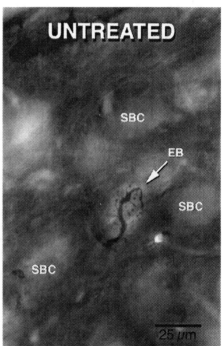

FIGURE 3.2. Animal models reveal that neurons of the auditory tract within the auditory brainstem undergo measurable change as a consequence of auditory deafferentation. Cochlear treatment with tetrodotoxin (TTX) to silence primary auditory afferents can be shown to alter the pattern of their contact with secondary neurons in the anteroventral cochlear nucleus (Niparko 1999). The left panel demonstrates a disrupted end-bulb configuration, as contrasted by the untreated side shown in the right panel. However, short-term deafferentation does not appear to alter secondary neuron size—a phenomenon observed in cases of chronic, progressive deafness. Endbulbs of Held (EB) represent the terminals of primary auditory afferent fibers upon target secondary neurons (spherical-bus cells: SBC) in the anteroventral cochlear nuclues.

before implantation. Outcomes also showed acclimatization in the form of significant improvements in performance over time after implantation. For adult patients, the level of performance measured shortly after implantation was, on average, about half the level measured eventually. Eighty percent of the adult subjects show significant performance improvements with time. On average, performance reached asymptote after 30 to 40 months of implant use, although individual differences in the rate and amount of improvement were large. These authors suggested that the accuracy of speech perception with implants by adults related to preoperative measures in three domains: (1) the number and physiological responsiveness of auditory ganglion cells and nerve fibers, indexed by measures of hearing sensitivity, duration of deafness, and age; (2) the responsiveness of the central nervous system, indexed by measures of cognitive and linguistic ability, and possibly also by age and duration of deafness; and (3) motivation in learning to use the implant.

These observations, paired with clinical results with cochlear implantation, indicate that an established, durable pathway for auditory processing is often present even in profound SNHL. Though there are important negative correlations between duration of deafness and performance (Gantz et al. 1993; Waltzman et al. 1995; Rubinstein et al. 1999), such correlations do not appear to be determinative. Even a prolonged period of deafness does not rule out the prospect for open-set speech understanding with a cochlear implant.

The greatest contrast in performance of open-set speech recognition exists between adults who acquired their deafness postlingually and those who acquired their deafness prelingually. Accordingly, the effect of prelingual deafness on the integrity of the central auditory pathway is a critical area of investigation (see Leake and Rebscher, Chapter 4; Hartmann and Kral, Chapter 6). Clinical evidence of improved performance with earlier implantation (Osberger et al. 2000) and the greater benefit with length of use in children (Miyamoto et al. 1994b; Cheng et al. 1999) suggest that learning effects on implant-mediated speech comprehension are large.

2.1.4 Controllable Interaction Between Applied Stimulus and Response Patterns of Target Neurons

Parkins (1985) postulated that the relationship between the pattern of electrical stimulation and responses of auditory neurons will dictate the effectiveness of information transfer from an implant system. Ultimately, response patterns within the auditory pathway depend on individual auditory nerve fiber excitability. Neuronal shrinkage and demyelination produce elevated thresholds and limit the maximal rate at which auditory nerve fibers can be driven. This change can affect not only the threshold for activation but also a fiber's adaptation (fatigability) and dynamic range of responsivity.

Spiral ganglion cell loss and shrinkage of remaining neurons likely account for reduced dynamic ranges (Kiang and Moxon 1972) of both individual neuronal and psychophysical (behavioral) responses. Whereas acoustic stimulation is associated with behavioral and individual fiber dynamic ranges of 120 dB and 20 to 40 dB, respectively, electrical stimulation is associated with dynamic ranges of behavioral responses of 10 to 40 dB (Zeng et al. 2002) and individual fiber responses of 7 to 10 dB, respectively (Javel et al. 1987; Parkins and Columbo 1987). These relatively constrained dynamic ranges require that compression be introduced into processing schemes. This design feature feature simulates normal cochlear transduction in order to accommodate the wide variance in sound intensities that are normally rendered both detectable and discriminable by the ear. Models of electrical stimulation of the auditory nerve that incorporate these parameters can be used to predict auditory nerve responses to changes in stimulus parameters and may be useful in designing strategies for implant activation (Finley et al. 1990; Javel 1990; Rubinstein et al. 1998).

The need for controllable interactions between the implant and target neurons underscores the importance of assessing surgical suitability. Clinical trials bear out the requirement of a stable interaction between the applied stimulus and target neurons. Substantially reduced benefit has been reported in association with compromised cochlear anatomy in clinical trials of adults (Cohen et al. 1993) and children (Geers et al. 2000). With normal cochlear anatomy the scala tympani offers a robust, shielded site to ensure electrode stability.

Electrode carriers placed within the scala tympani lie proximate to target peripheral dendrites and cell bodies of auditory nerve afferent fibers (Fig. 3.1). The design of the electrode array must be biocompatible and mechanically stable. It must allow for practical fabrication and facilitate atraumatic insertion. From a surgical perspective, the trauma of inserting the device is minimized through both effective design of the array and surgical technique.

Multichannel arrays are typically placed within scala tympani along the first turn and a varying portion of the second turn of the cochlea. The depth of insertion is determined by several variables: surgical technique, obstructing tissue within the cochlea, and electrode design (e.g., surface features and longitudinal stiffness imparted by the carrier and connectors within the core of the implanted array). Electrode design, particularly the longitudinal (curved vs. straight) and cross-sectional shape, and the stiffness of the carrier influence bending and the trajectory of the array tip during insertion. These properties influence the depth of insertion and the position of electrode contacts relative to neuronal fibers housed within the modiolar core of the cochlea. Current electrode carriers are typically inserted over a distance of 20 to 30 mm. Although the potential advantages of insertion beyond 20 mm into the cochlea are as yet uncertain, deep insertion provides, at least theoretically, the opportunity to differentially stimulate according to particularly the low-frequency sites, and, depending on electrode design, may enable wider electrode spacing with the goal of providing distinct pitch perception via individual electrodes.

Electrode positions relative to target neurons may influence electrode selectivity (see Abbas and Miller, Chapter 5). With closer approximation to the medial wall, stimulation of spiral ganglion cell subpopulations may be more selective (Lenarz et al. 2000), potentially reducing activation thresholds and, as a result of lower thresholds and enhanced focus of electrical fields, reduce crosstalk between electrodes (Finley et al. 2001). Recent designs of electrode arrays have attempted to improve approximation of electrode contacts with modiolar neurons by using shim-like positioners or by enhancing the curvature of the array carrier (Jolly et al. 2000). Initial clinical results with positioned electrodes are promising (Zwolan et al. 2001), though complication associated with positioners have prevented their contained use (Reefhuis et al., 2003).

2.2 Clinical Approaches to the Assessment of Implant Candidacy

To ensure complete assessment of candidacy, clinicians should consider the myriad factors likely to affect daily use and performance of the device. The importance of comprehensive assessment is underscored by several factors:

- The cochlear implant is a communication tool and does not cure disease of the cochlear transducers responsible for profound SNHL.
- Expectations, often defined by a patient's (Watson 1991) or family's (Beadle et al. 2000) psychological set, provide a critical framework for postoperative satisfaction with auditory rehabilitation (Ross and Levitt 1997).
- The multifaceted nature of communication disorders often necessitates more than one rehabilitative strategy, particularly in children whose prior communication mode and level vary and whose deficits in auditory processing, speech production, cognitive ability, and attention often need to be addressed.

Candidates should have the psychological makeup, motivation, and motivated support system to learn to use the device consistently to optimize their performance. Speech-recognition performance begins to asymptote only after at least 3 years of use, on average, for children (Cheng et al. 1999) and adults (Tyler and Summerfield 1996). Infants and toddlers demonstrate improvements in speech recognition over much more protracted periods of time (e.g., Miyamoto et al. 1994b; O'Donoghue et al. 2000).

2.3 Multidisciplinary Candidacy Assessment

The varied and interactive factors that potentially affect performance suggest the need for candidacy assessment by a multidisciplinary team that can address the possible challenges to a successful implant experience. Each candidate for cochlear implantation presents with a unique set of baseline capabilities, needs, and objectives. Although advanced levels of SNHL are common to this group, implant candidates differ in virtually every other descriptor. Age, age of onset, etiology, and pattern of progression of deafness; cognitive and educational level; communication mode; language competence; family and environment; sensory and motor skills; and personal motivation influence both candidacy decisions and the outcome of implantation. While medical and audiological criteria can usually be evaluated with objective measures, other important characteristics of an individual's candidacy extend beyond quantifiable baseline measures. A team approach to candidacy assessment optimizes information on which to render a final recommendations. A team approach also offers an efficient means of providing preoperative intervention or counseling, if needed, and an informational base on which to formulate the plan of (re)habilitation after device activation.

Hellman et al. (1991) offer a test protocol that comprehensively considers not only the implant and the implantation process, but also the habilitative services needed to optimize device use after surgery. The Children's Implant Profile (ChIP) encompasses an evaluation of 11 factors that nominally contribute to successful implant use in children. The ChIP is based on components of an earlier developed Diagnostic Early Intervention Program (DEIP) by Brookhouser and Moeller (1986). The ChIP evaluates candidacy on the basis of chronological age, deafness duration, medical or radiological anomalies, secondary handicaps (motoric abnormalities or learning disabilities), functional level of hearing (levels of detection or discrimination), speech and language abilities, family structure and support, expectations of level of benefit, educational environment, availability of support services needed to monitor implant use, and cognitive learning style. If evaluation generates "great concern" on any single attribute, limited implant success is anticipated. Thus any factor evaluated as being of "great concern" suggests that a remedy should be sought prior to implantation.

2.3.1 Audiology

Preliminary consideration of implant candidacy in deafness is based on an individual's baseline hearing and experience with amplification. As a first order of approximation of severity of SNHL, unaided, pure-tone thresholds are measured. Depending on the age of the candidate, pure-tone thresholds may be measured with simple behavioral or visual-reinforce audiometry.

Threshold evaluations can be performed also with auditory evoked potentials. Evoked potential evaluations are useful when supporting measurements are needed or when behavioral, volitional responses cannot be obtained as with infants and toddlers. Otoacoustic emission (OAE) and far-field evoked potential assessments have emerged as the tests of choice in such cases.

Otoacoustic emissions represent released sound from within the cochlea that is recordable within the ear canal (Kemp 1978). Their presence is consistent with normal or near-normal hearing. However, OAEs are not useful for estimating pure-tone thresholds as they are absent when thresholds exceed 35 to 50 dB HL (Lonsbury-Martin et al. 1991). Auditory brainstem responses (ABRs) reflect synchronized neural activity to the onset of the stimuli and can be used to estimate thresholds (Picton et al. 1994). They are most accurate when acoustic-transient stimuli are used, thus limiting their frequency specificity. A more recently developed modality—the steady-state evoked potential (SSEP) response—offers a frequency-specific measurement (Rickards et al. 1994).

Threshold determinations can be used to broadly predict benefit with amplification, though exceptions are not infrequent. An early classification scheme of children with advanced SNHL developed by Osberger et al. (1993) offers a useful conceptual model (Table 3.1). Bronze-category hearing-aid users derive no benefit from acoustic amplification and appear to respond to auditory stimuli on the basis of vibrotactile sensation alone. Silver-category hearing-aid users comprise an intermediate group. Gold-category hearing-aid users obtain substantial benefit from conventional amplification, revealing high levels of open set speech recognition with amplification. However, predicting aided benefit and implant candidacy based on threshold data alone fails to address the commonly encountered "borderline" candidate (e.g., silver-category hearing-aid users).

Measures to evaluate candidacy vis-à-vis functional benefit from amplification have employed an "oddball" paradigm in which the subject's response to an infrequent, random change in a featured stimulus (e.g., changing the voiced consonant from a stream of /ba/ stimuli to /ta/) is measured with far-field potentials, mismatched negativity (Picton 1995), or visual reinforced infant speech discrimination (Eilers et al. 1977). However, the clinical application of these discrimination procedures to children with

TABLE 3.1. Classification of children with sensorineural hearing loss (Osberger et al. 199)

Hearing-aid user class	Unaided thresholds @ 0.5 kHz/1 kHz/2 kHz (HL)
Gold	90 to 100 dB @ 2 of 3 frequencies (mean = 94 dB)
Silver	101 to 110 dB @ 2 of 3 frequencies (mean = 104 dB)
Bronze	>110 dB @ 2 of 3 frequencies (mean >110 dB)

minimal residual hearing has been slow to evolve, and intersubject and intrasubject variability is yet to be characterized.

An alternative approach to obtaining discrimination data in infants and toddlers uses criteria-referenced measures. The Meaningful Auditory Integration Scale (MAIS) consists of 10 questions, answered by caregivers, that assess device (hearing aid or implant) bonding, spontaneous alerting to sound, and the ability to derive meaning from auditory events (Robbins et al. 1991). Osberger et al. (2000) found that MAIS scores of bronze-category hearing-aid users improved dramatically with cochlear implantation to levels that matched or exceeded those demonstrated by gold-category hearing-aid users.

Extreme elevations in thresholds that are consistent with the most profound SNHL and that offer little potential for speech understanding with amplification provide an audiological indication for early implantation of infants and toddlers (Osberger 1997). Audiological criteria have been expanded to include children with some open-set speech recognition ability. Silver-category hearing-aid users (Osberger et al. 1991) and even well-selected gold-category hearing-aid users (Gantz et al. 2000) have been implanted, with initial reports of significant postoperative gains in open-set speech recognition. These children are generally older (often beyond 2 years), and discrimination data are more easily obtained.

Selected pathologies present a unique challenge in the assessment of implant candidacy from an audiologic viewpoint. In 1996 Starr et al. described a distinctive type of hearing deficit termed auditory neuropathy (AN). Patients with presumed AN have findings that suggest the outer hair cell function is normal (as revealed by the presence of OAEs) and that the site of lesion is retrocochlear, with possible inner hair cell or eighth nerve impairment (as revealed by an absent or abnormal ABR). Other findings include absent acoustic stapedial reflexes and word recognition abilities that are disproportionately poorer than would be expected from audiometric thresholds. Published reports suggest that approximately 0.3% of childhood candidates exhibit findings consistent with AN (Rance et al. 1999). Experience with cochlear implantation in children with AN is largely anecdotal at this stage, with published findings varying between observations of high (Trautwein et al. 2000; Shallop et al. 2001) to low (Miyamoto et al. 1999) implant benefit.

For adults, current criteria hold that speech understanding of up to 40% correct using words-in-sentence testing under best-aided conditions constitute audiologic candidacy. For example, mean speech recognition scores using similar measures after implantation far exceed the 40% level, and individuals with some preserved speech-recognition ability preoperatively often score substantially higher than they did in their preoperative condition (Tyler et al. 1989; Waltzman et al. 1995). Residual hearing as reflected in aided speech-recognition levels can serve as an important predictor of implant success (Rubinstein et al. 1999). When combined with duration of deafness, preoperative scores on tests of sentence recognition provide a predictive composite that accounts for 80% of the variance in word recognition achieved with a cochlear implant.

Electrical responsivity of the auditory nerve may be verified with transtympanic stimulation with behavioral responses in adults or averaged, far-field auditory potentials in children (Kileny et al. 1994). In the preoperative assessment of both childhood and adult cochlear implant candidates, transtympanic electrical stimulation may be helpful in verifying responsiveness and choosing the ear to be implanted.

Although the strict prognostic value of preoperative promontory testing is probably limited (Kileny et al. 1991, 1994; Blamey et al. 1992), the test is useful when asymmetries in the appearance of the cochlea are noted on imaging and for adult patients who are reticent about implantation owing to concerns about the responsiveness of their ear to a novel stimulus.

2.3.2 Otologic and Medical Assessment

Medical examination yields essential information about the health status of the candidate by identifying potential health concerns and determining suitability for a general anesthetic. An otologic evaluation, including history and physical examination, is essential to identifying structural changes in the temporal bone that may affect the surgical approach. Clinical evaluation of the vestibular system may help to predict whether implantation will produce vestibular symptoms.

Establishing the precise etiology of deafness can provide useful information in guiding the implantation process, particularly when awareness of etiology is likely to affect strategies of surgical implantation and postoperative rehabilitation. Cochlear ossification often requires a refined surgical plan (Balkany et al. 1988). A history of meningitis should prompt a discussion with the candidate of methods for implanting an ossified cochlea. Note should be taken of prior otologic surgeries because plans for device placement and stabilization can be affected. Cochlear otosclerosis and temporal bone fractures may be more likely to manifest adventitial facial nerve stimulation with activation of the implant (Niparko et al. 1991), thereby necessitating a considered choice of the electrode to be implanted and the processing program to be used later.

Profound SNHL associated with congenital absence of neural foramina (Jackler et al. 1987a), cochlear development, and profound loss due to acoustic tumors are rare disorders in which the etiology often rules out the option of cochlear implantation because of inadequate auditory innervation. For those patients with bilateral acoustic tumors (neurofibromatosis type 2) producing profound SNHL, auditory brainstem implants offer a viable option to restore auditory access (Briggs et al. 1994).

Medical assessment early in the candidacy process is needed to determine whether hearing loss is due to a disorder correctable by medical treatment or an otologic procedure short of cochlear implantation. Such a possibility is most frequent in young children, who may present with a hearing loss that arises from cochlear dysfunction combined with eustachian tube dysfunction. Eustachian tube dysfunction is a common disorder that can manifest suppurative otitis media and a consequent conductive hearing loss. Early management of eustachian tube dysfunction can facilitate accuracy in assessing the level of SNHL and response to amplification.

Luntz et al. (1996) reviewed their experience with 60 children younger than 18 years of age who received implants. While 74% of the children had one or more episodes of acute otitis media before implantation, only 16% had diagnoses of acute otitis media after implantation. Acute otitis media was also decreased after implantation, as observed by House et al. (1985). Reduced infection rates were thought to be due to several factors, including the natural tendency for otitis media to diminish with age, the use of intraoperative and perioperative antibiotics, and the effect of air cell removal as part of the mastoidectomy procedure. None of the children in these studies had inner ear or intracranial complications.

Poor general health status is rarely a contraindication to cochlear implantation. Candidacy for implantation should include assessment of the patient's fitness for a general anesthetic, the necessary mastoid surgery, and the effort required for device programming and rehabilitation. Although implantation under local anesthesia has been described, this approach constrains soft tissue dissection and the drilling of the retrosigmoid tissues to effectively position and stabilize the internal device.

Consideration should be given to conditions for which a patient may need future assessment with magnetic resonance imaging (MRI). Implantation of a magnet in the internal device may be contraindicated in these patients. A nonmagnetic modification of one commercially available device is available for patients whose medical or neurological condition mandates further studies using MRI (Heller et al. 1996).

2.3.3 Radiologic Assessment

High-resolution computed tomography (HRCT) scans of the temporal bone help to define the surgical anatomy and provide information about cochlear abnormalities that can aid the surgeon in surgical planning and counseling the patient (Fig. 3.3). Temporal bone CT scans reveal anatomical details important in treatment planning: the pattern of mastoid pneumatization, the position of vascular structures, middle ear anatomy, and position of the facial nerve (Woolley et al. 1997). Scans should also be examined for evidence of cochlear malformation, cochlear ossification, enlarged vestibular aqueduct, and other inner ear and skull base anomalies.

Although HRCT is the standard for evaluation of most aspects of temporal bone anatomy, it does have limitations, particularly in the assessment

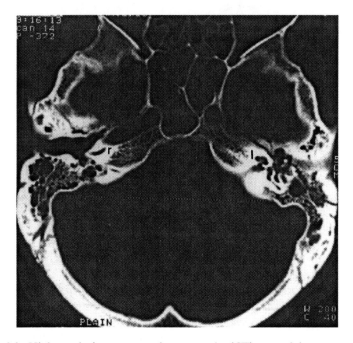

FIGURE 3.3. High-resolution computed tomography (CT) scan of the temporal bone helps to define the bony anatomy of the cochlea as an aid surgical planning. This axial CT scan from a patient with temporal bone fractures demonstrates two levels of cochlear anatomy (due to a slightly skewed plane of sectioning). The CT demonstrates patency of the basal turn on the right (r), and patency of the middle and apical turns on the left (l).

of cochlear patency. Computed tomography findings of cochlear patency generally correlate with surgical findings (Langman and Quigley 1996), but significant discrepancies have been reported (Jackler et al. 1987c; Wiet et al. 1990; Frau et al. 1994). The radiographic appearance of the cochlea should be considered in light of clinical information, particularly a history of meningitis or otosclerosis, which may directly affect the likelihood of complete insertion of the electrode array. Balkany and Dreisbach (1987) list four categories of cochlear patency: C0, normal cochlea; C1, indistinctness of the endosteum of the basal turn; C2, definite narrowing of the basal turn; and C3, bony obliteration of portions of the basal or middle turn or the entire cochlea.

Magnetic resonance imaging may be a useful adjunct to HRCT for assessment of implant candidacy (Harnsberger et al. 1987; Tien et al. 1992; Casselman et al. 1993). Whereas HRCT reveals detailed bony anatomy, MRI offers imaging of soft tissue structures such as the membranous

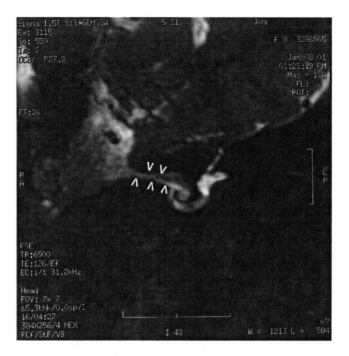

FIGURE 3.4. Whereas conventional computed tomography (CT) scans reveal details of the bony anatomy, magnetic resonance imaging (MRI) offers imaging of soft tissue structures such as the membranous labyrinth and neural substrate. This high-resolution MRI demonstrates the pattern of innervation of the right cochlea (arrowheads) in a case of a mild Mundini (cochlear) malformation.

labyrinth and neural substrate (Fig. 3.4), and MRI techniques make it possible to visualize the presence or absence of fluid within the cochlear turns, thus enabling evaluation of cochlear patency with greater resolution.

Tien et al. (1992) first reported the use of fast spin-echo (FSE) MRI techniques to image the inner ear. Fast spin-echo imaging has an advantage over both conventional spin-echo T1-weighted images, which lack tissue contrast among fluid, neural tissue, otic capsule septa, and surrounding temporal bone, and conventional T2-weighted images, which require considerably longer scanner times. The speed advantage of FSE allows the radiologist to obtain thin-section (2-mm) high-resolution T2-weighted images with excellent contrast in a fraction of the time needed for conventional spin-echo techniques. Images obtained with this technique can delineate the fluid-filled otic capsule and internal auditory canal. Nadol (1997) has reported a strong positive correlation between the diameter of the cochlear and vestibular nerves and the total spiral ganglion cell count, and suggests the theoretical prospects for MRI to be used in quantifying neuron survival.

2.3.4 Auditory Skills

An auditory skills assessment evaluates the child's ability to attend to and integrate sound with use of conventional amplification. A child may demonstrate residual hearing on an audiological evaluation but not have the necessary skills to make use of that hearing. An auditory skills assessment then determines the child's ability to use residual hearing to attend to speech and environmental sounds, integrate auditory perception with speech production, make meaningful associations with sounds ranging from single words to conversational contexts, and integrate hearing in the context of communication. The development of a child's residual hearing before implantation can provide a foundation for listening via the cochlear implant.

2.3.5 Language Assessment

Assessing language in a child seeking a cochlear implant often requires evaluation of the area that is most deficient in that child. A child may have no real language in either the signed or spoken form. Behavioral observation is employed to assess how a candidate communicates with others. In the case of young candidates it is important to establish the level of communicative intent and the prelinguistic and linguistic strategies used by the child to support this intent.

The assessment of prelinguistic communication behaviors includes eye contact and gaze patterns, gesture, pointing, vocalization, object and physical manipulation, turn-taking, imitation, and willingness to maintain engagement (Schopmeyer 2000a). The evaluation of linguistic communication examines receptive and expressive vocabulary in sign or speech, beginning syntax, use of grammatical markers, and narration and conversation.

Pragmatic development refers to the social use of language skills and applies to both linguistic and prelinguistic behaviors. Pragmatic skills include communicating for a variety of purposes—to request, comment, gain attention or information, protest, choose, and demonstrate social conventions such as greetings. Children who display appropriate social use of nonlinguistic and linguistic behaviors before implantation can bring social engagement and communicative intent to the postimplant rehabilitation process. Anomalous patterns of pragmatics may signal pervasive developmental and autistic disorders.

The development of prelinguistic behaviors and language pragmatics can result in the identification of noncognitive (e.g., cerebral palsy and low muscle tone) and cognitive (e.g., autism, pervasive developmental disorder, and Asberger syndrome) handicaps that require a special consideration of implant candidacy and postoperative rehabilitation. Assessments by a developmental pediatrician, a pediatric neurologist, and occupational and physical therapists may offer therapeutic options. Such an assessment screens for sensory and motor dysfunctions beyond those routinely associated with communication deficits (Rapin 1978). For example, appropriate treatment of attention disorders may be required to optimize benefit from audiological and rehabilitative services.

2.3.6 Cognitive and Psychological Assessment

The candidate's skills in nonverbal problem solving, attention, and memory affect the postimplant rehabilitation process and may provide predictive information regarding potential use and benefit. The psychological assessment includes measurement of verbal and nonverbal intelligence, visual-motor integration, attention, motor development, child behavior, and parental stress. The evaluator also counsels candidates or parents about expectations, noting whether there are realistic expectations of what an implant can provide given the particular attributes of the case.

A wide range of psychosocial attitudes should be addressed in evaluating family support of childhood implantation (Quittner et al. 1991). Parental attitudes toward their child's hearing loss and motivations for implantation may prove predictive of their commitment to the rehabilitation process and psychological adjustment to the sometimes slow pace of improvements in speech understanding and production. The preimplantation assessment should ascertain long-range prospects for access to oral-based education.

The process of communication change associated with cochlear implantation can have far-reaching psychological implications. A cochlear implant will most benefit individuals who possess sufficient motivation and surrounding support to encourage a program of postimplantation rehabilitation. Motivations and expectations should be discussed in detail with the candidate. For pediatric candidates, parental expectations and attitudes should be carefully determined. The very best efforts of an implant team can be thwarted by a patient's or family's frustrations based in unrealistic expectations and resulting in consequent disuse. Though it is certainly reasonable for candidates or their family to expect improved hearing, practical hearing gains often require periods of training, and current implant systems have limitations that will always be present. Personality traits that may hamper compliance should be looked for. Though cochlear implantation may be associated with improvement in some mood disorders (Proops et al. 1999), the implant itself cannot be expected to produce such benefits. Psychotic tendencies or a substantial burden of endogenous depression can limit compliance with any systematic plan of rehabilitation. Psychological assessment should also screen for other conditions, such as organic brain disease, that would compromise the implantation and rehabilitation processes.

2.3.7 Ophthalmology

Vision plays a critical role in the deaf child's development. Even after implantation, the child may rely on vision to begin to associate meaning with auditory inputs. The ophthalmological evaluation can identify visual abnormalities associated with congenital deafness including refractive errors, strabismus, adnexal anomalies, and cataracts (Siatkowski et al. 1994). Rubella retinopathy, Usher's syndrome, and Waardenburg's syndrome often manifest anomalies that are detectable on ocular examination for implant candidates of all ages.

2.3.8 Occupational Therapy

Children with cochlear implants need to integrate their new sense of hearing with their more developed sensory systems. As a result of auditory deprivation, some children display subtle motor or sensory delays that may impede the smooth acceptance and integration of new auditory stimulation. An evaluation by an occupational therapist may identify subtle vestibular, tactile, or proprioceptive deficits that could affect the child's ability to integrate auditory information, use language in a social context, engage with others, and gain volitional control over body movement, especially those of oral articulators.

2.3.9 Educational Placement

A key factor in optimizing benefit from the implant is appropriate educational placement and opportunity for flexibility within that placement as the auditory skills of the child evolve and the support requirement changes (Hellman et al. 1991; Koch et al. 1997). An appropriate school environment motivates the use of audition in learning, maximum attention to language development, and opportunities for communication. Essential features of an appropriate school placement include encouragement of spoken language, opportunities to interact verbally with adults and peers, appropriate support services, and school personnel who understand a cochlear implant and are willing to participate in a team approach. A school visit by a rehabilitation therapist from the implant team initiates the important collaboration with a child's teachers and therapists.

3. Surgery

Surgical techniques of cochlear implantation represent modifications of procedures used over the past century in managing chronic infections of the mastoid and middle ear. Surgery is performed in the conventional otological position with the use of aseptic precautions, microscopy (up to 160×), and a rotating drill.

The mastoid cortex and surrounding temporal skull are exposed by raising a "flap" of soft tissue, pedicled anteriorly and inferiorly (Fig. 3.5). Soft tissue flaps should accommodate placement of the implant at a safe

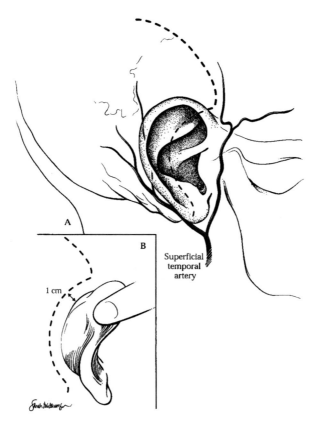

FIGURE 3.5. Artist's rendering of the surgical approach to cochlear implantation of the right ear. A: Blood supply to the soft tissues surrounding the ear. B: Usual location of the incision to expose the mastoid cortex by raising a postauricular skin flap.

margin from overlying incisions and in a location compatible with an ear-level processor.

A simple mastoidectomy is performed with the superior (middle fossa tegmen) and posterior (sigmoid sinus) cortical margins maintained intact (Figs. 3.6 and 3.7). The resulting bony ledge provides protection for connecting leads and a bony plateau for embedding the receiver-stimulator. The bony facial recess (between the corda tympani nerve and vertical segment of the facial nerve) is opened to visualize the incudostapedial joint and cochlear promontory. The facial recess is opened to maximize visualization via thinning of the bony canal wall, and adequate removal of air cells to enable systematic exposure of the horizontal semicircular canal, fossa incudus, and corda-facial angle. As the dimensions of the tympanic

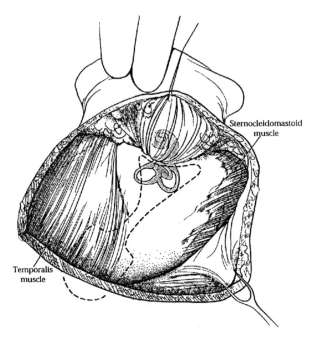

Sternocleidomastoid
muscle

Temporalis
muscle

FIGURE 3.6. With the patient in the surgical position, the mastoid cortex is opened and a bed created to accommodate the receiver-stimulator of the internal device. The area of bony dissection is outlined by the dashed line.

cavity are established at birth and completed by 6 months of age, the risk associated with the facial recess approach is no greater in children than in adults (Bielamowicz et al. 1988). The corda tympani nerve may be sacrificed if the facial recess is small or if the array to be implanted requires a generous exposure (Lalwani et al. 1998). Permanent taste disturbance is rare. The round window niche is opened exposing its tegmen and the round window membrane (Proctor et al. 1986).

The preferred approach to cochleostomy provides access to the scala tympani through the promontory, anterioinferior to the round window (Fig. 3.8). The electrode array is advanced under direct visualization, minimizing trauma to the membranous components of the scala, particularly the basilar membrane. Resistance to array insertion can produce buckling of the carrier-array. Buckling can induce spiral ligament and basilar membrane injury, and consequent neural loss. Full insertion of the array within the basal turn of the cochlea requires an insertion depth of 25 to 30 mm. Before insertion of the electrode array, a well is created behind the mastoid to accommodate the receiver-stimulator portion of the internal device (Fig. 3.6). The receiver-stimulator is stabilized with sutures to the bony cortex.

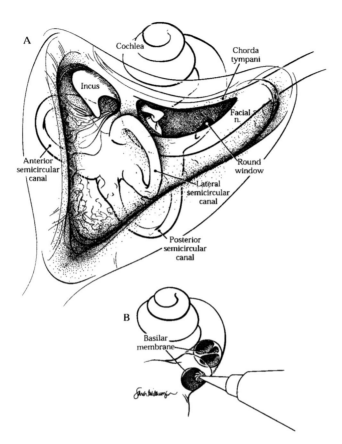

FIGURE 3.7. A: Simple mastoidectomy has been performed. Opening the middle ear from behind via a facial recess approach, anterior to the facial nerve, exposes the round window. B: Enlargement of the round window niche with a rotating bur opens the scala tympani beneath the basilar membrane.

The external processor and antenna are fitted 1 month after implantation. This delay allows time for the incision to heal and edema that may accompany all aspects of the implantation procedure to resolve. The internal device and external processor establish linkage via two antennae, one of which is magnetically retained behind the ear and the second of which is contained within the internal device.

The feasibility of revision implant surgery, including single-to multiple-channel conversion to enhance perception spectral cues, is now well recognized (Miyamoto et al. 1994a; Rubinstein et al. 1997). Minor modifications of the implant procedure when performed in conjunction with labyrinthectomy for vertigo have been described (Zwolan et al. 1993).

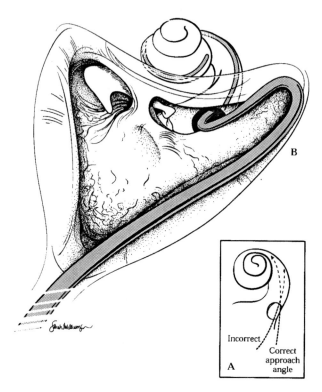

FIGURE 3.8. A: Employing a trajectory of electrode array insertion along the floor of the scala tympani allows for atraumatic insertion. B: Electrode array and connecting lead in proper position.

3.1 Implantation of Young Children

Implantation of infants and toddlers can be achieved with no greater risk of complication than that observed for adults (Hoffman 1997; Parisier et al. 1997; Waltzman and Cohen 1998; Tucci and Niparko 2000). Experience with surgical implantation of young children has been reported by several centers, and initial results are encouraging (Gantz et al. 1993; Lenarz et al. 1997; Waltzman and Cohen 1998). Although temporal bone growth has been shown to continue through adolescence, the anatomy of the facial recess is fully developed at birth (Bielamowicz et al. 1988; Eby and Nadol 1996). The most significant developmental changes are in the size and configuration of the mastoid cavity, which has been shown to expand in width, length, and depth from birth into the teenage years.

3.2 Surgical Implantation of Obstructed and Malformed Cochleas

Profound SNHL may be associated with intracochlear inflammatory reactions (labyrinthitis ossificans). Obstruction of each of the scalae has been associated with inflammation associated with meningitis, Cogan's syndrome, syphilis, chronic otitis media, malignant otitis externa, otosclerosis, and trauma (Jackler et al. 1987c). Labyrinthitis ossificans results in the formation of fibrous tissue, with the possibility of subsequent dystrophic calcification to produce bone within the normally fluid-filled scalae. The scala tympani, especially in the proximal portion of the basal turn, is the most common site of fibrous tissue and new bone growth, although Green et al. (1991) demonstrated that ossification due to meningogenic labyrinthitis extended further into the cochlea than did ossification due to other causes.

Labyrinthitis ossificans was previously considered a contraindication to implantation of long, multielectrode arrays (as discussed by Balkany et al. 1988, 1996). In most cases, however, ossification involves only the proximal (basal) portion of the cochlea, and total ossification of the cochlea is unusual (Green et al. 1991). Balkany et al. (1988) found that 14% of patients had cochlear ossification at the time of implant surgery. However, electrode insertion was complete in 14 of 15 patients as bony growth was confined to the most basal portion of the cochlea. In cases of otosclerosis Fayad et al. (1990) found some ossification of scala tympani that required drilling in 30%, but the extent of ossification did not exceed 5 mm. Performance with the implant was found to be similar to that of patients without ossification.

Preservation of spiral ganglion cells has been shown to be poor in patients with labyrinthitis ossificans, particularly if it is a consequence of bacterial meningitis (Nadol 1997). In general, the greater the degree of ossification, the lower the spiral ganglion cell counts. However, even in cases of severe bony occlusion, significant numbers of neurons remain. Because patients with as few as 10% of the normal complement of SGCs are known to demonstrate at least average performance with the implant (Linthicum et al. 1991), implantation is not necessarily contraindicated in patients with extensive ossification.

Surgical techniques used for implantation of the ossified cochlea utilize the usual transmastoid-facial recess approach (Balkany et al. 1988). Complete drill-out of the cochlea has been described by Gantz et al. (1988b) as an alternative option in implanting the obstructed cochlea. Limited experience with this approach indicates that some patients are capable of achieving performance levels that match those of patients implanted without bony obstruction. However, implant performance after cochlear drill-out has not been assessed in a large number of patients.

Congenital deafness is frequently associated with labyrinthine malformations, typically cochlear hypoplasia or a common cavity (Jackler et al.

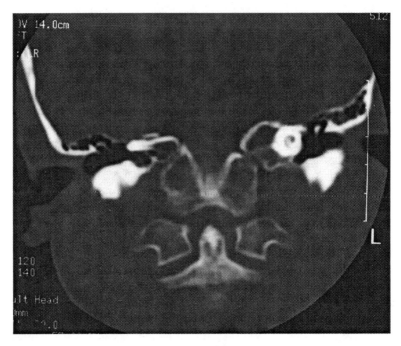

FIGURE 3.9. A significant percentage of children with deafness of congenital onset exhibit cochlear malformation, typically cochlear hypoplasia or a common cavity. This coronal CT demonstrates a mild Mundini malformation on the left, and complete aplasia of the labyrinth and cochlear absence on the right.

1987a) (Fig. 3.9). Hypoplastic cochleas are associated with poor definition of cochlear turns and partitions between the modiolus and internal auditory canal (IAC) and relatively low spiral ganglion cell populations (Schmidt 1985). When associated with a narrow IAC on preoperative CT scanning, innervation may be completely absent (Jackler et al. 1987b), as supported by electrical evidence of lack of excitability with preoperative testing with a transtympanic electrode (Tucci et al. 1995). Implantation of an ear with a narrow IAC is thus contraindicated unless electrical testing shows responsiveness.

  Mondini deformity (hypoplasia of the cochlea) is associated with shorter length with relatively low spiral ganglion cell populations and often has a thin bony partition between the modiolus and a widened IAC (Jackler et al. 1987a)—an anomaly that accounts for the cerebrospinal fluid leak that can occur during a cochleostomy for electrode insertion. Other anomalies involving the round window niche and facial nerve are observed with cochlear malformations (House and Luxford 1993; Tucci et al. 1995). Despite associated anomalies, implantation of a sufficient number of electrodes within a malformed cochlea is feasible and can provide a surprising degree of speech perception in many subjects (Tucci et al. 1995).

3.3 Surgical Results

Cochlear implantation entails risks inherent in extended mastoid surgery and those associated with the implanted device. Hoffman and Cohen (1995) characterized implant-related complications as major if they required revision surgery, and as minor if they resolved with minimal or no treatment, and reported 55 major (12%) and 32 minor (7%) complications. Hoffman and Cohen (1993) noted in later follow-up 220 (8%) major and 119 (4.3%) minor complications among 2,751 implantations. Though direct comparisons of complication rates between reports fail to include information on duration of device use, and observations vary in length and frequency of follow-up, there is a suggested substantial reduction in the incidence of major complications over the past 15 years.

Major complications that are strictly device-related involve partial or complete device failure. As the materials used to fabricate the internal device are expected to last more than 100 years, use-related failure per se is not expected; failure is attributed to either manufacturing flaws or trauma. Electrode array compression can occur with aggressive array insertion and may simply reduce the number of functional channels, and produce noxious sensations with stimulation and even complete device malfunction. The connecting lead between the receiver-stimulator and the electrode array is vulnerable to shearing if the device is not properly secured. Accordingly, embedding the device in bone and fixation with suture material is strongly advised. Device reliability is better than that of cardiac pacemakers, likely owing to design differences and the cochlear implant's position in a relatively immobile site. Device failures requiring replacement with a new device do not appear to lead to a compromise of the level of performance obtained with the second implant (Rubinstein et al. 1997).

Major complications also include facial nerve paralysis and implant exposure due to flap loss. Facial nerve injury is uncommon and unlikely to produce permanent, complete paralysis when appropriately managed. Loss of flap viability can lead to infection and device extrusion, necessitating coverage with a scalp flap or possibly device removal.

Approximately 70% of profoundly hearing-impaired individuals exhibit reduced vestibular function as indicated by tests of the vestibulocular reflex (Huygen et al. 1994). For an ear with residual vestibular function, implantation ablates vestibular function in about 50%, yielding potential postimplantation disequilibrium and vertigo. During the first week after surgery, patients often experience some degree of disequilibrium and unsteadiness. True vertigo that persists in the postoperative phase, however, is rare (Dobie et al. 1995). Extended periods of disequilibrium are also rare but when present are treatable with exercise regimens designed to elicit central compensation.

The risk of bacteria entering the cochlear scalae after implantation, producing labyrinthitis or meningitis and associated reactive fibrosis and destruction of neural elements, appears to be low. However, reports of meningitis months to years after cochlear implantation have raised vigilance for this potential complication (O'Donoghue et al. 2002). Franz et al. (1987) inoculated the bullae of implanted cats with streptococci to evaluate the risk of infection spreading from the middle ear into the implanted cochlea. Despite uniform inflammation of the round window niche, cochlear inflammation was absent. A protective seal had formed around the electrode as it entered either the round window or the cochleostomy, suggesting a protective role in preventing implant-surface colonization with bacteria (biofilm) (An and Friedman 1998) and further infection.

Users and physicians should be aware that most implants are not compatible with MRI. Although selected devices can be modified to enable MRI, such modifications require breaking the implant's hermetic seal.

3.4 Histological Assessment of Implantation of the Cochlea

Potential histopathological injury to the cochlear scalae can be classified as surgical or device-related (see Leake and Rebscher, Chapter 4). The histological results of cochlear implantation have been well studied (Kennedy 1987; Fayad et al. 1991; Zappia et al. 1991). Surgical trauma may induce cochlear fibrosis, neossification, and injury to the membranous cochlea including the loss of sensorineural elements and injury to the stria vascularis. In general, these changes appear limited, and though neuronal loss may ensue, loss appears to be focal. There is no histological evidence of reaction to extended electrical stimulation.

4. Outcomes

4.1 Factors Related to the Measurement of Auditory Performance

Measurement variables should be standardized as much as possible. Speech perception as reflected by speech-reading capability evaluates the effects of combining implant-derived percepts with visual cues based on lip and facial movement. Clinicians can choose between closed-set tests (e.g., forced choice of one answer from a list of four) and open-set tests of words and sentences. Closed-set tests and sentence tests (as scored by words correct) typically produce substantially higher percentages correct than do open-set tests and tests of single words.

Capabilities for understanding speech in noise can be assessed by altering the signal-to-noise ratio (i.e., intensity differences) of taped sentence material relative to background noise. Normal-hearing individuals can comprehend speech words when this ratio approaches zero, and sentences often down to a ratio of −5 dB. Implant recipients typically show degraded speech recognition when signal-to-noise ratios are lowered beyond +10 dB (e.g., Zeng and Galvin 1999).

Trends toward higher rates of open-set speech recognition with newer implant technology and longer implant experience have prompted more stringent assessments of receptive capability. Increasing the difficulty of a speech perception test has the effect of limiting the "ceiling effects" that result from testing with simple, everyday phrases. For the purpose of generating more meaningful comparative data, increasing test difficulty also tends to normalize distributions across populations, thereby permitting the use of more powerful statistical designs in searching for differences between groups.

4.2 Results in Adults

One goal of cochlear implantation is to improve speech perception. Whereas initial clinical series judged implant efficacy mostly on environmental sound perception and performance on closed-set tests, greater emphasis is now placed on measures of open-set speech comprehension. Speech-perception results from early clinical trials have served to guide the evolution of cochlear implantation.

Gantz et al. (1988a) provide early comparative data in their assessment of environmental sound and speech perception in a large cohort of non-randomized, single- and multiple-channel implant users. Multiple-channel implants provided significantly higher levels of performance on all measures. Cohen et al. (1993) performed the first prospective, randomized trial of cochlear implants through the U.S. Veterans Administration hospital system. The trial provided high-level, quality clinical data and convincing evidence of open-set speech recognition and the clear superiority of multiple- over single-channel designs. Moreover, the study was the first to report a low complication rate and high device reliability in postlingually deafened individuals, despite the fact that implantations were performed and monitored across multiple centers and with varying levels of prior implant experience.

Clinical observations in patients with current processors indicate that for patients with implant experience beyond 6 months, the mean score on open-set word testing approximates 25% to 40%, with a range of 0% to 100% (Cohen et al. 1993; Gantz et al. 1993; Waltzman et al. 1995; Rubinstein et al. 1999). Results achieved with the most recently developed speech-processing strategies reveal mean scores above 75% correct on words-in-sentence testing, again with a range of 0% to 100%. Though subjects perform substantially less well on single-word testing, these mean scores continue to improve as the speech-processing strategy evolves (Skinner et al. 1999). After implantation, speech recognition by telephone (Cohen 1997) and music appreciation are often observed. Again, this benefit seems to be achieved in larger percentages of patients with the use of more recently developed processing strategies (see Wilson, Chapter 2; Shannon et al., Chapter 8).

The high prevalence of SNHL among the elderly has prompted evaluations of the benefit of cochlear implantation in this age group (Horn et al. 1991; Facer et al. 1995; Kelsall et al. 1995; Francis et al. 2002). For recipients of the cochlear implant after the age of 65 years, open-set speech recognition scores are not as high as those reported in younger cohorts, potentially representing an effect of longer duration of deafness as opposed to age itself. Nonetheless, implant usage is high among elderly recipients, with nonuse observed in few patients.

4.3 Predictors of Benefit in Adults

Evaluation of the benefit of cochlear implantation in adults has largely focused on measuring gains in speech perception. The individual factors commonly employed to predict speech performance are categorized as subject variables and device variables (e.g., Gantz et al. 1988a, 1993; Kileny et al. 1991; Cohen et al. 1993; Miyamoto et al. 1994a; Waltzman et al. 1995; Rubinstein et al. 1999). The subject variables include age of onset, age of implantation, deafness duration, etiology, preoperative hearing, survival and location of spiral ganglion cells, patency of the scala tympani, cognitive skills, personality, visual attention, motivation, engagement, communication mode, and auditory memory. The device variables include processor, implant, electrode geometry, electrode number, duration and pattern of implant use, and the strategy employed by the speech-processing unit.

Generally, duration of implant use and duration of deafness account for a high degree of variance on speech-perception measures. Preoperative hearing, particularly with respect to speech recognition, age of implantation, cochlear patency, subject engagement with the therapeutic regimen, and processor type also carry relatively higher predictive values for speech understanding.

4.4 Testing Implant Performance in Children

Implant performance in children is assessed with a battery of audiological tests that can address a wide range of perceptual skills before and after implantation. Although substantial auditory gains are apparent in implanted children, the range of quantifiable improvement varies widely between children and depends heavily on the duration of use of the device, as well as preoperative variables. For this reason testing should survey a range of levels of speech recognition—from simple awareness of sound, to pattern perception (discrimination of time and stress differences of utterance), to closed-set (multiple choice) and open-set speech recognition (Tucci et al. 1995; Kirk et al. 1997; Kirk 2000).

4.5 Speech Comprehension Results in Children with Cochlear Implants

The era of pediatric cochlear implantation began with 3M/House, single-channel implants in 1980. Investigational trials with multiple-channel cochlear implants began with older children and adolescents (aged 10 through 17 years) in 1985 and with younger children (aged 2 through 9 years) in November 1986. Implantation of infants and toddlers younger than 2 years of age began in 1995 (NIH Consensus Conference 1995). Although clinical experience with cochlear implantation is considerably shorter in children than in adults, a larger body of clinical reports is now available.

To provide a systematic review of the relevant literature on speech recognition in children with cochlear implants, Cheng et al. (1999) performed a meta-analysis that surveyed peer-reviewed published reports. Of 1916 reports on cochlear implants published since 1966, 44 provided sufficient patient data to compare speech recognition results between published (n = 1904 children) and unpublished (n = 261) trials. Pooling results of these studies was hampered by the diversity of tests required to address the full spectrum of speech reception in implanted children. To address this impediment, an expanded format of the Speech Perception Categories (Geers and Moog 1987) was designed to integrate results across studies. Although this conversion introduces statistical constraints, study results could be compared and the impact of selected variables (e.g., age at implantation, duration of use, etiology, and age at onset of deafness) could be determined in larger populations. The main conclusions of this meta-analysis were that earlier implantation is associated with a greater trajectory of gain in speech recognition, differences in performance diminish in time between congenital and acquired etiologies, and there is a distinct *absence* of plateauing of speech-recognition benefits over time. More than 75% of the children with cochlear implants reported in peer-reviewed publications have achieved substantial open-set speech recognition after 3 years of implant use. Comparisons of published and unpublished data failed to show a publication bias (Fig. 3.10).

4.6 Language Acquisition in Children with Cochlear Implants

Preverbal communication behaviors underpin verbal language learning. Substantial gains in prelinguistic behaviors, including eye contact and turn-

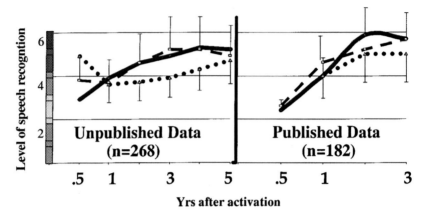

FIGURE 3.10. Data from a meta-analytic synthesis of literature on cochlear implants in children (Cheng et al. 1999) indicating level of speech recognition achieved over time (see text for information on speech recognition level). Earlier implantation is associated with a greater trajectory of gain in the highest level of (open-set) speech recognition. More than 75% of the children with cochlear implants achieve substantial open-set speech recognition after 3 years of implant use. Dotted line: implantation after 6 years of age; dashed line: implantation between 4 and 6 years of age; solid line: implantation under 4 years of age.

taking (Tait and Lutman 1994), and in verbal spontaneity (Schopmeyer et al. 2000a) are observed as early correlates of benefit, developing within 6 months of implantation in young children. Moreover, Tait et al. (2001) found that preverbal measures obtained 12 months after implantation are predictive of late performance on speech perception tasks. They also observed a significant association between the preverbal measure of "autonomy" obtained before implantation and later speech-perception performance. This latter finding has important theoretical implications for understanding language development and suggests that intervention that promotes autonomy in adult–child interaction may lead to improved outcomes. Such intervention can be introduced as soon as deafness is discovered.

Although challenging to characterize, the effects of cochlear implantation on receptive language skills and language production after implantation have been quantified. One approach is to assess language performance on standardized tests. The Reynell Developmental Language Scale evaluates both receptive and expressive skills independently (Robbins et al. 1997). These scales have been normalized on the basis of performance levels of hearing children over an age range of 1 to 8 years and have been used in populations of deaf children. Whereas deaf children without cochlear implants achieved language competence at half the rate of normal-hearing peers, implanted subjects exhibited language-learning rates that early accel-

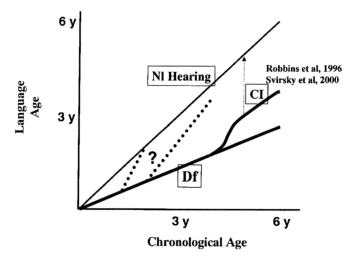

FIGURE 3.11. A schematic summary of language learning curves of hearing and deaf children. Whereas deaf children (DC) without cochlear implants (CI) achieved language) competence at half the rate of normal-hearing peers, implanted subjects exhibited language-learning rates that approached those of their normal-hearing peers (Robbins et al. 1997; Svirsky et al. 2000). Though improved rates of language learning was achieved with implantation, a gap in language level between implanted children and their hearing peers persists (dashed arrow)—apparently due to delays in language acquisition noted prior to implantation. The average age of implantation in this cohort was just under 4 years. It remains to be seen whether earlier implantation might limit early language delays to enable more age-appropriate language acquisition (dotted lines).

erate to approach those of their normal-hearing peers (Robbins et al. 1997; Svirsky et al. 2000). Though an improved rate of language learning was achieved with implantation, a gap in language level between implanted children and their hearing peers persists—apparently due to delays in language acquisition noted prior to implantation. The average age of implantation in this cohort was approximately 4 years. It remains to be seen whether earlier implantation might limit early language delays to enable more age-appropriate language acquisition (Fig. 3.11).

In evaluating the role of communication methodology, Robbins et al. (1997) noted that implantation improved language-learning rates for children in both oral and total communication settings based on the Reynall Developmental Language Scale. Geers et al. (2000), also assessing language skills in implanted children enrolled in oral and total communication (TC) settings, found that such groups did not differ in language level, though the oral group demonstrated significantly better intelligibility in their speech production.

Delays in receptive and expressive language among hearing-impaired children are well documented, although such delays are not strongly related to degree of hearing loss (e.g., Levitt et al. 1987; Osberger 1986). Performance on language measures can be influenced by the child's mode of communication such that results may not directly reflect the influence of auditory capacity and prosthetic intervention. Clinical findings, however, support the notion that some deaf children are, in fact, able to utilize the acoustic-phonetic cues provided by the implant in ways that way reduce the language gap between normal-hearing and deaf children. The ability to use these cues is closely related to cognitive processing (for details, see Pisoni and Cleary, Chapter 9).

4.7 Educational Placement and Support of Implanted Children

The hearing-impaired child is at substantial risk for educational under-achievement (Trybus and Karchmer 1977; Holt 1993). Educational achievement by the hearing-impaired child can be enhanced by verbal communication, and traditionally teaching children to talk has been more successful with children who have residual hearing adequate to perceive speech signals with amplification (Geers and Moog 1991). Improved speech perception and production provided by cochlear implants appear to offer the possibility of increased access to oral-based education and enhanced educational independence.

Koch et al.(1997) and Francis et al. (1999) tracked the educational progress of implanted children by using an educational resource matrix to map educational and rehabilitative resource utilization. The matrix was developed on the basis of observations that changes in classroom settings (e.g., into a mainstream classroom) are often compensated by an initial increase in interpreter and speech-language therapy. Follow-up of school-aged children with implants indicated that relative to age-matched hearing-aid users with similar levels of baseline hearing, implanted students are mainstreamed at a substantially higher rate, but this effect is not immediate and appears to require added rehabilitative support to be achieved. Within 5 years after implantation, the rate of full-time assignment to a mainstream classroom increases from 12% to 75%.

The educational resource matrix also offers a basis for assessing overall cost-benefit ratios of the cochlear implant in children in the United States. Although educational costs for all implanted students remained static or actually increased initially, ultimate achievement of educational independence for the majority of implanted children produced net savings that ranged from $30,000 to $100,000 per child, including the costs associated with initial cochlear implantation and postoperative rehabilitation.

5. Quality of Life and Cost-Effectiveness Assessment

As a relatively expensive health care intervention, the practice of cochlear implantation has been subjected to cost-effectiveness considerations. Studies of the cost-effectiveness of cochlear implants have assessed quality of life and health status to determine the utility gained from the multi-channel cochlear implant (Summerfield and Marshall 1995; Wyatt et al. 1996; Cheng et al. 2000). *Utility* is a concept basic to commerce that reflects the true value of a good or service. Cost-utility methods determine the ratio of monetary expenditure to change in utility as defined by a change in quality of life over a given period. The assessment of cost-utility was based on the following:

$$\text{Cost-Utility} = \text{Costs(in \$)} / \Delta(\text{Quality-Adjusted Life-Years})$$
$$= \text{Costs(in \$)} / \Delta(\text{Life-Years} \times \text{Health Utility})$$

Life-years is the mean anticipated number of years of implant experience based on the life expectancies of the implanted cohort. The change in health utility reflects the preimplant and postimplant difference in scores on survey instruments designed and validated to accurately reflect quality of life.

In the United States, England, and Canada, health interventions with a cost-utility ratio under $20,000 are generally considered to represent acceptable vale for money expended, i.e., they are "cost-effective" (Summerfield et al. 1997; Azimi and Welch 1998).

5.1 Quality of Life and Cost-Effectiveness Assessment in Adults

Costs per quality-adjusted life-year (QALY) for the cochlear implant in adult users were determined with use of cost data that account for the pre-, post-, and operative phases of cochlear implantation (Summerfield and Marshall 1995; Wyatt et al. 1996). Benefits were determined on the basis of functional status and quality of life. Results indicated that multiple-channel cochlear implants in adults are associated with cost-utility ratios in the range of $14,000 to $18,000/QALY, indicating a favorable position in terms of their cost-effectiveness relative to other medical and surgical interventions.

Hearing impairment is one of the most common clinical conditions affecting elderly people in the United States (Campbell et al. 1999). Hearing loss is so profound in 10% of the aged hearing-impaired population that little or no benefit is gained with conventional amplification (Havlik 1986). Assessing the effectiveness of cochlear implants in the elderly requires consideration of general health status and psychosocial factors, as well as impact on hearing per se. The social isolation associated with acquired hearing loss in the elderly (Weinstein and Ventry 1982) is accompanied by a significant decline in quality of life and an increase in emotional handicap (Mulrow et al. 1990). However, well-recognized age-related degeneration of the spiral ganglion (Otte et al. 1978; Ng et al. 2000) and progressive central auditory dysfunction (Welch et al. 1985; Stach et al. 1990) raise concerns about the efficacy of cochlear prostheses in the elderly. Though comparable gains in speech understanding have been reported for elderly implant recipients (Kelsall et al. 1995), the broader implications of these functional gains have only recently been characterized. The determination of both auditory efficacy and quality of life impact may help to guide clinical and resource decisions, particularly in light of the high costs associated with cochlear implantation as an unforced health intervention.

Reports of quality of life gains in elderly patients with cochlear implants have been favorable (Horn et al. 1991; Facer et al. 1995), but are based on questionnaires that are difficult to correlate with function and cost-utility. Chee et al. (2002) evaluated patients between the ages of 50 and 80 years who had cochlear implants with the Ontario Health Utilities Index Mark 3 (HUI 3) survey of health utility. There was a significant mean gain in health utility of 0.24 (SD 0.33) associated with cochlear implantation ($p < 0.0001$). Improvements in hearing and emotional health attributes were primarily responsible for this increase in the health-related quality of life measure. There was a significant increase in speech-perception scores at 6 months after surgery and a strong correlation between the magnitude of health utility gains and the postoperative enhancement of speech perception. Speech-perception gains were also correlated with improvements in emotional status and use of the implant. The authors concluded that cochlear implantation has a significant impact on the quality of life of older deaf patients and represents a cost-effective intervention in this population.

5.2 Quality of Life and Cost-Effectiveness Assessment in Children

Published cost-utility analyses of the cochlear implant in children have been limited by using either health utilities obtained from adult patients (Summerfield et al. 1997; O'Neill et al. 2000) or hypothetically estimated utilities of a deaf child (Lea 1991; Hutton et al. 1995; Lea and Hailey 1995; Carter and Hailey 1999). These studies yielded cost-utility ratios that fell out over a wide range ($3141 to $25,450/QALY). Utilities derived from adult patient surveys may not capture the impact of issues unique to childhood deafness (Cheng and Niparko 1999).

Cheng et al. (2000) surveyed parents of a cohort of 78 children with cochlear implants to determine direct and total cost to society per QALY. Parents of profoundly deal candidate children ($n = 48$) awaiting cochlear implantation served as a comparison group to assess the validity of recall. Parents rated their child's health state "now," "immediately before," and "one year before" the cochlear implant using the Time Trade-Off (TTO), Visual Analog Scale (VAS), and (HUI #3). Mean VAS scores increased from 0.59 to 0.86, TTO scores from 0.75 to 0.97, and HUI scores increased from 0.25 to 0.64. Discounted direct medical costs were $60,228, yielding cost-utility ratios of $9029/QALY using the TTO, $7500/QALY using the VAS, and $5197/QALY using the HUI. Including indirect costs, such as reduced educational expenses, the cochlear implant yielded a net savings of $53,198 per child.

When language- and education-related outcomes in children with cochlear implants are combined with parental perspectives of quality of life effects, overall cost-utility ratings can be generated (Cheng et al. 2000; O'Neill et al. 2000). These independent studies from different continents support the view that pediatric cochlear implantation, relative to other medical and surgical interventions, is highly cost-effective in profoundly hearing-impaired young children.

6. Auditory Training After Cochlear Implantation

A central clinical question relates to the potential benefit to be derived from dedicated sensory training, commonly referred to as "auditory training." Can such training enable a cochlear implant user to apply both latent skills and adaptive strategies to listening tasks that (1) offset at least partially the limitations of cochlear implant listening and prior deprivation and (2) capitalize on the device capabilities?

Much of the question of the potential benefit of auditory training centers on the ability to promote the acquisition of perceptual and productive skills to yield effective spoken language. In the case of an adult with little or no auditory experience in childhood, the "critical period" (Lennenberg 1967) of auditory language learning is past. In such cases, adaptive strategies using other sensory modalities may be emphasized. In contrast, early implanted deaf children may utilize adaptive capabilities to process sound information. In such cases, employing strategies that augment a child's ability to fully appreciate the stimuli that normally subserve auditory development would seem on empiric grounds to offer benefit.

Classic descriptions of auditory training strategy (Carhart 1947) stress the importance of training skills in auditory discrimination. Discrimination is based on acoustic characteristics that furnish critical cues in speech perception–cues based on pitch, loudness, overtones, and the way in which these parameters change from one instant to another. Auditory discriminations required in everyday life vary in the ease with which distinctions can be made, ranging from gross discriminations (e.g., between environmental and voiced sounds), to difficult speech discriminations that demand that every phonetic element be heard with precision (e.g., discrimination in noise or discriminations of personal names or technical terms.)

6.1 Auditory Training in Children with Cochlear Implants

Cochlear implants provide profoundly deaf children with a level of hearing sensitivity that enables them to perceive most of the sounds of spoken language, even when presented at low levels of intensity. Rehabilitative practices are designed to build on this level of sound awareness in practical increments (Mecklenberg 1990). In children who develop an auditorally based language system prior to deafness, the cochlear implant restores access to spoken language. However, in the case of children who have never heard or who lost their hearing before the establishment of spoken language, access to sound provided by the cochlear implant initially lacks meaningful associations. From an auditory training perspective, postlingually deafened children are often characterized as more like the postlingually deafened adult than like the prelingually deaf child. Rehabilitation with the postlingually deafened child emphasizes communication strategies and mapping new percepts onto an existing linguistic code. In contrast, the prelingually deaf child with a cochlear implant must use information from the implant to develop such a code de novo, with early auditory training necessarily emphasizing early constructs of spoken language.

Age of onset and degree of hearing loss, as well as amplification history (Osberger et al. 1991), will determine the quantity and quality of auditory stimulation received before implantation and, therefore, readiness to interpret acoustic information. A child experienced from an early age in using residual hearing through amplification is likely to adapt to the sounds provided by the implant more quickly than a child with limited experience in listening (Boothroyd et al. 1991). Moreover, the close scrutiny afforded by regular interaction with an auditor training therapist allows for the early detection of either the need for device reprogramming or failure of components of the implant system.

Components of auditory training in children with cochlear implants are designed to foster the development of auditory and speech skills in a manner that simulates patterns of acquisition of spoken language without SNHL, while addressing remedial needs that arise as a consequence of auditory deprivation (Robbins 2000). It follows that rehabilitative strategies related to auditory, speech, and language skills should be combined to achieve overall communicative competence. Ideally, rehabilitative strategies do not teach isolated auditory, speech, or language subskills. Subskills are the means to an end that is multimodal communicative competence; training in any one individual subskill should not be overweighted. Intuitively, programs are most effective when they take an integrated approach to rehabilitation and avoid rote, drill-oriented approaches.

Programs and methodologies that focus on the development of spoken language are generally felt to be indicated after cochlear implantation (Hellman et al. 1991). These include the oral-aural approach, which emphasizes lip-reading and speech; auditory-verbal, which emphasizes use of residual hearing through the elimination of visual cues; cued speech, which includes a system of hand shapes to clarify visually ambiguous or invisible mouth shapes of speech; and total communication, which is based on the simultaneous presentation of spoken language with signs. Total communication programs that fail to emphasize auditory input may exert an inhibitory effect on a child's use of capabilities gained from residual hearing or implant hearing to develop spoken language (Ross and Calvert 1984; Geers et al. 2000; Osberger et al. 2000).

The anticipated mode of communication to be used in the educational setting affects the expected benefit in developing spoken language skills. Programs relying exclusively on American Sign Language (ASL) are not considered to be appropriate placements for children after cochlear implantation (Hellman et al. 1991), as ASL is an exclusively manual system often referred to as the "native language of the deaf"; its idiotypic foundation bears few correlates to spoken English. Though considered a language in the full linguistic sense of the word (e.g., Goldin-Meadow and Mylander 1984), ASL does not have a spoken or written correlate. Anecdotal evidence suggests that ASL can be a tool for helping children to develop language skills before embarking on the cochlear implant process, and it can complement the rehabilitation process by providing a bridge to the use of spoken language, provided that the use of implant audition is not compromised by behavior and communication patterns that de-emphasize its importance.

Children born with profound hearing loss demonstrate characteristic difficulties with voice quality and speech intelligibility as a result of the inability to monitor their speech production auditorally (Carney et al. 1993; Tobey et al. 1994). These difficulties are rooted in abnormal use of both voice and articulators (tongue, teeth, and lips) and speech rhythm. It appears that with appropriate intervention and consistent expectation, children with cochlear implants can learn to monitor speech production to improve intelligibility (Tye-Murray 1994; Geers et al. 2000).

6.2 Rehabilitation of Adults

After implant activation, adult users must adjust to the novel signals transmitted by the speech processor. Recipients must learn to associate electrically elicited sound patterns with perceptions that were previously meaningful sound sensations. Rehabilitative needs differ in implant recipients depending on their auditory experience before the onset of deafness. For selected prelinguistically deafened adults, for example, auditory and speech (articulation) training may facilitate communication change. For the postlinguistically deafened patient, auditory training often focuses on more complex listening skills—understanding speech in noise, telephone use, and music appreciation (Edgerton 1985).

6.3 The Efficacy of Auditory Training in Cochlear Implant Rehabilitation

Several studies have systematically assessed sensory and perceptual skill development with training in implant listeners. Watson (1991) noted that representation of spectral-temporal features can become more salient through selective training. Dawson and Clark (1997) tested whether the ability to use place-coded vowel formant information would be subject to training effects in a group of congenitally deafened patients with limited speech perception ability. They found that significant gains in vowel perception occurred posttraining, suggesting the need for children to continue to have aural rehabilitation for a period after implantation. Svirsky et al. (2000) examined two possible reasons underlying longitudinal improvements in vowel identification by cochlear implant users: improved labeling of vowel sounds and improved electrode discrimination. Improvements in vowel identification were attributed to improved labeling, suggesting cortical learning effects were responsible for the observed changes, as opposed to enhanced electrode discrimination.

7. Cultural Considerations

Programs actively engaged in the habilitation or rehabilitation of profound deafness are inevitably drawn into discussions of the cultural implications of prosthetic intervention, particularly as they relate to childhood implantation (Balkany et al. 1996). Proponents of communication strategies such as signing hold that implants put recipients on an ill-defined middle ground between the hearing and deaf worlds. Successful users of implants wonder why anyone would not want to take advantage of an auditory pathway that is capable of responding to meaningful stimuli, providing cues to speech recognition that escape visual detection.

There are no correct positions in these controversies. Professionals should be guided by candidates' perceptions of how they can best participate in and relate to their environment or, in the case of young children, their parents' perceptions. In a legal opinion, Brusky (1995) offers support for the notion that the existing legal framework in the U.S. holds that parents rather than any other potential decision maker hold the principal decision-making responsibility for their children.

The decision to pursue implantation should be as fully informed as possible (Brusky 1995). Implant team personnel should make every attempt to ensure that patients and their families fully comprehend the implant procedure and postoperative rehabilitation, including the risks, benefits, and cultural alternative. Without informed consent, surgical intervention is subject to tort law. The informed consent doctrine safeguards an individual's right to self-determination and encourages intelligent decisions about medical care. Comprehensively informed consent provides the foundation for decision making that conforms with a patient's values, preferences, and needs.

8. Summary

Cochlear implants provide effective auditory rehabilitation for congenital as well as acquired deafness in people across the entire age spectrum. The implant outcomes depend heavily on case assessment prior to intervention. Children with early-onset deafness lack a base of auditory memory with which to pair implant-mediated percepts, and they may harbor other disabilities that may prevent instinctive language learning. Such conditions can produce a wide range of individual variability, particularly when intervention with a cochlear implant is delayed. Adults with cochlear implants exhibit a similarly wide range of results usually owing in large part to factors outside of the device per se. Be enabling simultaneous input of multiple perspectives, a multidisciplinary team is the most effective approach to assessing a candidate's needs and desires and the potential for an implant to meet them.

In the 1980s candidacy for a cochlear implant required total or near-total sensorineural hearing losses as characterized by a pure-tone average of 100 dB or greater, amplified thresholds that failed to reach 60 dB, and an absence of open-set speech recognition despite the use of powerful, best-fit hearing aids. Because clinical experience has indicated that the mean speech reception scores of implant recipients generally exceed the aided results of individuals with lesser impairments, the audiological criteria have been progressively relaxed over the past 20 years to include those with a range of pure-tone thresholds, focusing instead on observed receptive benefit with amplification. In children, initial reports suggest that implantation prior to the age of 3 years provides distinct advantages over later implantation in cases of early-onset deafness. Whether implantation of ever-younger infants may yield greater benefits relative to risks will require longitudinal follow-up. Postimplantation rehabilitation can be important for some adult implant recipients, but appears critical for children to optimize the usefulness of an implant. It is often assumed that effective interactive skills and language comprehension directly result from the sensitivity with which sound is perceived through a cochlear implant. However, hearing is not a sufficient condition for these higher skills, and there is a compelling rationale for offering auditory rehabilitation as an important adjunct to enhance fundamental skills in verbal communication for many implant recipients.

Overall, a 25-year clinical experience with cochlear implantation indicates the need for provision of service and an environment that facilitates a fundamental change in an individual's daily life in order to use this implantable technology to its fullest potential. Providing an individual with a cochlear implant entails high costs and intensive support. However, when properly supported, cochlear implantation can enhance communication capability to a degree that cost-effectiveness relative to other health care interventions is clear.

References

An Y, Friedman R (1998) Animal models of orthopedic implant infection. J Invest Surg 11:139–146.

Azimi N, Welch H (1998) The effectiveness of cost-effectiveness analysis in containing costs. J Gen Intern Med 13:664–669.

Balkany T, Dreisbach J (1987) Workshop: surgical anatomy and radiographic imaging of cochlear implant surgery. Am J Otol 8:195–200.

Balkany T, Gantz B, Nadol JB (1988) Multichannel cochlear implants in partially ossified cochleas. Ann Otol Rhinol Laryngol Suppl 135:3–7.

Balkany T, Gantz BJ, Steenerson RL, Cohen NL (1996) Systematic approach to electrode insertion in the ossified cochlea. Otolaryngol Head Neck Surg 114:4–11.

Beadle E, Shores A, Wood E (2000) Parental perceptions of the impact upon the family of cochlear implantation in children. Ann Otol Rhinol Laryngol Suppl 185:111–114.

Bielamowicz S, Coker N, Jenkins H, Igarashi M (1988) Surgical dimensions of the facial recess in adults and children. Arch Otolaryngol Head Neck Surg 114:534–537.

Blamey PJ, Pyman BC, Gordon M, Clark GM, et al. (1992) Factors predicting postoperative sentence scores in postlinguistically deaf adult cochlear implant patients. Ann Otol Rhinol Laryngol 101:342–348.

Boothroyd A, Geers A, Moog J (1991) Practical implications of cochlear implants in children. Ear Hear 12(suppl 4):81s–89s.

Briggs R, Brackmann D, Baser M, Hitselberger W (1994) Comprehensive management of bilateral acoustic neuromas. Arch Otolaryngol Head Neck Surg 120:1307–1314.

Brookhouser P, Moeller MP (1986) Choosing the appropriate habilitative track for the newly identified hearing-impaired child. Ann Otol Rhinol Laryngol 95(1 pt 1):51–59.

Brusky A (1995) comments. Making decisions for deaf children regarding cochlear implants: the legal ramifications of recognizing deafness as a culture rather than a disability. Wisconsin Law Rev 235–270.

Campbell VA, Crews JE, Moriarty DG, Zack MM, et al. (1999) Surveillance for sensory impairment, activity limitation, and health-related quality of life among older adults—United States, 1993–1997. MMWR CDC Surveill Summ 48:131–156.

Carhart R (1947) Auditory training. In: Davis H, ed. Hearing and Deafness. New York: Rinehart & Co., pp. 276–299.

Carney a, Obsberger MJ, Carney E, Robbins A, et al. (1993) A comparison of speech discrimination with cochlear implants and tactile aids. J Acoust Soc Am 94:2036–2049.

Carter R, Hailey D (1999) Economic evaluation of the cochlear implant. Int J Tech Assess 15:520–530.

Casselman JW, Kuhweide, Deimling M, Ampe W, et al. (1993) Constructive interference in steady state-3DFT MR imaging of the inner ear and cerebellopontine angle. AJNR 14:47–57.

Cheng A, Niparko J (1999) Cost utility of the cochlear implant in adults. Arch Otolaryngol Head Neck Surg 125:1214–1218.

Cheng A, Grant G, Niparko J (1999) A meta-analysis of the pediatric cochlear implantation. Ann Otol Laryngol Rhinol 177:124–128.

Cheng A, Rubin H, Powe N, Mellon N, Francis H, Niparko J (2000) A cost-utility analysis of the cochlear implant in children. JAMA 284:850–856.

Cohen NL (1997) Cochlear implant soft surgery: fact or fantasy? Otolaryngol Head Neck Surg 117:214–216.

Cohen NL, Waltzman S, Fisher S, The Department of Veterans Affairs Cochlear Implant Study Group (1993) A prospective, randomized study of cochlear Implants. N Engl J Med 328:233–237.

Dawson P, Clark G (1997) Changes in synthetic and natural vowel perception after specific training for congenitally deafened patients using a multichannel cochlear implant. Ear Hear 18:488–501.

Dobie R, Jenkins H, Cohen N (1995) Surgical results. Ann Otol Rhinol Larygol 104(suppl 165):6–8.

Eby TL, Nadol JB (1986) Postnatal growth of the human temporal bone: implications for cochlear implants in children. Ann Otol Rhinol Laryngol 95:356–382.

Edgerton B (1985) Rehabilitation and training of postlingually deaf adult cochlear implant patients. Semin Hear 6:65–89.

Eilers R, Wilson W, Moore J (1977) Developmental changes in speech discrimination in infants. J Speech Hear Res 20:766–780.

Facer G, Peterson A, Brey R (1995) Cochlear implantation in the senior citizen age group using the Nucleus 22-channel device. Ann Otol Rhinol Laryngol 166(suppl):187–190.

Fayad J, Moloy P, Linthicum FH (1990) Cochlear otosclerosis: does bone formation affect cochlear implant surgery? Am J Otol 11:196–200.

Fayad J, Linthicum F, Otto S, Galey F, et al. (1991) Cochlear implants: histopathologic findings related to performance in 16 human temporal bones. Ann Otol Rhinol Laryngol 100:807–811.

Finley C, Wilson B, White M (1990) A finite-element model of bipolar field patterns in the electrically stimulated cochlea. A two-dimensional approximation. In: Miller J, Spelman F, eds. Cochlear Implants: Models of the Electrically Stimulated Ear. New York: Springer-Verlag, pp. 339–376.

Finley C, van den Honert C, Wilson B, Miller R, Carteel L, Smith D, Niparko J (2001) Factors contributing to the size, shape, latency and distribution of intracochlear evoked potentials. Abstracts of the 2001 Asilomar Conference on Cochlear Implants, pp. 39–40.

Francis H, Koch M, Wyatt J, Niparko J (1999) Trends in educational placement and cost-benefit considerations in children with cochlear implants. Arch Otolaryngol Head Neck Surg 125:499–505.

Francis HW, Chee N, Yeagle J, Cheng A, Niparko JK (2002) Impact of cochlear implants on the functional health status of older adults. Laryngoscope 112:1482–1488.

Franz BK, Clark GM, Bloom DM (1987) Effect of experimentally induced otitis media on cochlear implants. Ann Otol Rhinol Laryngol 96:174–177.

Frau CN, Luxford WM, Lo W, Berliner KI, et al. (1994) High-resolution computed tomography in evaluation of cochlear patency in implant candidates: a comparison with surgical findings. J Laryngol Otol 108:743–748.

Gantz BJ, Tyler RR, Knutson JF, Woodworth G, et al. (1988a) Evaluation of five different cochlear implant designs: audiologic assessment and predictors of performance. Laryngoscope 98:1100–1106.

Gantz BJ, McCabe BF, Tyler R (1988b) Use of multichannel cochlear implants in obstucted and obliterated cochleas. Otolaryngol Head Neck Surg 98:72–81.

Gantz B, Woodworth G, Knutson J, Abbas P, Tyler R (1993) Multivariate predictors of audiological success with multichannel cochlear implants. Ann Otol Rhinol Laryngol 102:909–916.

Gantz B, Rubinstein J, Tyler R, Teagle H et al. (2000) Long-term results of cochlear implants in children with residual hearing. Ann Otol Rhinol Laryngol Suppl 185:33–36.

Geers A, Moog J (1987) Predicting spoken language acquisition in profoundly deaf children. J Speech Hear Dis 52:84–94.

Geers A, Moog J (1991) Evaluating the benefits of cochlear implants in an education setting. Am J Otol 12(suppl):116–125.

Geers A, Nicholas J, Tye-Murray N, Uchanski R, et al. (2000) Effects of communication mode on skills of long-term cochlear implant users. Ann Rhinol Otol Laryngol 109:89–92.

Goldin-Meadow S, Mylander C (1984) Gestural communication in deaf children: the effects and noneffects of parental input on early language development. Monogr Soc Res Child Dev 49:1–151.

Green JD, Marion MS, Hinojosa R (1991) Labyrinthitis ossificans: histopathologic consideration for cochlear implantation. Otolaryngol Head Neck Surg 104:320–326.

Harnsberger HR, Dart DJ, Parkin JL, Smoker W, et al. (1987) cochlear implant candidates: assessment with CT and MRI imaging. Radiology 164:53–57.

Havlik R (1986) Aging in the eighties: impaired senses for sound and light in persons age 65 years and over. Preliminary data from the supplement on aging to the National Health Interview Survey: United States, January–June 1984. National Center for Health Statistics, DHHS.

Heller J, Brackmann D, Tucci D, Nyenhuis J, et al. (1996) Evaluation of MRI compatibility of the modified Nucleus auditory brainstem and cochlear implants. Am J Otol 17:724–729.

Hellman S, Chute P, Kretschmer R, Nevins ME, Parisier S, Thurston L (1991) The development of a children's implant profile. Am Ann Deaf 136:77–81.

Hinojosa R, Marion M (1983) Histopathology of profound sensorineural deafness. Ann N Y Acad Sci 405:459–484.

Hoffman RA (1997) Cochlear implant in the child under two years of age: skull growth, otitis media, and selection. Otolaryngol Head Neck Surg 117:217–219.

Hoffman RA, Cohen NL (1995) Complications of cochlear implant surgery. Ann Otol Rhinol Laryngol Suppl 166:420–422.

Holt J (1993) Stanford achievement test, 8th ed. Am Ann Deaf 138:172–175.

Horn KL, McMahon NB, McMahon Dc, Kewis JS, et al. (1991) Functional use of the of Nucleus 22-channel cochlear implant in the elderly. Laryngoscope 101:284–288.

House JR, Luxford WM (1993) Facial nerve injury in cochlear implantation. Otolaryngol Head Neck Surg 109:1078–1082.

House WF, Luxford WM, Courtney B (1985) Otitis media in children following cochlear implant. Ear Hear 6:24S–26S.

Hutton J, Politi C, Seeger T (1995) Cost-effectiveness of cochlear implantation of children. A preliminary model for the UK. Adv Otorhinolaryngol 50:201–206.

Huygen P, Broke P, Mens L, Spies T, et al. (1994) Does intracochlear implantation jeopardize vestibular function? Ann Otol Rhinol Laryngol 103:609–614.

Jackler R, Luxford W, House W (1987a) Congenital malformations of the inner ear: a classification based on embryogenesis. Laryngoscope 97(suppl 40):2–14.

Jackler R, Luxford W, House W (1987b) Sound detection with the cochlear implant in five ears of four children with congenital malformations of the cochlea. Laryngoscope 97(suppl 40):15–17.

Jackler R, Luxford W, Schindler R, McKerrow W (1987c) Cochlear patency problems in cochlear implantation. Laryngoscope 97:801–805.

Javel E (1990) Acoustic and electrical encoding of temporal information. In: Miller J, Spelman F, eds. Models of the Electrically Stimulated Cochlea. New York: Springer-Verlag, pp. 247–290.

Javel E, Tong Y, Shepherd R, Clark G (1987) Responses of cat auditory nerve fibers to biphasic electrical current pulses. Ann Otol Rhinol Laryngol 96(suppl 128):26–30.

Jolly C, Gstottner W, Hochmair-Desoyer I, Baumgartner W, Hamzavi J (2000) Principles and outcome in perimodiolar positioning. Ann Otol Rhinol Laryngol Suppl 185:20–23.

Kelsall Dc, Shallop JK, Burnelli T (1995) Cochlear implantation in the elderly. Am J Otol 16:609–615.

Kemp DT (1978) Stimulated acoustic emission from the human auditory system. J Acoust Soc Am 64:1386–1391.

Kennedy D (1987) Multichannel intracochlear electrodes: mechanism of insertion trauma. Laryngoscope 97:42–49.

Kerr A, Schuknecht H (1968) The spiral ganglion in profound deafness. Acta Otolaryngol (Stockh) 65:586–598.

Kiang N, Moxon E (1972) Physiological considerations in artificial stimulation of the inner ear. Ann Otol Rhinol Laryngol 81:714–730.

Kileny P, Zimmerman-Phillips S, Kemink J, Schmaltz S (1991) Effects of preoperative electrical stimulability and historical factors on performance with multichannel cochlear implants. Ann Otol Rhinol Laryngol 100:563–568.

Kileny P, Young K, Niparko J (1994) Acoustic and electrical assessment of the auditory pathway. In: Jackler R, Brackmann D, eds. Neurotology. St. Louis: Mosby, pp. 261–282.

Kirk K (2000) Challenges in the clinical investigation of cochlear implant outcomes. In: Niparko J, Kirk K, Mellon N, Robbins A, Wilson B, eds. Cochlear Implants: principles and Practices. Philadelphia: Lippincott Williams & Wilkins, pp. 225–258.

Kirk KI, Diefendorf A, Pisoni D, Robbins A (1997) Assessing speech perception in children. In: Mendel L, Danhauer J, eds. Audiologic Evaluation and Management and Speech Perception Assessment. San Diego: Singular, pp. 101–132.

Koch M, Wyatt JR, Francis H, Niparko J (1997) A model of educational resource use by children with cochlear implants. Otolaryngol Head Neck Surg 117:174–179.

Lalwani AK, Larky JB, Wareing MJ, Kwast K, et al. (1998) The Clarion multi-strategy cochlear implant surgical techniques, complications, and results: a single institutional experience. Am J Otol 19:66–70.

Langman AW, Quigley SM (1996) Accuracy of high-resolution computed tomography in cochlear implantation. Otolaryngol Head Neck Surg 114:38–43.

Lea A (1991) Cochlear implants. Health Technology Series, No. 6. Canberra: Australian Institute of Health.

Lea A, Hailey D (1995) The cochlear implant. A technology for the profoundly deaf. Med Prog Technol 21:47–52.

Lenarz T, Battmer R, Bertram B (1997) Cochlear implantation in children under the age of two. In: Abstracts of Vth International conference on Cochlear Implants. New York: NYU Postgraduate Medical School, p. 45.

Lenarz T, Kuzma J, Weber B, Reuter G, et al. (2000) New Clarion electrode with positioner: insertion studies. Ann Otol Rhinol Laryngol Suppl 185:16–18.

Levitt H, McGarr N, Geffner D (1987) Development of language and communication skills in hearing-impaired children. ASHA Monographs, Number 26. Rockville, MD: American Speech-Language-Hearing Association, pp. 1–158.

Linthicum F, Galey F (1983) Histologic evaluation of temporal bones with cochlear implants. Ann Otol Rhinol Laryngol 92:610–613.

Linthicum F, Fayad J, Otto S, Galey F, et al. (1991) Cochlear implant histopathology. Am J Otol 12:245–311.

Lonsbury-Martin B, Whitehead M, Martin G (1991) Clinical applications of otoacoustic emissions. J Speech Hear Res 34:964–981.

Luntz M, Hodges AV, Balkany T, Dolan-Ash S, Schloffman J (1996) Otitis media in children with cochlear implants. Laryngoscope 106:1403–1405.

Miyamoto R, Osberger M, Cunningham L, Kirk K, et al. (1994a) Single-channel to multichannel conversions in pediatric cochlear implant recipients. Am J Otol 15:40–45.

Miyamoto R, Osberger M, Todd S, Robbins A, et al. (1994b) Variables affecting implant performance in children. Laryngoscope 104:1120–1124.

Miyamoto RT, Kirk KI, Renshaw J, Hussain D (1999) Cochlear implantation in auditory neuropathy. Laryngoscope. 109(2 pt 1):181–185.

Moore J, Niparko J, Miller M, Linthicum F (1997) Effect of profound hearing loss on a central auditory nucleus. Am J Otol 15:588–595.

Mulrow CD, Aguilar C, Endicott JE, Velez R, et al. (1990) Association between hearing impairment and the quality of life of elderly individuals. J Am Geriatr Soc 38:45–50.

Nadol J (1984) Histological considerations in implant patients. Arch Otolaryngol 110:160–163.

Nadol J (1997) Patterns of neural degeneration in the human cochlea and auditory nerve: implications for cochlear implantation. Otolaryngol Head Neck Surg 117:220–228.

Nadol J, Young Y-S, Glynn R (1989) Survival of spiral ganglion cells in profound sensorineural hearing loss: implications for cochlear implantation. Ann Otol Rhinol Laryngol 98:411–416.

Ng M, Niparko JK, Nager GT (2000) Inner ear pathology in severe to profound sensorineural hearing loss. In: Niparko JK, Kirk KI, Mellon NK, Robbins AM, et al., eds. Cochlear Implants: Principles and Practices. Philadelphia: Lippincott Williams & Wilkins, pp. 57–100.

NIH Consensus Development Statement (1988) Cochlear Implants. Bethesda, MD: U.S. Department of Health and Human Services PHS, 7:2, 1–9.

NIH Consensus Development Statement (1995) Cochlear implants in adults and children. JAMA 274:1955–1961.

Niparko J (1999) Activity influences on neuronal connectivity within the auditory pathway. Laryngoscope 109:1721–1730.

Niparko J, Oviatt D, Coker N, Sutton L, et al.(1991) Facial nerve stimulation with cochlear implants. Otolaryngol Head Neck Surg 104:826–830.

O'Donoghue G, Nikolopoulos T, Archbold S (2000) Determinants of speech perception in children after cochlear implantation. Lancet 5(356):466–468.

O'Donoghue G, Balkany T, Cohen N, Lenarz T, Lustig L, Niparko J (2002) Meningitis and cochlear implantation. Otol Neurotol 23:823–824.

O'Neill C, O'Donoghue GM, Archbold SM, Normand C (2000) A cost-utility analysis of pediatric cochlear implantation. Laryngoscope 110:156–160.

Osberger MJ (1986) Language and learning skills of hearing-impaired students. ASHA Monographs, Number 23. Rockville, MD: American Speech-Language-Hearing Association, pp. 3–107.

Osberger MJ (1997) Cochlear implantation in children under the age of two years: candidacy considerations. Otolaryngol Head Neck Surg 117:145–149.

Osberger M (2001) Candidacy and performance trends in children. 8th Symposium on Cochlear Implants in Children, Los Angeles.

Osberger MJ, Miyamoto RT, Zimmerman-Phillips S, Kemink J, et al. (1991) Independent evaluation of the speech perception abilities of children with the Nucleus 22-channel cochlear implant system. Ear Hear 12(4 suppl):66S–80S.

Osberger MJ, Maso M, Sam L (1993) Speech intelligibility of children with cochlear implants, tactile aids or hearing aids. J Speech Hear Res 36:186–203.

Osberger MJ, Kalberer A, Zimmerman-Phillips S, Barker MJ, Geier L (2000) Speech perception results in children using the Clarion Multi-Strategy Cochlear Implant. Ann Otol Rhinol Laryngol Suppl 185:75–77.

Otte J, Schuknecht HF, Kerr A (1978) Ganglion cell populations in normal and pathological human cochleae: implications for cochlear implantation. Larygoscope 88:1231–1246.

Parisier SC, Chute PM, Popp AL, Hanson MB (1997) Surgical techniques for cochlear implantation in the very young child. Otolaryngol Head Neck Surg 117:248–254.

Parkins C (1985) The bionic ear: principles and current status of cochlear prostheses. Neurosurgery 16:853–865.

Parkins CW, Colombo J (1987) Auditory-nerve single-neuron thresholds to electrical stimulation from scala tympani electrodes. Hear Res 31:267–285.

Picton T (1995) The neurophysiological evaluation of auditory discrimination. Ear Hear 16:1–5.

Picton T, Durieux-Smith A, Moran L (1994) Recording auditory brainstem responses from infants. Int J Pediatr Otorhinolaryngol 28:93–110.

Proctor B, Bollobas B, Niparko J (1986) Anatomy of the round window niche. Ann Otol Rhinol Laryngol 95:444–446.

Proops D, Donaldson I, Cooper H, Thomas J, et al. (1999) Outcomes from adult implantation, the first 100 patients. J Laryngol Otol Suppl 24:5–13.

Quittner A, Steck J, Rouiller R (1991) Cochlear implants in children: a study of parental stress and adjustment. Am J Otol 12:95–104.

Rapin I (1978) Consequences of congenital hearing loss—a long-term view. J Otolaryngol 7:473–483.

Reefhuis J, Honein MA, Whitney CG, Chamany S, et al. (2003) Risk of bacterial meningitis in children with cochlear implants. N Engl Med. 349(5):435–445.

Rickards FW, Tan LE, Cohen LT, Wilson OJ, Drew JH, Clark GM (1994) Auditory steady-state evoked potential in newborns. Br J Audiol 28:327–337.

Robbins AM (2000) Rehabilitation after cochlear implantation. In: Niparko J, Kirk K, Mellon N, Robbins A, Wilson B, eds. Cochlear Implants: Principles and Practices. Philadelphia: Lippincott Williams & Wilkins, pp. 323–362.

Robbins AM, Renshaw JJ, Berry S (1991) Evaluating meaningful auditory integration in profoundly hearing-impaired children. Am J Otol 12(suppl):144–150.

Robbins AM, Svirsky M, Kirk KI (1997) Children with implants can speak, but can they communicate? Otolaryngol Head Neck Surg 117(3 pt 1):155–160.

Ross M, Calvert D (1984) Semantics of deafness revisited: total communication and the use and misuse of residual hearing. Audiology 9:127–145.

Ross M, Levitt H (1997) Consumer satisfaction is not enough: hearing aids are still about hearing. Semin Hear 18:7–11.

Rubinstein JT, Parkinson WS, Lowder MW, Gantz BJ, et al. (1997) Single-channel to multichannel conversions in adult cochlear implant subjects. Am J Otolaryngol 19:461–466.

Rubinstein JT, Parkinson WS, Lowder MW, Gantz BJ, et al. (1998) Single-channel to multichannel conversions in adult cochlear implant subjects. Am J Otol 19:461–466.

Rubinstein J, Parkinson W, Tyler R, Gantz B (1999) Residual speech recognition and cochlear implant performance: effects of implantation criteria. Am J Otol 20:445–452.

Ryugo DK, Pongstapom T, Huchton DM, Niparko JK (1997) Ultrastructural analysis of primary endings in deaf white cats: morphologic alterations in endbulbs of Held. J Comp Neurol 385:230–244.

Schmidt J (1985) Cochlear neuronal populations in developmental defects of the inner ear: implications for cochlear implantation. Acta Otolarygol 99:14–20.

Schopmeyer B, Mellon N, Dobaj H, Niparko J (2000a) Emergence of spontaneous language in children with cochlear implants. In: Waltzman S, Cohen N, eds. Proceedings of the Vth International Cochlear Implant Conference. New York: Thieme, pp. 298–303.

Schopmeyer B, Mellon N, Dobaj H, Grant G, Niparko J (2000b) Use of Fast ForWord to enhance language development in children with cochlear implants. Ann Otol Rhinol Laryngol Suppl 185:95–98.

Schuknecht HF (1989) Pathology of Presbycusis. In: Goldstein J, Kashima C, Koopmann C, eds. Geriatric Otorhinolaryngology. Philadelphia: BC Dekker, pp. 40–45.

Shallop J, Peterson A, Facer G, Fabry L, Driscoll C (2001) Cochlear implants in five cases of auditory neuropathy: postoperative findings and progress. Laryngoscope (4 pt 1):555–562.

Skinner MW, Fourakis MS, Holden TA, Holden LK, et al. (1999) Identification of speech by cochlear implant recipients with the multipeak (MPEAK) and spectral peak (SPEAK) speech coding strategies II: consonants. Ear Hear 20:443–460.

Stach BA, Spretnjak ML, Jerger J (1990) The prevalence of central presbycusis in a clinical population. J Am Acad Audiol 1:109–115.

Starr A, Picton TW, Sininger Y, Hood LJ, Berlin C (1996) Auditory neuropathy. Brain 119:741–753.

Summerfield A, Marshall D (1995) Cochlear implantation in the U.K. 1990–1994. Nottingham: Medical Research Council Institute of Hearing Research, pp. 199–236.

Summerfield A, Marshall D, Archbold S (1997) Cost-effectiveness considerations in pediatric cochlear implantation. Am J Otol 18(suppl 6):S166–S168.

Svirsky M, Robbins A, Kirk K, Pisoni D, et al. (2000) Language development in profoundly deaf children with cochlear implants. Psychol Sci 11:153–158.

Tait M, Lutman M (1994) Comparison of early communicative behavior in young children with cochlear implants and with hearing aids. Ear Hear 15:352–362.

Tait M, Lutman ME, Nikolopoulos TP. (2001) Communication development in young deaf children: review of the video analysis method. Int J Pediatr Otorhinolaryngol 61:105–112.

Tien RD, Felsberg GJ, Macfall J (1992) Fast spin-echo high-resolution MR imaging of the inner ear. Am J Radiol 159:395–398.

Tobey E, Geers A, Brenner C (1994) Speech production results: speech feature acquisition. In: Geers A, Moog J, eds. Volta Review Effectiveness of Cochlear Implants and Tactile Aids for Deaf Children: The Sensory Aids Study at the Central Institute for the Deaf. Washington, DC: AG Bell Association, pp. 109–130.

Trautwein PG, Sininger YS, Nelson R (2000) Cochlear implantation of auditory neuropathy. J Am Acad Audiol 11:309–315.

Trybus R, Karchmer M (1977) School achievement scores of hearing impaired children: national data on achievement status and growth patterns. Am Ann Deaf 122:62–69.

Tucci D, Niparko J (2000) Cochlear implant surgery. In: Niparko J, Kirk K, Mellon N, Robbins A, Wilson B, eds. Cochlear Implants: Principles and Practices. Philadelphia: Lippincott Williams & Wilkins, pp. 198–224.

Tucci D, Telian S, Zimmerman S, Zwolan T, et al. (1995) Cochlear implantation in patients with cochlear malformations. Arch Otolaryngol Head Neck Surg 121:833–838.

Tye-Murray N (1994) Cochlear Implants and Children: A Handbook for Parents, Teachers and Speech and Hearing Professionals. Washington, DC: Alexander Graham Bell Publishing.

Tyler R, Summerfield Q (1996) Cochlear implantation: relationships with research on auditory deprivation and acclimatization. Ear Hear 17:38S–52S.

Tyler R, Moore B, Kuk F (1989) Performance of some of the better cochlear-implant patients. J Speech Hear Res 32:887–911.

Waltzman SB, Cohen NL (1998) Cochlear implantation in children younger than 2 years old. Am J Otol 19:158–162.

Waltzman S, Fisher S, Niparko J, Cohen N (1995) Predictors of postoperative performance with cochlear implants. Ann Otol Rhinol Laryngol 104(suppl 165):15–18.

Watson C (1991) Auditory perceptual learning and the cochlear implant. Am J Otol 12(suppl):73–79.

Weinstein BE, Ventry IM (1982) Hearing impairment and social isolation in the elderly. ASHA 25:593–599.

Wiet RJ, Pyle GM, O'Connor CA, Russell E, et al. (1990) Computed tomography: how accurate a predictor for cochlear implantation? Laryngoscope 100:687–692.

Woolley AL, Oser AB, Lusk RP, Bahadori RS (1997) Preoperative temporal bone computed tomography scan and its use in evaluating the pediatric cochlear implant candidate. Laryngoscope 107:1100–1106.

Wyatt JR, Niparko JK, Rothman ML, de Lissovoy GV (1996) Cost utility of the multichannel cochlear implants in 258 profoundly deaf individuals. Laryngoscope 106:816–821.

Ylikoski J, Savolainen S (1984) The cochlear nerve in various forms of deafness. Acta Otolaryngol (Stockh) 98:418–427.

Zeng FG, Galvin JJ (1999) Amplitude compression and phoneme recognition in cochlear implant listeners. Ear Hear 20:60–73.

Zeng FG, Grant G, Niparko J, Galvin JJ, et al. (2002) Speech dynamic range and its effects on cochlear implant performance. J Acoust Soc Am 111:377–386.

Zwolan T, Sheperd N, Niparko J (1993) Labyrinthectomy with cochlear implantation. Am J Otolaryngol 14:220–224.

Zwolan T, Kileny PR, Smith S, Mills D, Koch D, Osberger MJ (2001) Adult cochlear implant patient performance with evolving electrode technology. Otol Neurotol 22:844–849.

4
Anatomical Considerations and Long-Term Effects of Electrical Stimulation

PATRICIA A. LEAKE and STEPHEN J. REBSCHER

1. Introduction

There is a consensus that achieving high levels of speech intelligibility with auditory prostheses requires the use of several independent channels of stimulation, each engaged in the transmission of limited bandwidth information (see Zeng, Chapter 1; Wilson, Chapter 2; Shannon et al., Chapter 8). Several types of multichannel cochlear prostheses have been designed with the objective of selectively exciting discrete sectors of the auditory nerve by placing individual electrodes at several sites along the cochlear spiral. Most of these devices utilize the tonotopic organization of the primary afferent auditory neurons within the scala tympani, one of the spiral fluid channels of the cochlea, which is accessible from the middle ear via entrance from the round window or through an opening created just anterior to the window (see Niparko, Chapter 3). As described by Wilson in Chapter 2, several multichannel devices have been developed and used in clinical populations. There are substantial differences between the basic components for each of these multichannel cochlear implants in current use:

1. The mechanical characteristics of intracochlear electrode arrays are substantially different for each of the devices (see Rebscher et al. 2001 for review).
2. Electrode geometries and configurations differ greatly among devices.
3. Strategies for encapsulation of receiving electronics and for demultiplexing and controlling levels of applied signals are idiosyncratic.
4. The external speech processors differ significantly.
5. As a result of these differences, we must assume that the distributed stimulation patterns of auditory nerve excitation differ greatly among the various implant systems.

It is important to note that, despite these substantial differences, impressive levels of auditory-only speech recognition have been enjoyed by the majority of subjects implanted with each of these multichannel systems. This chapter reviews the structure of the cochlea, some aspects of the electrode–neural interface, pathological alterations in the cochlea and central auditory system resulting from deafness, some of the long-term consequences of electrical stimulation observed in animal models, and suggests some areas for future research and development.

2. Anatomy of the Cochlea and Electrode–Neural Interface

2.1 Structure of the Cochlea

The cochlear prosthesis is intended for individuals with sensorineural hearing impairment, in whom the cochlea is dysfunctional. (For a detailed review of the normal anatomy and function of the cochlea, the reader is referred to Dallos et al. 1996, *The Cochlea*, Volume 8 in this series of the *Springer Handbook of Auditory Research*.) In a subject with normal hearing, sound energy passes through the external ear canal to the tympanic membrane (ear drum) and is transmitted in the middle ear via a system of three small bones, the ossicles, to the fluid-filled inner ear within the petrous temporal bone. Sound vibrations are detected within the auditory portion of the inner ear, the cochlea, by specialized mechanoreceptors, the hair cells of the organ of Corti. The organ of Corti is a long, coiled neuroepithelial ridge sitting on the flexible basilar membrane. The basilar membrane attaches medially to a delicate bony shelf (the osseous spiral lamina), which spirals around and projects outward from the central bony core or modiolus of the cochlea. Laterally, the basilar membrane is anchored to the bone of the otic capsule by the spiral ligament. Above the organ of Corti, the delicate Reissner's membrane bounds the triangular fluid compartment called the scala media, which contains endolymph, a high-potassium fluid that is critical to normal cochlear function. These structures (organ of Corti, basilar membrane, osseous spiral lamina, spiral ligament, and Reissner's membrane) together are termed the cochlear partition, and they divide the long, coiled cochlea into two additional fluid-filled chambers, the scala vestibuli and scala tympani, both of which contain perilymph, a fluid similar in composition to cerebrospinal fluid. The scala vestibuli and scala tympani are actually continuous at the apex of the cochlea, where a small opening called the helicotrema connects them. Both of these chambers are surgically accessible, but cochlear implants are usually implanted in the lower chamber, the scala tympani, which is accessible via entrance from the round window or via access from a cochleostomy positioned near the round window (Fig. 4.1).

The vibration of the innermost ossicle, the stapes, in response to sound produces displacement of the fluids of the cochlea, resulting in a traveling wave along the basilar membrane. The point of maximum displacement along the length of the cochlea is determined largely by the impedance

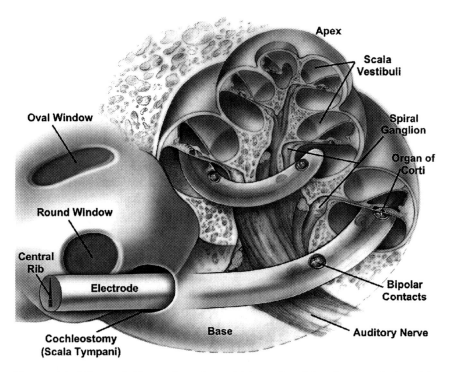

Apex

Scala Vestibuli

Oval Window

Spiral Ganglion

Organ of Corti

Round Window

Central Rib

Electrode

Bipolar Contacts

Cochleostomy (Scala Tympani)

Base

Auditory Nerve

FIGURE 4.1. Schematic illustration of a multichannel cochlear implant designed in the early 1980s at the University of California at San Francisco. Eight bipolar pairs of hemispherical platinum-iridium electrodes, 350 µm in diameter, were oriented in a roughly radial orientation and positioned under the osseous spiral lamina, on either side of the habenula perforata. The goal of this design was to effect spatially selective stimulation of the distal processes of the spiral ganglion neurons at eight different locations along the frequency gradient of the cochlea.

characteristics of the basilar membrane. High frequencies produce maximum displacement at the base of the cochlea, where the basilar membrane is narrow and stiff, and progressively lower frequencies produce maximum displacement at more apical locations, where the basilar membrane and structures of the organ of Corti are wider and more compliant. The cochlear partition is thus tonotopically organized or "tuned" by a systematic and precise regional gradation in its compliance. In the normal cochlea, the organ of Corti contains specialized mechanoreceptors or "hair" cells, consisting of a single row of inner hair cells (IHCs) and three rows of outer hair cells (OHCs). The IHCs transduce movements of the vibrating basilar membrane into neural signals, responding highly selectively (i.e., are tuned) to a narrow range of frequencies at low intensities. Specifically, when the basilar membrane moves in response to incoming sounds, the stereocilia on top of the hair cells are displaced. This displacement of the stereocilia causes ion-specific channels to open or close in the membranes of the hair cells and results in depolarization of the membrane potential of the IHCs.

Depolarization of the hair cells ultimately causes release of neurotransmitter, which is delivered via chemical synapses to the primary afferent auditory neurons of the cochlear spiral ganglion, thus initiating transmission of information to the central nervous system. Each IHC is innervated by 10 to 15 such neurons, whose discharge rates are proportional to the mechanical movement of the basilar membrane at that point. The exquisitely tuned frequency selectivity of this transduction process is due to the nonlinear amplification provided by the OHC. The neural population thus encodes a precise spatial map of frequency (cochlear place), which is maintained and re-represented at each successive level within the central auditory system. Frequency information sent to the brain in terms of this spatial selectivity (i.e., the position along the cochlear spiral of the neurons that are stimulated) has been termed the "place code." In addition, however, individual auditory neurons also encode frequency information in the rate and timing of their bursts of activity, which is termed the "temporal code." The relative importance of these place and temporal codes in determining the pitch characteristic and other perceptual attributes of sounds is currently an area of intensive debate and investigation by auditory psychophysicists and physiologists involved in cochlear prosthesis coding (see McKay, Chapter 7).

The parent cell bodies of the auditory neurons, the spiral ganglion cells, are situated in an irregular spiral channel within the bone of the modiolus. Their peripheral dendrites, the so-called radial nerve fibers, pass through the osseous spiral lamina and fan out in a spectacular spiral array representing the range of human hearing from approximately 20 Hz to 20 kHz. Thus, the longitudinal array of spiral ganglion cell somata and their peripheral axons are directly accessible for frequency-patterned electrical stimulation via multiple electrode arrays positioned in the scala tympani.

Several types of multichannel cochlear prosthesis electrodes have been designed to bypass the hair cell transduction process in individuals with sensorineural hearing loss and to selectively activate discrete sectors of the auditory nerve at several sites along the cochlear spiral, utilizing its precise tonotopic organization. Most of these devices employ a flexible cylindrical insert made of silicon rubber and containing multiple wires, each of which terminates in a stimulating electrode contact on the surface of the insert. These electrode arrays are designed to position several stimulating electrodes either directly under the basilar membrane to stimulate the array of surviving peripheral axons of spiral ganglion neurons, or facing the modiolus to directly activate the spiral ganglion cell somata.

2.2 Histopathology of the Human Cochlea in Profound Sensorineural Hearing Loss

The success of multichannel cochlear implants relies on the likelihood that a significant fraction of the auditory neurons are intact and are widely distributed within the spiral ganglion (i.e., distributed across the frequency range) of the cochlea in most candidate subjects with severe to profound sensorineural deafness. In fact, studies of cochlear histopathology in subjects with various deafness etiologies suggest that degeneration of the spiral ganglion is progressive after deafness. But this is a very slow process, and a substantial fraction of the auditory neurons may survive for many years even after profound hearing loss. Studies by Hinojosa et al. in the early 1980s evaluated 65 temporal bones from patients profoundly deaf due to a representative distribution of etiologies (Hinojosa and Lindsay 1980; Hinojosa and Marion 1983). They concluded that although most of the ears exhibited severe or total loss of hair cells, only 20% had ganglion cell counts below 10,000 (less than ≈30% of normal).

Table 4.1 summarizes data on spiral ganglion survival reported in several histophathological studies of human temporal bones from profoundly deaf subjects (Hinojosa and Lindsay 1980; Hinojosa and Marion 1983; Nadol 1984, 1997; Hinojosa et al. 1987; Otte et al. 1978; Fayad et al. 1991). For subjects considered appropriate candidates for a cochlear implant, the mean survival of spiral ganglion cells varied from 41% to 77% of normal, as reported in the various studies. The most comprehensive of these studies analyzed 93 temporal bones from 66 patients who were documented to have suffered profound sensorineural hearing loss during life (Nadol et al. 1989; Nadol 1997). For this large and representative group, the mean survival of spiral ganglion cells was estimated to be about 47% of normal. Moreover, the specific etiology of deafness accounted for more than half (57%) of the variance in spiral ganglion cell counts in this large group. Spiral ganglion cell survival was the highest in subjects deafened by aminoglycoside toxicity or sudden idiopathic deafness. Substantially lower spiral ganglion cell counts were found in the diagnostic categories of congenital or genetic deafness and bacterial meningitis. Individuals deafened by postnatal viral labyrinthitis had the most severe cochlear pathology and neural degeneration. It is important to note that degeneration of the spiral ganglion is progressive over time. Thus, longer durations of hearing loss and deafness are correlated with more severe neural degeneration (Nadol et al. 1989; Nadol 1997). Further, in most deafness etiologies, degeneration of the spiral ganglion is more severe in the basal half of the cochlea (i.e., where most stimulating electrodes of cochlear implants are positioned) as compared with the apical half (Nadol 1997).

In general, these studies of pathological alterations in the human cochlea following profound sensorineural hearing loss indicate that the population of spiral ganglion neurons remains relatively intact for a time even following severe hair cell loss. In a few years, however, secondary nerve degeneration occurs (for review see Johnsson et al. 1981), usually in a basal-to-apical progression. Characteristically, the first degenerative change seen in the auditory neurons is the loss of the distal processes of the spiral

TABLE 4.1. Data on spiral ganglion survival derived from histopathological studies.

Reference	Etiology of Deafness	Number of Subjects	Hearing Loss (dB SPL)	Number of Ganglion cells (Total)*	Ganglion survival (% normal)
Hinojosa and Marion 1983	Disease/Genetic	14	Profound	16,084 ± 5,672	47.3
Hinojosa and Marion 1983	Ototoxicity	1	≈90dB	19,608	57.8
Suzuka and Schuknecht 1988	Disease/Genetic	3	Profound	23,839 ± 832	70.9
Suzuka and Schuknecht 1988	Ototoxicity/Trauma	10	≈90dB	25,899 ± 7,672	77.0
Hinojosa et al. 1991	Disease/Genetic	6	Profound	18,196 ± 8,307	54.1
Fayad et al. 1991	Cochlear Implant	13	Profound	12,175 ± 4,720	40.6
Nadol et al. 1989; 1997	Varied Pathology	66	Profound	13,444 ± 5,436	47.3

*Mean number of ganglion cells ± Std. Dev.

ganglion neurons within the osseous spiral lamina (Ylikoski 1974; Egami et al. 1978; Otte et al. 1978; Johnsson et al. 1981). This process occurs first in regions where the supporting elements of the organ of Corti have degenerated, and the presence of supporting cells of the organ of Corti can delay neural degeneration even after total hair cell loss. However, it should be noted that all of the human temporal bone studies have reported great variability in the relative status of survival of ganglion cells and organ of Corti among individuals with similar histories, as emphasized by Hinojosa and Marion (1983).

Finally, one aspect of cochlear pathology that is of particular concern for cochlear implantation is the presence of new bone formation, or labyrinthitis ossificans, within the cochlea. This is a common finding in temporal bones from individuals deafened by bacterial meningitis (Eisenberg et al. 1984; Balkany et al. 1988). Labyrinthitis ossificans presents a potentially difficult mechanical impediment to insertion of a cochlear implant electrode, although it is now considered only a relative, rather than absolute, contraindication to cochlear implantation (Balkany et al. 1988; Gantz et al. 1988; Nadol 1997).

2.3 Histopathology of Cochlear Implantation

A potential major factor determining long-term survival of cochlear neurons (and the consequent long-term efficacy of a cochlear implant, particularly in pediatric implant recipients) is the degree of trauma to the cochlea that occurs during surgical insertion of an electrode array. A number of studies have evaluated temporal bones from cochlear implant recipients and have demonstrated significant trauma to the spiral ligament, basilar membrane, and osseous spiral lamina that occurs predictably near the site of insertion and at a location about 8 to 15 mm from the base, in the ascending segment of the basal cochlear turn (Shepherd et al. 1985; Kennedy 1987; Clark et al. 1988; O'Leary et al. 1991; Zappia et al. 1991; Nadol 1997). Another common finding in these studies of human temporal bones after cochlear implantation is evidence of ectopic new bone formation around the electrode array. Nadol (1997) suggests that several factors may contribute to such bone formation, including the original etiology of deafness, bone dust introduced into the cochlea during surgery, and damage to the cochlear blood supply resulting from direct insertion trauma to the lateral cochlear wall or basilar membrane. Finally, with regard to the effects of implantation on the survival of spiral ganglion neurons, the results from temporal bone studies are mixed. Some studies have suggested a reduction in the number of spiral ganglion cells surviving in the implanted cochlea as compared with the opposite ear (Zappia et al. 1991; Marsh et al. 1992), but others have not observed a significant effect (Clark et al. 1988; Linthicum et al. 1991; Nadol et al. 1994; Nadol 1997).

3. Central Auditory Degeneration

3.1 Studies of the Human Central Auditory System

In addition to the condition of the cochlea and survival of the spiral ganglion neurons, the relative integrity or extent of degeneration of the central
auditory pathway is likely to affect the benefit that can be obtained from a
cochlear implant (see Hartmann and Kral, Chapter 6; Pisoni and Cleary,
Chapter 9). Severe degeneration of the brain structures that normally
process the sensory input from the ear is likely to limit the success of a
cochlear implant in restoring hearing. Although there have been relatively
few detailed studies of the changes occurring in the human central auditory
pathways following deafness, Moore and colleagues (1994, 1997) have
provided some informative data on this topic. First, profound hearing loss
(audiometric responses poorer than 90–100 dB HL) has been shown to
result in a severe reduction (>50%) in cell size of neurons within the posterior division of the ventral cochlear nucleus (Moore et al. 1994). Smaller
neuronal cell size in the cochlear nucleus was correlated with fewer surviving cochlear spiral ganglion neurons, and also showed a moderate correlation with longer duration of profound deafness. In addition, pathology
was more severe in two cases of Scheibe degeneration in which deafness
has a genetic basis, as compared to cases of acquired pathology with equivalent ganglion cell populations or periods of hearing loss.

In a subsequent study, Moore et al. (1997) assessed degenerative changes
by measuring cell size at three levels of the brainstem auditory pathway
(anterior division of the cochlear nucleus, medial superior olivary nucleus,
and inferior colliculus) in several human subjects with profound bilateral
adult-onset deafness. Their results clearly indicate that profound hearing
loss affects all these levels of the central auditory system, and further that
neuronal changes tend to be quite similar across all levels of the brain in a
given individual. In contrast, large intersubject variability again was
observed in this study, and the findings confirm the suggestion that the
number of surviving ganglion cells and duration of deafness both contribute
to the extent of central degenerative change. Finally, with regard to the application of a cochlear implant, it is encouraging to note that even in the worst
cases examined, there were many viable neurons present in all of the central
auditory nuclei for up to three decades after onset of bilateral deafness.

3.2 Animal Studies of the Effects of Auditory Deprivation and Developmental Critical Periods

Numerous studies have been conducted in a variety of laboratory animal
models, evaluating the consequences of hearing loss in both the peripheral
and the central auditory system. It is well known that in virtually all causes
of deafness, degeneration of the cochlear hair cells results in subsequent sec-

ondary degeneration of the primary afferent spiral ganglion neurons and their central axons, which form the auditory nerve. Numerous studies have documented specific details of the sequence and time course of pathology following hearing loss in guinea pigs, chinchillas, cats, and monkeys (Kohenen 1965; Stebbins et al. 1969; Ylikoski 1974; Spoendlin 1975; Bohne 1976; Kiang et al. 1976; Hawkins et al. 1977; Liberman and Kiang 1978; Leake and Hradek 1988). Following hearing loss induced by ototoxic drugs or by exposure to high-intensity sounds, the sequence of pathological alterations in the spiral ganglion neurons reported in experimental animals appears to be very similar to that reported in the human cochlea as outlined above. This degeneration is progressive over many months to years, depending on the cause of deafness, but it is important to point out that the progression of pathological alterations usually is substantially more rapid in experimental animal models than in the human inner ear and central auditory system.

As mentioned previously, another potentially important issue for the successful application of a cochlear implant is the relative integrity of the central auditory system after deafness. One of the most intriguing and medically important questions, for which there is presently no clear answer, is, *Why* do some postsynaptic neurons live while others die after loss of their afferent input? (For review, see Rubel and Fritszch 2002.) One important factor determining the relative extent of immediate and long-term degenerative changes induced by auditory deprivation or deafferentation is the time of occurrence of the deprivation. Many studies have emphasized the existence of a critical period of development during which deprivation or deafferentation at an early age causes substantially more severe changes than the same manipulation in adult animals. For example, several investigators have removed the cochlea (usually including the ganglion cells) in experimental animals at various ages and examined the cochlear nucleus (CN) weeks or months later (Levi-Montalcini 1949; Powell and Erulkar 1962; Parks 1979, Trune 1982a,b; Nordeen et al. 1983). Large reductions in neuron size, nuclear volume, and neuron number have been observed following removal of the cochlea when performed in young animals. In contrast, changes seen after deafferentation in mature animals are almost always less severe and do not include a reduction in number of neurons in the CN.

Convincing evidence that the effects of deafferentation or acoustic deprivation on CN neuronal survival and atrophy are much more severe in neonates than in adults has been provided from studies in several different species, including chicks (Born and Rubel 1985), mice (Webster and Webster 1977, 1979; Trune 1982a,b; Webster 1983), rats (Coleman and O'Connor 1979; Coleman et al. 1982; Blatchley et al. 1983), gerbils (Nordeen et al. 1983; Hashisaki and Rubel 1989), guinea pigs (Dodson et al. 1994), chinchillas (Fleckeisen et al. 1991), and ferrets (Moore and Kowalchuk 1988). Moreover, reports have demonstrated that the temporal boundaries or "window" for these critical periods of enhanced susceptibility to deprivation-induced pathology in animals may be quite sharp (Tierney et al. 1997; Mostafapour et al. 2000; Rubel and Fritzsch 2002). For example, Tierney et al. found that in gerbil pups between the seventh (P7) and ninth (P9) postnatal day there is an abrupt change in the response of CN neurons following deafferentation. Removal of the cochlea prior to 7 days of age results in extensive cell death, with loss of 45% to 88% of the CN neurons, whereas this same manipulation at 9 days of age or later causes no consistent cell death. A similar dramatic decrease in susceptibility to deprivation is seen in the mouse during the first 10 days postnatal (Mostafapour et al. 2000).

Transneuronal changes at higher levels of the auditory system also appear to be more severe when auditory deprivation occurs at an early age, as compared to the degenerative changes reported when adult animals are deafened (Powell and Erulkar 1962; Jean-Baptist and Morest 1975; Feng and Rogowski 1980). Other research has shown that neonatal unilateral cochlear ablation can result in pronounced modifications in the anatomical organization of neural projections from the contralateral CN to the superior olivary complex and inferior colliculus (Moore and Kitzes 1985; Moore and Kowalchuk 1988; Nordeen et al. 1983; Moore et al. 1994; Russell and Moore 1995; Kitzes 1996). Thus, auditory deprivation during early development clearly can induce profound changes in the central nervous system, and there is evidence for the existence of "critical periods" in auditory system development. That is, some of the interactions that occur between the hair cell and spiral ganglion neurons or between the spiral ganglion neurons and their targets in the cochlear nucleus occur throughout life (and thus cause pathological changes after auditory deprivation even in the adult), whereas others apparently are limited to a specific time, a "critical period," during development. The molecular mechanisms underlying these trophic relationships are just beginning to be elucidated (Rubel and Fritzsch 2002).

Finally, it is important to recognize that these studies have been conducted in a wide variety of species, and in many different models of auditory deprivation and deafness. Thus, the precise implications of this research for defining the nature and timing of critical periods and the role of early auditory deprivation for later structural and functional development of the central auditory system, as would apply in a young deaf child who receives a cochlear implant, are currently unknown, although general hypotheses are emerging based on this body of work (Ponton et al. 1996; Svirsky et al. 2000).

4. Animal Studies of the Consequences of Cochlear Implantation, Chronic Electrical Stimulation, and Plasticity

In addition to the impact of auditory deprivation on survival of cochlear neurons and in causing central auditory degenerative changes, an equally interesting and important question is the potential impact of aberrant input activity on these neuronal populations and their long-term functional capacities. As will be described in subsequent chapters (see Abbas and Miller, Chapter 5; Hartmann and Kral, Chapter 6), electrical stimulation of the cochlea generally elicits auditory nerve activity that is much more highly synchronized and temporally invariant than normal responses of the auditory nerve to acoustic signals. Given the increasing number of individuals who are now receiving cochlear implants, and the enormous range of intersubject variability in benefit from the implant, achieving a better understanding of the anatomical and functional consequences of implantation and the highly abnormal input delivered by a cochlear implant are important goals of research in animal models. Moreover, because so many children, including children who are congenitally deaf, are now receiving cochlear implants at a very young age, it is of particular interest to better understand the anatomical consequences of implantation of a prosthesis and the functional effects of its abnormal inputs upon the deafened, developing auditory system.

4.1 Factors Influencing Survival of Cochlear Spiral Ganglion Neurons

4.1.1 Deafness Etiology and Duration of Deafness Determine Survival of Cochlear Spiral Ganglion Neurons

As mentioned previously, both the cause and duration of deafness are known to be critical factors in determining the extent of peripheral and central auditory system pathology. This is important to bear in mind as we now consider the numerous animal studies of cochlear implants. In various studies, several different methods of inducing deafness have been employed, the age and duration of deafness have varied widely, and studies have been conducted in several different species of animals.

Many animal studies of cochlear implants have employed subjects deafened by ototoxic drugs, because the resulting cochlear pathology is bilaterally symmetrical. This is an advantage for examining the effects of unilateral insertion of a cochlear implant and chronic stimulation, using within-subject paired comparisons between ears. Moreover, in studying the central nervous system responses to electrical stimulation, studies conducted in bilaterally deafened animals are more relevant than studies of unilaterally deafened animals, because it is likely that maintenance of normal input from an intact ear will help to maintain more normal function. Thus, studies conducted after unilateral deafening may not reveal the full extent of alterations that would be expected to occur with clinical cochlear implants, which are almost always used in individuals with bilateral hearing loss.

However, several different ototoxic drug protocols have been employed, and the time course and extent of pathology induced may vary significantly among them. A number of studies have been conducted in cats after neonatal deafening by systemic administration of the highly ototoxic aminogly-

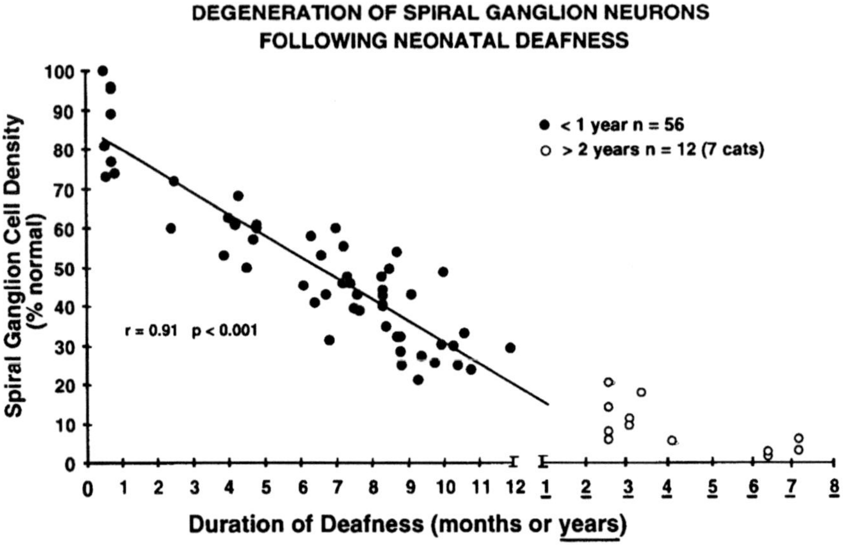

**DEGENERATION OF SPIRAL GANGLION NEURONS
FOLLOWING NEONATAL DEAFNESS**

FIGURE 4.2. Survival of spiral ganglion (SG) neurons is shown as a function of duration of deafness for cats deafened immediately after birth by administration of an ototoxic drug, neomycin sulfate. The data are presented as mean SG cell density expressed as percent of normal. SG survival is strongly correlated with duration of deafness, although there is considerable individual variability. Severe degeneration is seen in long-term deafened animals that were examined after periods exceeding 2 years, with a mean SG density of 9% of normal for this group. (The data for long-deafened subjects are shown as open symbols with duration indicated in years rather than months.)

coside antibiotic, neomycin sulfate (60 mg/kg IM), given over the first 16 to 21 days after birth. Cats are born deaf due to the immaturity of their auditory system (for review see Walsh and Romand 1992), and the ototoxic drug destroys the cochlear hair cells and induces a profound hearing loss prior to the age when adult-like hearing sensitivity would normally develop at about 21 days postnatal (Leake et al. 1997). Thus, these animals have no normal auditory experience and are a model of congenital or very early-acquired bilateral profound hearing loss. The degeneration of hair cells results in subsequent secondary degeneration of the spiral ganglion neurons and their central axons of the auditory nerve (Ylikoski 1974; Spoendlin 1975; Hawkins et al. 1977; Otte et al. 1978; Johnsson et al. 1981). Degeneration is progressive and continues for many months to years (Leake and Hradek 1988), although initial ganglion cell loss is seen as soon as 3 weeks postnatal after neonatal deafening (Leake et al. 1997). Figure 4.2 illustrates the time course of spiral ganglion (SG) cell degeneration in neonatally deafened cats. Although the data illustrate that there is considerable variation among individual subjects in the extent of neural degeneration for a specific duration of deafness, decreasing SG survival is strongly correlated to duration of deafness. Moreover, cochlear pathology is highly symmetrical in the two cochleas of individual animals (Leake et al. 1987, 1997). The consistent bilateral symmetry of cochlear pathology and the relatively rapid, progressive neuronal degeneration allow the systematic study of the effects of unilateral electrical stimulation using within-animal paired comparisons.

Another common method of inducing deafness via ototoxic drugs in animals was first introduced by West et al. (1973). This technique consists of administration of a bolus of kanamycin (400 mg/kg) given by subcutaneous injection, followed 30 minutes later by the loop diuretic, ethacrynic acid (15–25 mg/kg), which is administered by intravenous infusion until hearing loss occurs. This technique is highly effective in producing a very rapid and profound hearing loss within a couple of hours and destruction of essentially all hair cells within a week (Webster and Webster 1981). This acute deafening procedure has the advantage that it seldom causes kidney damage, in contrast to systemic aminoglycosides alone, which must be administered for many days to cause hearing loss and frequently causes nephrotoxicity in adults. However, comparisons across studies suggest that the time course of degeneration of the SG neurons may be more rapid with coadministration of kanamycin and ethacrynic acid than with aminoglycosides alone. This method has been used in number of studies of guinea pigs (Lousteau 1987; Hartshorn et al. 1991; Miller and Altschuler 1995; Li et al. 1999), and was adapted by Xu et al. (1993) for studies in cats deafened at 10 days to 1 month of age (Matsushima et al. 1991; Shepherd et al. 1994; Araki et al. 1998) and studies in cats deafened as adults (Moore et al. 2002).

Other studies have been conducted in animals with genetic defects that result in hearing loss, such as congenitally deaf white cats (Hartmann et al. 1997; Heid et al. 1997, 1998; Klinke et al. 1999; Kral et al. 2001) and dalmations (Niparko and Finger 1997). Such models may have an inherent advantage in more accurately modeling some forms of human deafness, but also have the disadvantage of often exhibiting great intersubject variability in the degree of hearing loss.

4.1.2 Chronic Electrical Stimulation Promotes Survival of the Cochlear Spiral Ganglion Neurons

Initial morphological studies of the effects of surgical implantation and chronic electrical stimulation focused largely on issues of safety and damage from the cochlear implant. Therefore, they used relatively short-term implantation and stimulation and emphasized evaluation of the possible deleterious effects of cochlear implantation (see Leake et al. 1990 for review). A number of subsequent studies reported that chronic electrical stimulation of the cochlea over time actually may promote the survival of SG neurons after deafness. That is, restoration of functional activation through effective electrical stimulation can be neurotrophic to the auditory nerve.

Studies of the effects of electrical stimulation have been conducted both in guinea pigs deafened as young adults (Lousteau 1987; Harshorn et al. 1991; Miller et al. 1991; Miller and Altschuler 1995) and in neonatally deafened cats (Leake et al. 1991, 1992). Results suggested that chronic stimulation with intracochlear bipolar electrodes and simple electrical signals can partially prevent or delay the degeneration of the SG neurons that otherwise occurs after deafness. Increased SG cell densities were observed over relatively large regions of the cochlea that appeared to be grossly related to the location of the stimulating electrodes. It should be noted, however, that other investigators studying the effect of chronic electrical stimulation on SG neuronal survival reported conflicting results. Shepherd et al. (1994) and Araki et al. (1998) found no difference in SG survival after chronic stimulation in cats deafened at 1 month of age by coadministration of kanamycin and ethacrynic acid, although the latter study did find a significant increase in the size of SG cells after chronic stimulation. Moreover, a study by Li et al. (1999) demonstrated an increase in ganglion cell density after chronic monopolar stimulation in guinea pigs, but these authors concluded that the increase resulted from narrowing of Rosenthal's canal apparently induced by stimulation, rather than an increase in the actual number of surviving neurons. These apparently conflicting results have led to some controversy as to whether or not stimulation by a cochlear implant can provide sufficient trophic support of SG neurons to significantly affect survival after deafness.

As suggested by Araki et al. (1998), there are several methodological differences among the various studies that could contribute to the disparate results, including differences in deafening procedures, in stimulation periods, in levels and types of applied stimuli, and in the use of monopolar vs. bipolar stimulation modes. Recent studies in neonatally deafened cats have helped to elucidate some of the specific factors that are critical in eliciting protective effects and promoting survival of the auditory neurons. These experiments examined the effects of intracochlear electrical stimulation that was delivered over prolonged periods (mean, 35 weeks) and high-frequency modulated signals (Leake et al. 1999, 2000a). Stimulation induced a marked increase in SG cell density, with more than 20% of the normal cell density maintained in the stimulated ears above that seen in the control deafened ears (Figs. 4.3 and 4.4). In addition, higher-frequency modulated electrical stimulation appeared to be more effective than simple low-frequency 30-Hz stimulation in preventing neural degeneration (Fig. 4.5). Because stimulation was continued over longer periods in this study, along with using higher frequency complex electrical signals, the *relative* extent to which signal properties and duration of stimulation contributed to the marked increase in neural survival is unclear. However, these results clearly demonstrated the importance of both factors, total duration of stimulation period and characteristics of the electrical signals applied, which in combi-

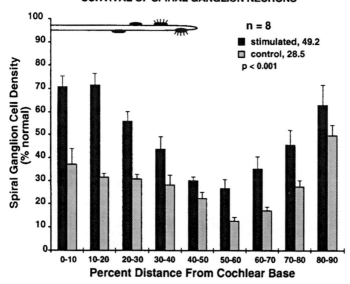

FIGURE 4.3. Summary graph illustrating the differences in SG survival between the stimulated and control ears in a group of eight cats that were neonatally deafened, implanted unilaterally (at 6–9 weeks of age), and received chronic stimulation with higher frequency, amplitude-modulated electrical signals for periods of 8 to 9 months. Mean SG cell density in the stimulated (black) and control deafened (shaded) ears is expressed as percent of normal and shown for different sectors of the cochlea from base to apex. Trauma caused by surgical insertion of the electrode in several animals resulted in a reduction in survival in the stimulated ears in the 40% to 50% cochlear segment (near the tip of the electrode as shown in the diagram at the top). In all other cochlear regions, notably higher mean SG density is observed in the stimulated cochleas as compared with the same regions in the contralateral control ears. Averaged over all sectors, SG cell density was more than 20% higher in the stimulated ears, a difference that was highly significant ($p < .001$; Student's paired t-test). (From Leake et al. 1999, with permission from *Journal of Comparative Neurology*, John Wiley & Sons.)

nation induce very substantial neurotrophic effects with cochlear implant stimulation.

This report suggested that one of the chief factors underlying the failure to elicit trophic effects of stimulation, as reported in several other studies, is the total duration of applied stimulation. A minimum of 3 months of stimulation was required in neonatally deafened cats in order to observe a consistent effect, and even greater effects were demonstrated with longer periods, averaging approximately 8 months (Leake et al. 1999). Most other

studies have used much shorter intervals of stimulation. It should be emphasized that it is the *total time period over which stimulation is delivered* that
appears to be critical in eliciting a trophic effect on neuronal survival, rather
than the total number of hours of stimulation. Studies in both guinea pigs
(Hartshorn 1991; Miller and Altschuler 1995) and cats (Leake et al. 1991,
1992) suggest that electrical stimulation for just an hour or two per day may
be enough to promote survival of the SG neurons, if stimulation is distrib-

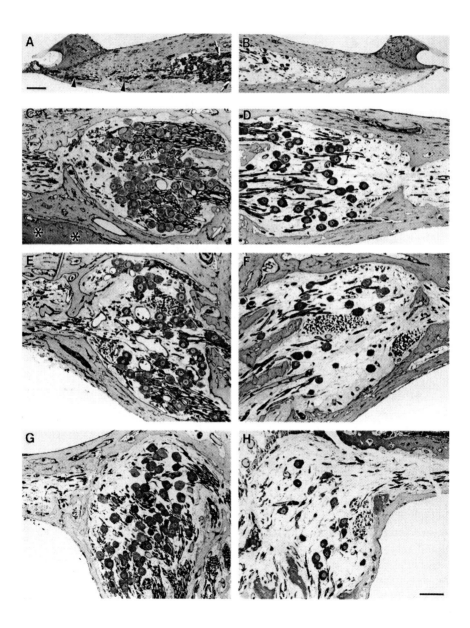

uted over a period that is long enough to allow sufficient degeneration to occur in the contralateral unstimulated ears to show a significant difference between sides. Thus, for example, Araki et al. (1998) used long daily periods of electrical stimulation in kittens, emphasizing that if SG survival is to be related to human cochlear implantation, daily stimulation periods of 12 hours or more should be chosen to approximate normal usage. The total number of hours of stimulation in that study was approximately 1000, but because animals received stimulation for 16 hours/day, 7 days/week, the stimulation was distributed over a period of 49 to 67 days (mean, ≈2 months). This stimulation resulted in slightly larger SG cell size in the stimulated region of the cochlea, but no significant increase in neuronal survival. In contrast, Leake et al. (1992) observed significantly increased neural survival after stimulating neonatally deafened cats for only 1 to 4 hours/day, 5 days/week. However in this study, the ≈1000 hours of stimulation was distributed over much longer total time periods of 70 to 161 days in individual subjects (mean, 3 months). Thus, the stimulation periods in the two sets of experiments were actually substantially different, and the results are consistent with the notion that a small effect on cell size is seen initially, and significant effects on SG density and number are observed after stimulation is continued for longer periods. Further, as mentioned previously, much greater trophic effects have been demonstrated in subsequent studies with

◀──

FIGURE 4.4. These histological sections illustrate the differences in SG cell survival observed in deafened cats following several months of temporally challenging electrical stimulation by a unilateral cochlear implant. Examples of SG density in the stimulated cochleae (left column) are compared to the same regions of the contralateral control deafened ears (right column) in four different subjects. **A,B:** These sections taken 3 mm from the cochlear base (10–20% region) are representative of the difference in SG cell survival observed in this cochlear region for the entire group as illustrated in Figure 4.3. SG density is ≈75% of normal in the stimulated cochlea and ≈31% on the control side. Also note the density of radial nerve fibers in the osseous spiral lamina (arrowheads) and central axons passing into the modiolus (arrows) on the stimulated side and the paucity of such fibers in the control deafened ear. These low-magnification images illustrate the orientation of the three pairs of higher magnification micrographs below. **C,D:** Sections taken ≈5 mm from the base in another subject after chronic electrical stimulation. SG density is 68% of normal in the stimulated cochlea and 38% in the control cochlea. **E,F:** A region of more moderate difference in SG density was seen ≈7 mm from the base (30–40% region), where survival after stimulation is about 55% or normal (E), compared to approximately 25% in the unstimulated ear. **G,H:** Sections from a more apical cochlear region (17 mm, 70–80% from the base), again illustrating markedly higher SG density in the stimulated cochlea (68% of normal) as compared to 26% of normal in the paired region from the control deafened ear. Scale bars = 100 μm in A,B; 50 μm in H (applies to C–H). (From Leake et al. 1999, with permission from *Journal of Comparative Neurology*, John Wiley & Sons.)

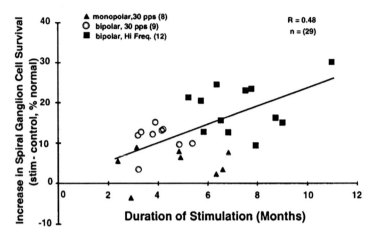

FIGURE 4.5. The extent of the survival-promoting effect of electrical stimulation on spiral ganglion neurons varies with different stimulation protocols. The overall survival in the stimulated ear less survival in opposite ear is expressed as percent of normal. Data are shown for individual subjects in three different experimental groups as a function of duration of stimulation. Subjects that received stimulation using a monopolar electrode near the round window (triangular symbols) clearly showed less increase in SG survival than other groups. In the bipolar stimulation groups, higher frequency stimulation appeared to be more effective than 30-pps stimulation, for the few animals with similar duration of stimulation, but there is a significant trend for greater increase in SG survival to be correlated with longer duration of stimulation ($R = 0.48$).

even longer overall duration of stimulation period of ≈8 months (Leake et al. 1999, 2000a).

It also has been reported that stimulation using a single wire monopolar electrode positioned near the round window elicited much more modest (but still significant) protection or maintenance of neurons than bipolar stimulation (Leake et al. 1995, 2000a). This suggests that these two different types of stimulation vary in their efficacy in maintaining the SG neurons. Thus, the lack of a trophic effect of electrical stimulation in guinea pigs reported by Li et al. (1999) may be related to both the use of a monopolar electrode and the limited stimulation period of 8 weeks. However, an additional variable that must be considered potentially relevant is that the guinea pig study was conducted in young-adult animals, whereas all the previously mentioned studies in cats were conducted in animals deafened and stimulated early in life.

Most of the studies cited above evaluated the effects of unilateral chronic electrical stimulation by using various morphometric methods to estimate the density of SG cell somata. All measures of density would be influenced by differences in cell size as well as number of cells. Further, as pointed out by Li et al. (1999), density would also be influenced by any alteration in the area of Rosenthal's canal. In the most recent study examining neonatally deafened cats, SG cell size, density, cell number, and area of Rosenthal's canal were all evaluated in the same subjects after prolonged periods of chronic electrical stimulation (Leake et al. 1999). Data demonstrated that stimulation induced a marked increase in SG cell density and further showed that the increase in density resulted from a substantial increase in the absolute number of surviving neurons, accompanied by a slightly larger mean cell size in the stimulated cochleas. Moreover, ultrastructural evaluation of the surviving SG neurons suggested that many neurons in the stimulated cochlea retain relatively normal morphological characteristics, as compared to the greater degree of pathological alteration seen in the SG in the control deafened ears (Fig. 4.6).

4.1.3 The role of Neurotrophins

Neurotrophins are a family of growth factors involved in the differentiation and survival of specific populations of neurons and glial cells (Dobrowsky and Carter 1998). The neurotrophin gene family includes nerve growth factor (NGF), brain-derived nerve growth factor (BDNF), neurotrophin-3 (NT-3) and neurotrophin-4/5 (NT-4/5). Mice with knockout mutations, which lack the genes for production of both BDNF and NT-3, exhibit a complete lack of SG neurons by the late stages of embryogenesis, indicating that these neurotrophins are essential for normal development and maintenance of SG neurons (Ernfors et al. 1994, 1995; Farinas et al. 1994; Fritzsch et al. 1997). Further, a number of studies have reported that exogenous administration of neurotrophins (Schindler et al. 1995; Shah et al. 1995; Zheng et al. 1995; Ernfors et al. 1996; Staecker et al. 1996; Miller et al. 1997) and other neurotrophic factors such as glial cell line-derived neurotrophic factor (GDNF) (Ylikoski et al. 1998; Yagi et al. 2000) can protect SG neurons and promote their survival after various types of insult causing deafness.

Studies of SG neurons in tissue culture preparations conducted by Green and colleagues (Hegarty et al. 1997; Hansen et al. 2001) have investigated the mechanisms by which depolarization promotes survival of SG neurons. They have shown that at least three distinct mechanisms are involved, operating in parallel and additively, and that one of these mechanisms is an autocrine neurotrophin response. The neurotrophins BDNF and NT-3 are both expressed by SG neurons, and exogenous BDNF and NT-3 (presumably normally supplied by hair cells or glia) both promote survival of SG neurons in culture. These neurotrophins signal through the Trk family of protein-tyrosine kinase receptors, and blocking the binding of BDNF and NT-3 to their respective receptors, TrkB and TrkC, partially inhibits the

trophic effect of depolarization in a manner consistent with their involvement in an autocrine mechanism (Hansen et al. 2001). Specifically, expression of BDNF and NT-3 in SG neurons does not require depolarization,
although expression of both these neurotropins also contributes to the
survival-promoting effects of depolarization.

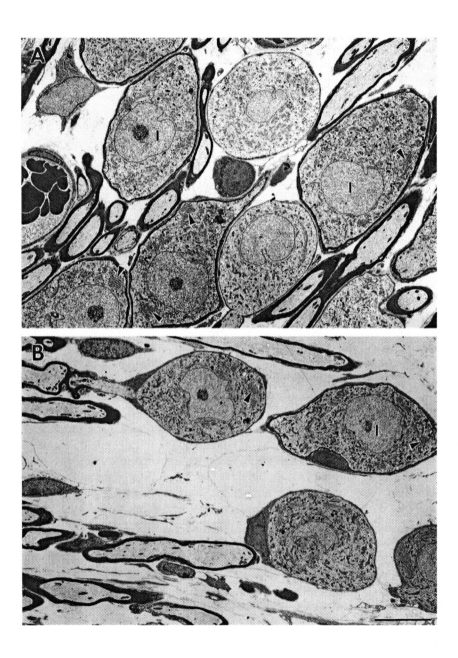

One hypothesis in interpreting in vivo studies suggesting that higher frequency modulated signals are more effective in promoting SG survival is that such signals may be more similar to the natural inputs to the auditory neurons (i.e., compared to simple low-frequency signals used in earlier experiments). Thus, higher frequency modulated stimulation is more effective in driving the mechanisms that underlie the trophic effects of depolarization and promotes marked improvements in SG survival (Leake et al. 1999, 2000a). However, even with the most effective electrical stimulation protocols examined to date, it is clear that neural survival in the deafened, stimulated ears does not approach normal levels.

Thus, there has been recent interest in the potential development of methods to induce increased expression of neurotrophins prior to or in conjunction with electrical stimulation by an implant in order to further promote optimal survival of SG neurons. Most of the experimental in vivo studies mentioned previously, which have reported that exogenous administration of neurotrophic factors promotes survival of SG neurons, have employed osmotic infusion pumps for direct administration of neurotrophins into the inner ear. This method has obvious limitations for human application (e.g., the pump reservoirs require frequent refilling, and neurotrophins maintain their bioactivity for quite limited periods). Other investigators are exploring gene therapy as a possible means of inducing increased expression of neurotrophic factors by the cells within the inner ear and thereby promoting SG neuronal survival. Studies investigating cochlear gene therapy have been initiated using a variety of viral vectors and various rodent animal models (Geshwind et al. 1996; Raphael et al. 1996; Weiss et al. 1997; Lalwani et al. 1998; Staecker et al. 1998; Yagi et al. 2000). Important challenges with this approach are the timing/regulation of gene transfer, limited duration of expression, and development of nonimmunogenic/nonpathogenic vectors, all of which are likely to be critical factors in the success of any procedure that might be proposed for eventual clinical application (Yagi et al. 2000).

FIGURE 4.6. Transmission electron micrographs show spiral ganglion neurons more than 8 months after neonatal deafening. Sections were taken near the cochlear base. **A**: In the stimulated ear, neural survival is ≈80% of normal, and many of the type I cells (I) appear to have relatively normal cytoplasm and myelin. Other cells, however, exhibit pathological alterations such as thinning of the myelin around the outside of the cell body, shrinkage, and compaction of the Nissl bodies (arrowheads). The type I neurons normally receive input from the inner hair cells (IHCs) and thus are considered of primary relevance to the function of a cochlear implant. Type II neurons, not shown here, receive convergent input from outer hair cells (OHCs) and their function is not well understood. **B**: In the control deafened cochlea the cell density is less than 30% of normal and most remaining cells show marked pathological alterations. Scale bar = 10 µm. (From Leake et al. 1999, with permission from *Journal of Comparative Neurology*, John Wiley & Sons.)

4.1.4 Insertion Trauma Causes Significant Degeneration of Spiral Ganglion Neurons

Research in animal models has suggested that mechanical damage to cochlear structures due to direct trauma caused by insertion of the electrode into the scala tympani is another factor that can significantly affect SG survival in the implanted cochlea. In studies of the effects of many months of chronic stimulation delivered by a cochlear implant, an important additional finding was that trauma caused by insertion of intracochlear electrodes, even slight damage to the osseous spiral lamina or basilar membrane, markedly decreased SG neural survival (Leake et al. 1999, 2000a). In these studies, the apical end of the electrode array caused some degree of insertion trauma in the cochlear region 40% to 50% from the base in many of the implanted cochleas (Fig. 4.7). In these damaged areas, the trophic effects of electrical stimulation in maintaining increased neural survival seen in other cochlear regions were largely offset by the effects of the insertion trauma (see Fig. 4.3, 40% to 50% cochlear region). Moreover, the degree of damage caused by the implanted electrode appeared to be related to the extent of SG cell loss seen in histological sections. The clinical relevance of these observations is emphasized by results in studies of human cochlear implant subjects. Advanced imaging techniques (submillimeter computed tomography, CT) have demonstrated marked intersubject variation in the intracochlear position of implanted electrodes and a high incidence (about one third of a group of 20 subjects) of insertions that would produce mechanical trauma, including compression of arrays, twisting, and intrascalae excursions of electrodes (Ketten et al. 1998) (see sections 4.2.1 and 4.2.2, below). The data from animal studies clearly suggest that such trauma would compromise SG neural survival in the damaged areas and would offset possible neurotrophic effects induced by chronic electrical stimulation. It seems clear that an important priority in the design of future generations of clinical cochlear implants should be the development of intracochlear electrodes with improved mechanical characteristics that minimize the likelihood of such mispositioning of electrodes and consequent trauma.

4.2 Anatomical Effects of Electrical Stimulation and Plasticity in the Central Auditory System

4.2.1 Effects of Electrical Stimulation in the Cochlear Nucleus

A few studies have examined the morphological effects of chronic electrical stimulation in the CN of deafened animals. Histological studies of the CN complex in neonatally deafened, chronically stimulated cats have demonstrated profound degenerative changes in the CN, changes that are progressive for many months after deafening (Hultcrantz et al. 1991; Lustig

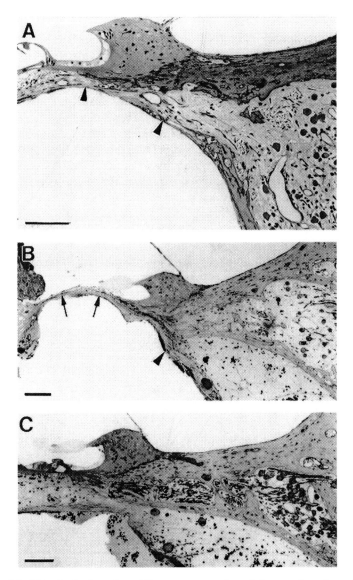

Figure 4.7. Histological sections illustrate examples of the results of mechanical trauma to the cochlea caused by insertion of an electrode array. **A**: Slight damage to the distal aspect of the osseous spiral lamina is seen here, in the region adjacent to the apical tip of the implanted electrode (arrowheads). **B**: A more severe fracture of the osseous spiral lamina occurred in this cochlea (arrowhead), in which the basilar partition is grossly displaced upward into the scala media (arrows). **C**: This section was taken from the same specimen as the preceding example, and from a region just 0.5 mm apical to the area shown in B. There is no induced trauma at this site, and SG survival is much better. Scale bars = 100 μm (From Leake et al. 1999, with permission from *Journal of Comparative Neurology*, John Wiley & Sons.)

et al. 1994; Osofsky et al. 2001). As compared to data from normal adults, the cochlear nuclei of neonatally deafened animals showed (1) marked shrinkage in the volume of the CN, (2) a significant reduction in the density (number of cells/unit area) of spherical cells within the anteroventral cochlear nucleus (AVCN), and (3) a significant reduction in the mean cross-sectional area of AVCN spherical cells. These degenerative changes are consistent with many previous studies showing that neonatal deafening results in profound adverse effects within the cochlear nucleus, as discussed previously (see above, section 4.1.2). Comparisons between the stimulated and control CN in these animals did not reveal any significant differences in either nuclear volume or spherical cell density due to chronic stimulation. However, in measurements of the cross-sectional area of spherical cells in the AVCN, a modest but significant increase (about 6%) was observed in the stimulated CN (Hultcrantz et al. 1991; Lustig et al. 1994; Osofsky et al. 2001) as compared to the opposite side. Figure 4.8 presents CN data from chronic stimulation experiments using higher frequency modulated stimuli, in which a mean increase in SG neural survival of more that 21% was demonstrated. These subjects showed CN data that were virtually identical

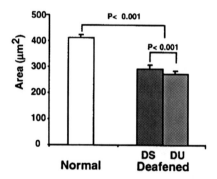

FIGURE 4.8. Cross-sectional areas of a class of neurons called spherical cells from the rostral area of the anteroventral cochlear nucleus (AVCN) are shown for normal cats and neonatally deafened animals after several months of unilateral electrical stimulation. Data demonstrate a marked reduction in cell size as a consequence of early deafness and as compared to normal. Neurons in the AVCN on the deafened control side had a mean area that was 66% of normal. By comparison, cells in the AVCN on the side of the stimulated ear were significantly larger at 71% of normal. This 5% increase induced by stimulation is quite modest, considering that these animals showed substantial increases in SG cell survival (>20%) in the stimulated cochleae as compared to the control deafened ears. (These spherical cell data are taken from six animals for which SG data are shown in Fig. 4.3.) Thus, CN changes do not appear to parallel the extent of SG maintenance induced by stimulation.

to data reported previously for animals with more modest differences in SG survival (Osofsky et al. 2001). Thus, the CN still showed only a modest effect of stimulation in preventing or reversing pronounced degenerative changes after deafening, despite marked differences induced in SG survival. One possible explanation for this finding is the delay that occurred before chronic stimulation was initiated (mean age of 7 weeks).

During normal development of the CN in cats, the neurons of the AVCN undergo an early growth phase with rapid increase in nuclear and cytoplasmic cross-sectional areas during the first 4 weeks of postnatal life (Larsen 1984). This is followed by a second, longer period of development during which cells grow more gradually and reach mature sizes by about 12 weeks postpartum. Therefore, in the studies described above, one possibility is that electrical stimulation was initiated too late in development to prevent or reverse the profound consequences of neonatal deafness. These findings provide additional evidence for the existence of a critical period, after which degenerative changes in the CN caused by deafness are largely irreversible, even when the input provided by electrical stimulation is sufficient to maintain markedly improved SG survival.

With respect to this issue, it should be noted that Matsushima et al. (1991) reported a similar study in cats that were deafened at 1 month of age and then chronically stimulated. These investigators observed somewhat greater differences in CN cell size, suggesting that chronic electrical stimulation may be more effective in preventing degenerative changes in the CN when deafness occurs at 4 weeks of age rather than at birth. Thus, results again suggest that age at time of deafening is critical in determining the extent to which the CN is sensitive to stimulation-induced "protective" effects. However, given the relative paucity of data currently available, this is clearly an area requiring additional research.

4.2.2 Functional Consequences of Chronic Electrical Stimulation and Plasticity

4.2.2.1 Alterations in Central Auditory System Spatial Representations

In addition to anatomical studies, a number of electrophysiological studies have examined the functional consequences of chronic electrical stimulation of the auditory nerve on neuronal responses in the central auditory system. Experiments have shown that stimulation delivered by a cochlear implant can result in significant changes both in the spatial selectivity of activation and in temporal processing within the auditory midbrain (inferior colliculus, IC) of neonatally deafened cats (Snyder et al. 1990, 1991, 1995; Vollmer et al. 1999; Leake et al. 2000a,b). Studies conducted in animals that were deafened, implanted as adults, and studied acutely as controls have shown that stimulation with intracochlear bipolar electrodes elicits spatially selective patterns of activation, which reflect the well-known and highly precise tonotopic organization of the IC. That is, electrodes at a basal location selectively activate the IC at a relatively superficial location, and more apical implant channels activate progressively deeper sites in the IC, corresponding to the dorsal-to-ventral gradient of high-to-low frequencies within the central nucleus of the IC.

Neonatally deafened subjects that are raised without an implant and examined as adults with no prior experience with electrical stimulation (unstimulated group) exhibit spatial selectivity that is similar to that of normal controls (Fig. 4.9). This finding suggests that the frequency organization within the midbrain develops relatively normally and is unaltered despite the complete lack of normal auditory input during development in these animals.

In contrast, when such neonatally deafened animals receive chronic electrical stimulation, spatial "maps" in the midbrain exhibit significant changes. Specifically, the area within the IC excited by the chronically activated electrodes is significantly expanded and on average is almost double the area for the identical stimulus configuration in either unstimulated deaf littermates or acutely deafened adult controls. These results indicate that the developing central auditory system is capable of substantial plasticity. The initially restricted area excited by the stimulated cochlear neurons expands over time as the central auditory system adapts to the only available afferent input. However, this expansion actually represents a significant degradation in the cochleotopic organization (frequency selectivity) of the central auditory system. It is important to note that chronic stimulation on two adjacent bipolar intracochlear channels of the cochlear implant can be effective in maintaining more normal selectivity of central representations of stimulated cochlear sectors. Thus, competing inputs elicited by electrical stimulation on adjacent channels may prevent the expansion and degradation of frequency selectivity seen after single-channel stimulation (Fig. 4.8).

Additional animal studies, conducted in the auditory cortex of both neonatally deafened cats (Raggio and Schreiner 1999) and congenitally deaf cats (Klinke et al. 1999), also have reported that long-term deafness alters spatial selectivity of electrical stimulation. Further, chronic electrical stimulation of the cochlea may result in an expansion of the cortical area activated by the cochlear implant (Klinke et al. 1999; Kral et al. 2001). Recent results have led to the suggestion that there is a critical or sensitive period of approximately 5 months in the congenitally deaf cat, after which cortical plasticity (as evidenced by expansion of activation areas) is more limited (Kral et al. 2001).

Taken together, findings from all these experiments indicate that in animals deafened early in life, electrical stimulation from a cochlear implant can induce marked functional plasticity and reorganization within the central auditory system. Central representations in these animals can vary substantially, because they are markedly influenced by intersubject vari-

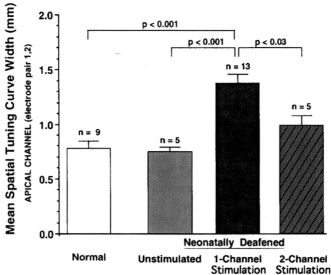

**SPATIAL REPRESENTATIONS
IN THE INFERIOR COLLICULUS**

FIGURE 4.9. This summary graph illustrates the mean spatial selectivity in the auditory midbrain for electrical stimulation by a standard cochlear implant channel. Spatial tuning curve (STC) widths are presented for four experimental groups of cats. (STC widths are calculated at 6 dB above minimum threshold for stimulation of the apical channel of the implant, and measurements are averaged for several recording penetrations in each subject.) STC width in normal control subjects was 0.78 mm, and the mean in the neonatally deafened, unstimulated group was virtually identical at 0.74 mm. The single-channel intracochlear stimulation group had a mean STC width of 1.39 mm, indicating an expansion of the area excited to almost double that seen in normal control animals and in neonatally deafened, unstimulated animals naïve to electrical stimulation. In contrast, the 2-channel stimulation group maintained narrower, more selective STC widths (mean of 1.00 mm) that were not significantly different from normal. (from Leake et al. 2000b, with permission from *Hearing Research*, Elsevier Science B.V., Amsterdam, The Netherlands.)

ables like threshold, extent of neural degeneration, and individual stimulation history.

Research in other sensory systems (particularly the visual system) has demonstrated that the initial input activity during development initiates a *critical period*, after which organizational changes driven by aberrant or distorted initial inputs are often largely irreversible. If the auditory system operates under similar mechanisms, then the initial input from the cochlear implant may be critically important. If the changes in the auditory midbrain demonstrated in the single-channel stimulation experiments described above were irreversible, they would clearly limit the possibility for selective multichannel stimulation. Unfortunately, this may be an actual problem in very young children using the cochlear implant, because fitting a processor and setting channel loudness levels is so difficult. If one channel is set too loud, it may dominate the input and produce a type of distortion similar to that described for single-channel stimulation (Leake et al. 2000b). Given the limited data currently available, it will be important in the future to further examine the effects of various formats of multichannel stimulation in animal experiments and to determine whether distortions induced by initial stimulation in young deafened animals are irreversible later in life.

The implications of these results in animal experiments are potentially very important for clinical pediatric application of cochlear implants. The marked intersubject variability and the severe (and possibly irreversible) distortions in the central cochleotopic (frequency) organizations seen in animals stimulated during maturation emphasizes the importance of the initial fitting of cochlear implants in the naive, developing auditory system. These results suggest that there may be specific ways of introducing stimulation in a young deaf child that might optimize setting up appropriately distinct central representations of individual channels of the cochlear implant (Leake et al. 2000). For example, this might be accomplished by introducing the channels one at a time and training implant recipients to discriminate among channels, rather than simply turning on all the channels simultaneously (Dawson and Clark 1997).

The specific mechanisms underlying the alterations or distortions in central auditory representations induced by electrical stimulation in animal studies have not been elucidated. It is not known at present whether these functional alterations result from modifications of synaptic efficacy or actual changes in neural connections, and if so, at what levels of the central auditory system these changes occur. It has been shown that neonatal cochlear ablation can induce dramatic modifications in the anatomical projections from the contralateral cochlear nucleus to both the superior olivary complex and the IC (Nordeen et al. 1983; Moore and Kitzes 1985; Moore and Kowalchuk 1988; Irvine and Rajan 1994; Russell and Moore 1995; Harrison et al. 1996; for review see Kitzes 1996). Such a phenomenon could account for the functional changes seen in the IC with chronic electrical stimulation. Unfortunately, however, no morphological studies have yet been conducted that directly examine the effects of electrical stimulation on the organization of central auditory connections after deafness.

4.2.2.2 Electrical Stimulation Effects on Temporal Response Properties of Central Auditory Neurons

In addition to evidence for plasticity in central auditory spatial representations induced by electrical stimulation as described above, electrophysiological studies also have analyzed effects on temporal properties of single neurons in the IC responding to electrical stimuli. Many temporal characteristics of both IC (Snyder et al. 1995) and cortical (Raggio and Schreiner 1994; Schreiner and Raggio 1996, 1999) neuronal responses to electrical signals are similar to responses to acoustic signals. For example, in the IC of control animals deafened and studied as adults, all major response types are identified, and first spike latencies and phase-locking capacities appear to be very similar (Snyder et al. 1995, 2000). However, quantitative analysis of response patterns (peristimulus time histograms, PSTHs) in cats deafened at a young age revealed significant changes in the temporal responses of midbrain neurons. Specifically, the temporal resolution of IC neurons, the ability of these neurons to phase lock to or follow repetitive signals, is altered both by severe sensory deprivation during development and by controlled, temporally stereotyped electrical stimulation. When frequency transfer functions for all IC neurons were analyzed quantitatively for adult-deafened "normal" control animals, the average maximum following (phase-locking) frequency is about 100 pulses per second (pps). Neonatally deafened, unstimulated cats, studied at prolonged intervals after deafening showed a significant decrease in the temporal resolution of IC neurons to an average of 86 pps (Snyder et al. 1995).

In contrast, chronically stimulated cats showed either maintenance of normal temporal resolution or an increase in temporal resolution, depending on the temporal properties of the electrical signals delivered by the implant (Vollmer et al. 1999). Animals stimulated only with a simple low-frequency signal (30 pps) were found to have maintained normal temporal resolution, (mean maximum following frequency of 109 pps). However, higher frequency modulated and in some subjects behaviorally relevant electrical stimulation resulted in a marked, highly significant increase in temporal resolution (mean maximum following frequency of 134 pps) (Fig. 4.10). These changes in temporal resolution were restricted to neurons in the central nucleus of the IC, while neurons in the external nucleus showed much lower temporal following and were not significantly modified following chronic stimulation.

Thus, experience with these electrical stimuli in neonatally deafened animals can profoundly alter temporal response properties of central auditory neurons, and the magnitude of these effects is dependent on the specific temporal properties of the signals delivered by the implant. These frequency-dependent effects of chronic stimulation in increasing the capacity of the midbrain neurons to resolve relatively fast temporal events may be important in understanding differences among performance of some cochlear implant subjects and in understanding how subjects improve over time (see Mckay, Chapter 7). Does this ability to follow electrical pulse trains at higher-than-normal frequencies underlie the success of recent sophisticated continuous interleaved sampling (CIS) speech processor

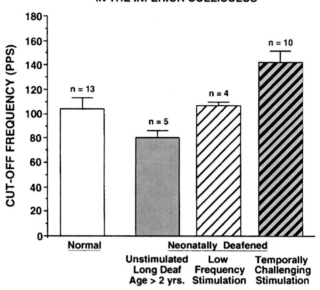

**TEMPORAL RESOLUTION OF NEURONS
IN THE INFERIOR COLLICULUS**

FIGURE 4.10. The mean maximum following (phase-locking) frequencies for neurons in the central nucleus of the inferior colliculus are shown for normals and three groups of neonatally deafened cats. Normal controls were adult cats that were acutely deafened just prior to study, and their mean following frequency was 104 pulses per second (pps). Unstimulated, neonatally deafened animals studied after prolonged periods of deafness showed a clear reduction in temporal resolution to a mean value of 81 pps. In contrast, neonatally deafened animals that received chronic low-frequency stimulation (30 pps pulses, 200 μsec/phase) exhibited maintenance or restoration of normal temporal resolution with a mean maximum cutoff frequency of 107 pps. Moreover, cats that received prolonged stimulation with higher frequency, temporally challenging signals (e.g., 300 pps amplitude modulated at 30 pps; or analog cochlear implant signal processor) showed a pronounced increase in frequency following capacity with a mean cutoff frequency of 142 pps. (From Leake et al. 2000a, with permission from Thieme Medical Publishers, Inc., Georg Thieme Verlag Stuttgart, Germany.)

designs, which utilize amplitude modulation of high-frequency pulse trains? Is poorer speech recognition capability of implant subjects related to an inability of their central auditory system to entrain to higher frequencies (e.g., due to specific deafness pathology) and a consequent inability to encode the fine structure of the electrical signals? These issues should be addressed in future studies by systematic evaluation of the functional effects of various parameters of chronic electrical stimulation in appropriate deaf animal models.

5 The Electrode/Neural Interface in Human Subjects

5.1 What Are the Target Neural Structures for Electrical Stimulation by a Cochlear Implant?

As described previously, in the normal organ of Corti the hair cells send signals to the central nervous system via chemical synapses formed by the distal processes of the spiral ganglion neurons. Within the organ of Corti, all the nerve fibers are unmyelinated. Soon after they pass through the basilar membrane into the osseous spiral lamina, the radial nerve fibers become myelinated, and this is assumed to be the normal spike generation site of the spiral ganglion neurons. However, in profoundly deaf individuals in whom all or most of the cochlear hair cells have degenerated, it can be expected that the spiral ganglion will have undergone pathological alteration, the extent of which depends on the etiology and duration of deafness as discussed above. In individuals with minimal pathology, the target site for cochlear implant electrodes to selectively activate SG neurons at lowest threshold might be within the osseous spiral lamina (e.g., with bipolar pairs of electrodes centered around the habenula, the presumed spike-initiation site). On the other hand, if the distal processes have largely degenerated in individuals with more severe pathology, the target site for electrical stimulation could be SG cell somata within the cochlear modiolus, suggesting that stimulating electrodes would be more efficient if positioned along the inner radius of the implant within the scala tympani.

When cochlear prostheses were first introduced, only individuals who were profoundly deaf and received no benefit whatsoever from hearing aids were considered to be candidates for an implant. Moreover, many subjects receiving these early devices had been deaf for extended periods, and thus were likely to have relatively severe cochlear pathology and degeneration of the spiral ganglion and auditory nerve, as indeed was demonstrated by initial temporal bone studies. Thus, the ganglion cell bodies within the modiolus were probably the primary target for stimulation in this group of cochlear implant users. However, the selection criteria for cochlear implant recipients have been gradually relaxed over time, and many individuals with substantial residual hearing are now receiving cochlear implants. In these subjects, it is likely that substantial populations of the peripheral axons of the SG neurons, called the radial nerve fibers, within the osseous spiral lamina will be intact and available for activation by electrical stimulation. Theoretically, the increased longitudinal distance around the spiral at a more peripheral site would spread out the cochlear frequency range and thus might allow more selective, non-overlapping stimulation of adjacent implant channels. Modeling studies predict that the presence of radial nerve fibers will sharpen selectivity of stimulation because the dendrites will respond with lower thresholds to a stimulating electrode in close proximity (for review, see Rebscher et al. 2001). However, there is currently no direct evidence indicating whether or not there is significant advantage in stimulation of the peripheral axons, as compared to direct activation of the cell bodies of neurons within the modiolus. Complicating factors such as shunting of current through the adjacent fluid of the scala if electrodes are not directly in contact with the structures they are designed to stimulate, and the difficulty in modeling the spread of current through the highly non-isotropic structure of the cochlea, make this a very difficult question to address experimentally (see Abbas and Miller, Chapter 5).

In this context, it is noteworthy that modifications in designs of intra-cochlear electrodes have been recently introduced, which are intended to place the stimulating contacts closer to the modiolus within the scala tympani. These second-generation clinical devices have been called "peri-modiolar" electrodes, and the Nucleus Contour™ and Advanced Bionics HiFocus™ electrodes are two examples (see below). Both of these new electrodes have individual stimulating contacts positioned in a single lon-gitudinal row along the inner radius of the carrier in the scala tympani. The implicit assumption of this design modification is that the target neural structures are the SG cells within the modiolus. The basis for this assump-tion is unclear, however, because the current population of implant candi-dates, many of whom have some residual hearing, are more likely to have surviving radial nerve fibers and thus might benefit from electrode config-urations designed to selectively stimulate the radial nerve fibers.

5.2 Evolution of Cochlear Implant Electrode Designs

As mentioned previously, insertion trauma was well documented in early cochlear implant subjects who had received electrodes consisting of one to five ball-tipped wires introduced through the round window (Johnsson et al. 1979, 1982; Fayad et al. 1991; O'Leary et al. 1991). Specific observations included the upward displacement of the basilar membrane and organ of Corti toward the scala vestibuli, fracture of the osseous spiral lamina or tearing of the spiral ligament as a result of upward pressure, and in many cases electrodes actually ruptured the basilar partition and were positioned within the scala vestibuli (Kennedy 1987).

To reduce the probability of such insertion trauma, multichannel elec-trodes introduced in the early 1980s incorporated two basic strategies. One approach was to mold the shape of the silicone electrode carrier to follow the spiral of the cochlea and to control the flexibility of the electrode array with a central stiffening structure designed to prevent upward deviation of the electrode tip (Loeb et al. 1983; Rebscher et al. 1999). This design, orig-inally developed at the University of California at San Francisco, evolved into the Clarion™ spiral electrode produced by Advanced Bionics Corpo-ration. A second approach was to make electrode arrays that were quite small in diameter and very flexible. Examples include the Nucleus 22 Banded™ and Med-El Combi 40™ electrodes.

When these devices were introduced, several investigations evaluated trauma following trial insertions into human cadaver temporal bones (e.g., O'Reilly 1981; Shepherd et al. 1985; Webb et al. 1988). Since that time a large number of profoundly deaf individuals have received these devices and many have used them for decades, and there have been several reports of postmortem evaluations of temporal bones of implant recipients (Zappia et al. 1991; Marsh et al. 1992; Nadol et al. 1994) as discussed previously. In addition, several reports have described the development of high-resolution radiographic methods (submillimeter spiral CT scans) to examine electrode position in living cochlear implant recipients (Skinner et al. 1994, 2002; Ketten et al. 1998). All of these studies have concluded that significant trauma is associated with cochlear implantation in a substantial proportion of cases. Rupture of the electrode through the basilar membrane and into the scala vestibuli was relatively common with these devices. For example, Ketten et al. (1998) reported clear excursions of electrode arrays into the scala media or scala vestibuli in 6 of 20 living subjects with Nucleus™ electrodes. These studies also revealed that even when insertion was atraumatic, these electrode arrays seldom achieved what is considered to be optimum positioning within the scala tympani, i.e., near either the modiolus or osseous spiral lamina.

Consequently, a new generation of "perimodiolar" multichannel electrodes has recently been developed, and several new electrode designs have been introduced into clinical application throughout the world. These perimodiolar electrodes are designed to increase overall system efficiency and subject performance by locating stimulating contacts closer to target neurons in the modiolus. However, because these novel electrodes represent very different design strategies than those of prior clinical devices, there is concern that the goal of placing an electrode nearer the modiolus may increase either the probability of insertion trauma or its severity. As a result, there has been a renewed interest in temporal bone studies to evaluate the precise positioning of intracochlear electrodes, to examine the incidence and severity of insertion trauma and to apply this information to further improve devices.

Earlier histological studies of temporal bones from deceased cochlear implant recipients or cadaver temporal bones cited above were limited by several factors. First, a number of different techniques were used to evaluate different devices, making comparison of results across studies difficult. Second, the earlier histology protocols generally required removal of wire electrodes prior to sectioning of the cochlea, introducing the possibility that withdrawal of the electrode may have caused additional trauma and often making reliable identification of the location of electrodes in situ impossible. These limitations have been largely overcome in contemporary studies by the use of plastic embedding and cutting techniques that were originally developed to evaluate bone and mineral specimens (Gstoettner et al. 1997, 1999; Roland et al. 2000; Tykocinski et al. 2001). These methods permit sectioning of temporal bones with electrodes in situ, so the position of the array and its relationship to any trauma or pathology can be documented accurately. In addition, when the surface is highly polished, these plastic-embedded specimens can provide excellent resolution permitting detailed examination of specific cochlear structures along the electrode path (Leake et al. 2003) (Fig. 4.11).

Several of these recent temporal bone studies have assessed intracochlear position and/or insertion trauma with the newly introduced perimodiolar electrodes using these plastic-embedding techniques. In general, these studies found that the newer electrode designs were successful in positioning stimulating contacts closer to the modiolus than earlier versions. However, there was no systematic improvement in the occurrence of trauma between the first- and second-generation devices. In fact, in at least some instances, the newer designs appear to produce more trauma than their predecessors. Leake et al. 2003 in press a,b directly compared two of these new perimodiolar electrodes with the previous devices from the same manufacturers, and Figure 4.12 illustrates typical trauma observed with two of the electrodes evaluated.

The mechanisms of damage observed in these studies can be divided into two distinct categories. The first type of trauma occurs when the electrode tip deviates upward through either the spiral ligament or basilar membrane and comes to rest in the scala media or scala vestibuli. This is the most frequently reported injury with straight, flexible electrodes such as the Nucleus 22[TM] banded array, the Med-El Combi 40[TM], and the new Cochlear Contour[TM] spiral electrode, which is inserted on a straight stylet. The mean frequency of penetration among the studies cited above is approximately 35% (n = 138) for these electrodes as a group, which is quite similar to the frequency of this finding reported by Ketten et al. (1998) in high-resolution imaging studies of 20 living Nucleus[TM] implant recipients.

A second type of injury was observed in recent studies of the Advanced Bionics HiFocus[TM] electrode with attached positioner. With this device, significant trauma was observed when the electrode was inserted to its full insertion depth (Fig. 4.12). However, when inserted to approximately two thirds of its full length (mean depth of 17.1 mm or 372 degrees), the electrode was positioned completely within the scala tympani and was largely atraumatic. At least in this particular study, it appeared that the dimensions of the HiFocus array were too large to allow full insertion in most of the temporal bone specimens studied (n = 8). In contrast, Fayad et al. reported test insertions of the original Clarion Spiral electrode in 10 temporal bones, with no evidence of excursion of the electrode from the scala tympani (Fayad et al. 2000), and a more recent study (Leake et al. 2003) reported only relatively minor damage in approximately one third of temporal bones implanted with the Spiral[TM] electrode. The findings to date suggest that these devices, which both incorporate a central stiffening structure designed to resist vertical deviation of the electrode upward into the scala vestibuli,

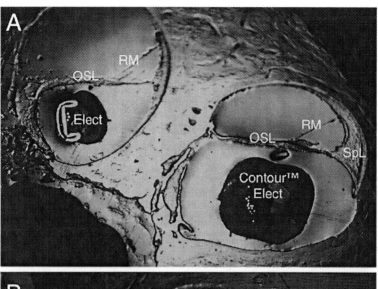

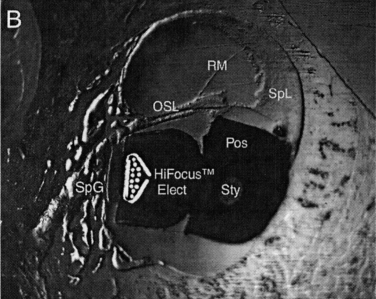

FIGURE 4.11. These micrographs show mid-modiolar sections through plastic-embedded temporal bone specimens after trial insertion of perimodiolar electrodes. **A**: A Cochlear Contour™ electrode is seen well positioned in the basal and middle cochlear turns. **B**: An Advanced Bionics HiFocus™ electrode with positioner (Pos). To insert this electrode, the positioner is loaded on a stylet (Sty) and extruded from the stylet into the scala tympani by an actuator. In this example, the electrode achieves excellent perimodiolar positioning with the stimulating contact directly adjacent to the spiral ganglion (SpG). The resolution achieved with plastic embedding is apparent in the detail seen in the osseous spiral lamina (OSL), Reissner's memebrane (RM), and spiral ligament (SpL).

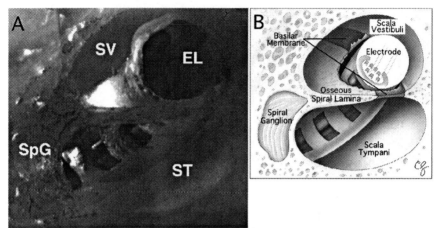

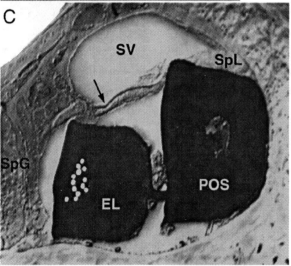

FIGURE 4.12. Insertion trauma observed with second-generation electrodes varied with the type of device. **A,B:** The smaller and very flexible Cochlear Contour™ array often penetrated the cochlear partition and was partially positioned in the scala vestibuli, as illustrated in a representative temporal bone specimen (**A**) and schematic drawing (**B**), with approximately the same frequency of occurrence as the banded Nucleus 22 electrode. **C:** The Advanced Bionics HiFocus™ array seldom deviated into the scala vestibuli due to its vertical rib structure. However, this electrode was associated with damage related to the overall dimensions of the two-part design, as shown in this image in which the electrode (EL) has fractured the osseous spiral lamina (arrow) and the spiral ligament (SpL) has been grossly dislocated upward into the scala vestibuli (SV) by the positioner (POS). The electrode and positioner also appear wedged and flattened against the bone of the modiolus and otic capsule, respectively, raising a concern about possible pressure erosion.

do effectively reduce the occurrence of this form of trauma. However, the HiFocus electrode with positioner is likely to be inserted to a point where it is too large for the dimensions of the scala tympani in many individuals.

Together, these reports describe two different types of trauma associated with different electrode designs. Moreover, the findings also suggest that specific aspects of each design have desirable attributes that in the future might be integrated into an optimized array, e.g., a cross-sectional profile that is scaled to fit atraumatically into almost every cochlea, and mechanical characteristics that effectively prevent vertical deviation of the array and also position stimulating contacts close to the modiolus. It should be noted that since the completion of the described studies at UCSF, Advanced Bionics Inc. has withdrawn the electrode positioner from clinical use with the HiFocus™ electrode. Cochlear Corporation has modified the tip of the Contour™ electrode and the suggested technique for insertion.

5.3 Fuure Directions for Optimizing Cochlear Implant Designs

In the future, additional studies using modern imaging techniques will be invaluable in relating specific data on placement of electrodes in living implant recipients (and the effects of mispositioning, twisting, or kinking of electrodes, etc.) to various measures of perception and performance. This approach may offer the most direct and empirical method for better defining the importance of various factors, such as precise electrode position and depth of insertion, for the optimum function of an implant—as exemplified by the findings of Ketten et al. (1988) and Skinner et al. (2002). Moreover, even higher resolution imaging protocols are now becoming available with multirow detector spiral CT technology (Skinner, personal communication). The potential for providing a wealth of information relating details of individual anatomy/pathology and electrode position to individual performance with the cochlear implant is an extremely promising area of future investigation.

As discussed above, the functional consequences of traumatic damage to the cochlea and auditory nerve degeneration are difficult to predict because of the many variables affecting the ultimate performance of individual implant recipients. For this reason, the importance of insertion trauma and even the consequences of severe neural loss are still areas of intensive debate (Incesulu and Nadol 1998). However, we suggest that preservation of the entire residual population of spiral ganglion neurons that is present in the cochlea of each implant recipient at the time of surgery is an important goal for future device development. One obvious reason for this mandate is that a number of subjects with significant residual hearing have now received special short cochlear implant electrodes with the specific goal of combining a hearing aid to utilize residual low-frequency hearing with a cochlear implant in the basal cochlea. Thus, the goal of atraumatic insertion of electrodes is likely to become increasingly important as sophisticated systems exploiting residual hearing evolve. Moreover, since so many very young children are now receiving implants, it is likely that reimplantation for the purpose of device replacement or upgrade will become increasingly common. As implant electrodes and signal processors evolve further in the future, it seems highly likely that more sophisticated devices will be able to take advantage of a more intact residual neural population to provide improved benefit to the recipient.

Finally, there is another implication of findings from the temporal bone studies of the new perimodiolar electrodes cited above that should not be overlooked. Specifically, when these electrodes deviated from their intended placement within the scala tympani (especially when there is excursion into the scala vestibuli or kinking or twisting of the array), the consequence was not only trauma to the cochlea, but also a very significant mispositioning of contacts away from the cochlear modiolus. Such mispositioning will clearly diminish or eliminate the intended performance gains of the perimodiolar designs. Moreover, such idiosyncratic positioning also has been shown to lead to the development of dense connective tissue or the deposit of new bone adjacent to or surrounding the electrode array and aberrant psychophysical findings (Ketten et al. 1998; Skinner et al. 2002). Each of these factors would be likely to further degrade performance of the device below what it would have been if the electrode have been ideally positioned in a given subject.

As outlined above, recent studies have identified several features of current intracochlear electrodes that appear to contribute to reduced trauma and improved positioning. Ideally, an optimized electrode array in the future would incorporate both a) the inherent advantages of a small, smoothly tapering cylindrical profile that would fit atraumatically into almost every temporal bone (and could be removed easily if replacement was required); and 2) the mechanical characteristics provided by a central rib or other integral structural feature that effectively prevents upward deviation of the electrode tip into the scala media or scala vestibuli, while at the same time positioning the stimulating contacts near the modiolus and osseous spiral lamina.

6. Summary

Modern multichannel cochlear prosthesis electrodes have been designed to selectively activate discrete sectors of the auditory nerve at several sites along the cochlear spiral in individuals with sensorineural hearing loss. Most of these devices employ a flexible cylindrical carrier of silicon rubber that is inserted into the scala tympani and positions individual stimulating contacts near the spiral ganglion neurons. The success of multichannel cochlear implants depends on the likelihood that a significant fraction of the auditory neurons may survive for many years even after profound hearing loss. However, in addition to the condition of the auditory periphery, the relative integrity or extent of degeneration of the central auditory pathway is likely to affect the benefit that can be obtained from a cochlear implant.

Studies conducted in animals have provided compelling evidence that the anatomical effects of auditory deprivation and deafferentation are much more severe when hearing loss occurs early in development as compared to the extent of pathology when deafness occurs later in life. In addition, well-controlled studies of the effects of stimulating the auditory nerve via a cochlear implant have demonstrated that several months of electrical stimulation is neurotrophic and promotes significantly increased survival of the cochlear spiral ganglion neurons. Less impressive effects have been reported in the cochlear nuclei, at least in animals deafened at a young age, in which the severe effects of auditory deprivation are only modestly ameliorated by chronic electrical stimulation. In contrast to these trophic influences of electrical stimulation, direct trauma to cochlear structures, which may occur at the time of insertion of the electrode array, is associated with a marked decrease in neuronal survival in the damaged regions. The clinical relevance of these findings in animal studies is emphasized by results in recent studies of living human cochlear implant recipients, in which high-resolution imaging techniques have demonstrated a high incidence of insertions that would produce mechanical trauma (i.e., about one third of a group of 20 subjects had compression, twisting, and/or intrascalar excursions of electrode arrays). Recent temporal bone studies of the newer "perimodiolar" cochlear implant designs have suggested specific desirable attributes of these designs that in the future might be integrated into further optimized arrays.

The functional consequences of electrical stimulation of the auditory nerve also have been studied in animal models, both at the level of the auditory midbrain (IC) and in auditory cortex. Results indicate that the fundamental tonotopic or frequency organization of the central auditory system develops relatively normally and is maintained despite severe bilateral auditory deprivation, even when deafness occurs early in life. However, chronic electrical stimulation delivered exclusively by a single bipolar channel of a cochlear implant in young, early-deafened animals can result in significant expansion of the central representation of that channel, and consequent distortion of the tonotopic (frequency) organization of the IC. In contrast, relatively normal (or even sharpened) spatial selectivity may be maintained by highly controlled, competing stimulation from multiple channels. Moreover, the temporal resolution of neurons in the IC may be significantly altered by experience with specific electrical signals delivered by an implant.

These results suggest that the synchronized neural activity elicited by a cochlear implant can exert a powerful influence on signal processing, particularly in the deafened, developing auditory system. Central auditory representations and temporal processing may be markedly altered by specific experience with electrical signals. These findings suggest that the initial fitting of a cochlear implant in the naive, developing auditory system of a young child may be extremely critical. Especially for such pediatric implant recipients, it may be highly advantageous to introduce stimulation in a manner that encourages the *discrimination of inputs from individual stimulation channels*. However, the exact implications of research conducted in animals for defining the nature and timing of "critical periods" as would apply in a young deaf child who receives a cochlear implant are still unclear, although general hypotheses are emerging based on this body of work.

Acknowledgments. Some of the research findings reviewed herein was supported by contract N01-DC02108 and grant R01-DC00160 to Dr. Leake from the National Institute on Deafness and Other Communication Disorders, National Institutes of Health.

References

Araki S, Atsushi K, Seldon HL, Shepherd RK, Funasaka S, Clark GM (1998) Effects of chronic electrical stimulation on spiral ganglion neuron survival and size in deafened kittens. Laryngoscope 108:687–695.

Balkany T, Gantz B, Nadol JB Jr. (1988) Multichannel cochlear implants in partially ossified cochleas. Ann Otol Rhinol Laryngol Suppl 97:3–7.

Blatchley BJ, Williams JE, Coleman JR (1983) Age-dependent effects of acoustic deprivation of spherical cells of the rat anteroventral cochlear nucleus. Exp Neurol 80:81–93.

Bohne B (1976) Safe level for noise exposure. Ann Otol Rhinol Laryngol 85:711–724.

Born DE, Rubel EW (1985) Afferent influences on brainstem auditory nuclei of the chicken: neuron number and size following cochlea removal. J Comp Neurol 231:435–445.

Clark GM, Shepherd RK, Franz BK-H (1988) The histopathology of the human temporal bone and auditory central nervous system following cochlear implantation in a patient. Acta Otolaryngol Suppl (Stockh) 448:1–65.

Coleman JR, O'Connor P (1979) Effects of monaural and binaural sound deprivation on cell development in the anteroventral cochlear nucleus of rat. Exp Neurol 64:533–566.

Coleman JR, Blatchley BJ, Williams JE (1982) Development of the dorsal and ventral cochlear nuclei in rat and effects of acoustic deprivation. Dev Brain Res 4:119–123.

Dallos P, Popper A, Fay RR, eds. (1996) The Cochlea. Springer Handbook of Auditory Research. New York: Springer-Verlag.

Dawson PW, Clark GM (1997) Changes in synthetic and natural vowel perception after specific training for congenitally deafened patients using a multichannel cochlear implant. Ear Hear 18:488–501.

Dodson HC, Bannister LH, Douek EE (1994) Effects of unilateral deafening on the cochlear nucleus of the guinea pig at different ages. Brain Res 80:261–267.

Dubrosky RT, Carter BD (1998) Coupling of the p75 neurotrophin receptor to sphingolipid signaling. Ann N Y Acad Sci 845:32–45.

Egami T, Sando I, Sobel J (1978) Noise-induced hearing loss: a human temporal bone case report. Ann Otol Rhinol Laryngol 87:868–874.

Eisenberg LS, Luxford WM, Becker TS, House WF (1984) Electrical stimulation of the auditory system in children deafened by meningitis. Otolaryngol Head Neck Surg 92:700–705.

Ernfors P, Lee K-F, Jaenisch R (1994) Mice lacking brain-derived neurotrophic factor develop with sensory deficits. Nature 368:147–150.

Ernfors P, van de Water T, Loring J, Jaenisch R (1995) Complementary roles of BDNF and NT-3 in vestibular and auditory development. Neuron 14:1153–1164.

Ernfors P, Duan ML, El-Shamy WM, Banlon B (1996) Protection of auditory neurons from aminoglycoside toxicity by neurotrophin-3. Nat Med 2:463–467.

Fariñas I, Jones KR, Backus C, Wang X-Y, Reichardt LF (1994) Severe sensory and sympathetic deficits in mice lacking neurotrophin-3. Nature 369:658–661.

Fayad J, Linthicum FH, Otto SR, Galey FR, House WF (1991) Cochlear implants: histopathologic findings related to performance in 16 human temporal bones. Ann Otol Rhinol Laryngol 100:807–811.

Fayad J, Luxford W, Linthicum FH (2000) The clarion electrode positioner: temporal bone studies. Am J Otol 21:226–229.

Feng AS, Rogowski BA (1980) Effects of monaural and binaural occlusion on the morphology of neurons in the medial superior olivary nucleus of the rat. Brain Res 189:530–534.

Fleckeisen C, Harrison R, Mount R (1991) Effects of total cochlear haircell loss on integrity of cochlear nucleus. Acta Otolaryngol Suppl (Stockh) 489:23–31.

Fritzsch B, Silos-Santiago I, Bianchi LM, Farinas I (1997) The role of neurotrophic factors in regulating the development of inner ear innervation. Trends Neurosci 21:159–164.

Gantz BJ, McCabe BF, Tyler RS (1988) Use of multichannel cochlear implants in obstructed and obliterated cochleas. Otolaryngol Head Neck Surg 98:72–81.

Geshwind MD, Hartnick CJ, Liu W, Amat J, Van De Water TR, Federoff HJ (1996) Defective HSV-1 vector expressing BDNF in auditory ganglia elicits neurite outgrowth: model for treatment of neuron loss following cochlear degeneration. Hum Gene Ther 7:173–182.

Gstoettner W, Plenk H, Franz P (1997) Cochlear implant deep electrode insertion: extent of insertional trauma. Acta Otolaryngol (Stockh) 117:274–277.

Gstoettner W, Franz P, Hamzavi J (1999) Intracochlear position of cochlear implant electrodes. Acta Otolaryngol (Stockh) 119:229–233.

Hansen MR, Zha X-M, Bok J, Green SH (2001) Multiple distinct signal pathways, including an autocrine neurotrophic mechanism contribute to the survival-promoting effect of depolarization on spiral ganglion neurons in vitro. J Neurosci 21(7):2256–2267.

Harrison RV, Ibrahim D, Stanton SG, Mount RJ (1996) Reorganization of frequency maps in Chinchilla auditory midbrain after long-term basal cochlear lesions induced at birth. In: Salvi RJ, Fiorino F, Henderson D, Colletti V, eds. Auditory System Plasticity and Regeneration. New York: Thieme Medical Publishers, pp. 238–255.

Hartmann R, Shepherd RK, Heid S, Klinke R (1997) Response of the primary auditory cortex to electrical stimulation of the auditory nerve in the congenitally deaf white cat. Hear Res 112:115–133.

Hartshorn DO, Miller JM, Altschuler RA (1991) Protective effect of electrical stimulation in the deafened guinea pig cochlea. Otolaryngol Head Neck Surg 104:311–319.

Hashisaki GT, Rubel EW (1989) Effects of unilateral cochlea removal on anteroventral cochlear nucleus neurons in developing gerbils. J Comp Neurol 283:465–473.

Hawkins JE Jr, Stebbins WC, Johnsson L-G, Moody DB, Muraki A (1977) The patas monkey as a model for dihydrostreptomycin ototoxicity. Acta Otolaryngol 83:123–129.

Hegarty JL, Kay AR, Green SH (1997) Trophic support of cultured spiral ganglion neurons by depolarization exceeds and is additive with that by neurotrophins or cAMP and requires elecation of $[CA^{2+}]i$ within a set range. J Neurosci 17:1959–1970.

Heid S, Jahn-Siebert TK, Klinke R, Hartmann R, Langner G (1997) Afferent projection pattern in the auditory brainstem in normal cats and congenitally deaf white cats. Hear Res 110:191–199.

Heid S, Hartmann R, Klinke R (1998) A model for prelingual deafness, the congenitally deaf white cat: population statistics and degenerative changes. Hear Res 115:101–112.

Hinojosa R, Lindsay JR (1980) Profound deafness: associated sensory and neural degeneration. Arch Otolaryngol 106:193–209.

Hinojosa R, Marion M (1983) Histopathology of profound sensorineural deafness. Ann N Y Acad Sci 405:459–484.

Hinojosa R, Blough RR, Mhoon EE (1987) Profound sensorineural deafness: a histopathologic study. Ann Otol Rhinol Laryngol 128(suppl 96):43–46.

Hultcranz M, Snyder RL, Rebscher SJ, Leake PA (1991) Effects of neonatal deafening and chronic introacochlear electrical stimulation on the cochlear nucleus in cats. Hear Res 54:272–180.

Incesulu A, Nadol JB (1998) Correlation of acoustic threshold measures and spiral ganglion cell survival in severe to profound sensorineural hearing loss: implications for cochlear implantation. Ann Otol Rhinol Laryngol 107:906–911.

Irvine DRF, Rajan R (1994) Plasticity of frequency organization in inferior colliculus of adult cats with unilateral restricted cochlear lesions. Abstr Assoc Res Otolaryngol 17:21.

Jean-Baptist M, Morest DK (1975) Transneuronal changes of synaptic endings and nuclear chromatin in the trapezoid body following cochlear ablations in cats. J Comp Neurol 16:111–133.

Johnsson L-G (1974) Sequence of degeneration of Corti's organ and its first-order neurons. Ann Otol Rhinol Laryngol 83:294–303.

Johnsson L-G, House WF, Linthicum FH (1979) Bilateral cochlear implants: histological findings in a pair of temporal bones. Laryngoscope 89:759–762.

Johnsson L-G, Hawkins JE Jr, Kingsley TC, Black FO, Matz GJ (1981) Aminoglycoside-induced inner ear pathology in man, as seen by microdissection. In: LernerSA, Matz GJ, Hawkins JE Jr, eds. Amino-Glycoside Ototoxicity. Boston: Little, Brown, pp. 389–408.

Johnsson L-G, House WF, Linthicum FH (1982) Otopathological findings in a patient with bilateral cochlear implants. Ann Otol Rhinol Laryngol 91(suppl 91):74–89.

Kennedy DW (1987) Multichannel intracochlear electrodes: mechanism of insertion trauma. Laryngoscope 97:42–49.

Ketten DR, Skinner MW, Wang G, Vannier MW, Gates GA, Neely JG (1998) In vivo measures of cochlear length and insertion depth of Nucleus cochlear implant electrode arrays. Ann Otol Rhinol Laryngol Suppl 107:1–16.

Kiang NYS, Liberman MC, Levine RA (1976) Auditory-nerve activity in cats exposed to ototoxic drugs and high-intensity sounds. Ann Otol Rhinol Laryngol 85:852–768.

Kitzes L (1996) Anatomical and physiological changes in the brainstem induced by neonatal ablation of the cochlea. In: Salvi RJ, Henderson D, eds. Auditory System Plasticity and Regeneration. New York: Thieme Medical Publishers, pp. 256–274.

Klinke R, Kral A, Heid S, Tillein J, Hartmann R (1999) Recruitment of the auditory cortex in congenitally deaf cats by long-term cochlear electrostimulation. Science 285:1729–1733.

Kohenen A (1965) Effect of some ototoxic drugs upon the pattern and innervation of cochlear sensory cells in the guinea pig. Acta Otolaryngol Suppl 208.

Kral A, Hartmann R, Tillein J, Heid S, Klinke R (2001) Delayed maturation and sensitive periods in the auditory cortex. Audiol Neuro-Otol 6:346–362.

Lalwani AK, Walsh BJ, Reilly PG, Zolotukhin S, Muzczka N, Mhatre AN (1998) Long-term in vivo cochlear transgene expression mediated by recombinant adeno-associated virus. Gene Ther 5:277–281.

Larsen SA (1984) Postnatal maturation of the cat cochlear nuclear complex. Acta Otolaryngol Suppl 417:1–43.

Leake PA, Hradek GT (1988) Cochlear pathology of long term neomycin induced deafness in cats. Hear Res 33:11–34.

Leake PA, Snyder R, Schreiner CE (1987) Cochlear pathology of sensorineural deafness in ctas: coadministration of kanamycin and aminooxyacetic acid. Ann Otol Rhinol Laryngol 96:48–50.

Leake PA, Kessler DK, Merzenich MM (1990) Application and safety of cochlear prostheses. In: Agnew WF, McCreery DB, eds. Neural Prostheses. Englewood Cliffs, NJ: Prentice Hall, pp. 253–296.

Leake PA, Hradek GT, Rebscher SJ, Snyder RL (1991) Chronic intracochlear electrical stimulation induces selective survival of spiral ganglion neurons in neonatally deafened cats. Hear Res 54:251–271.

Leake PA, Snyder RL, Hradek GT, Rebscher SJ (1992) Chronic intracochlear electrical stimulation in neonatally deafened cats: effects of intensity and stimulating electrode Location. Hear Res 64:99–117.

Leake PA, Snyder RL, Hradek GT, Rebscher SJ (1995) Consequences of chronic extracochlear electrical stimulation in neonatally deafened cats. Hear Res 82:65–80.

Leake PA, Kuntz AL, Moore CM, Chambers PL (1997) Cochlear pathology induced by aminoglycoside ototoxicity during postnatal maturation in cats. Hear Res 113:117–132.

Leake PA, Hradek GT, Snyder RL (1999) Chronic electrical stimulation by a cochlear implant promotes survival of spiral ganglion neurons in neonatally deafened cats. J Comp Neurol 412:543–562.

Leake PA, Snyder RL, Rebscher SJ, Hradek GT, et al. (2000a) Long term effects of deafness and chronic electrical stimulation of the cochlea. In: Waltzman SB, Cohen N, eds. Cochlear Implants. New York: Thieme Medical Publishers, pp. 31–41.

Leake PA, Snyder RL, Rebscher SJ, Moore CM, Vollmer M (2000b) Plasticity in central representations in the inferior colliculus induced by chronic single- vs. two-channel electrical stimulation by a cochlear implant after neonatal deafness. Hear Res 147:221–241.

Leake PA, Rebscher SJ, Wardrop PJC, Whinney DJD (2003) Optimizing Multichannel Cochlear Implant Electrodes, Assoc. Res. Otolaryngol. Abs.: Vol. 26, p. 194.

Levi-Montalcini R (1949) The development of the acousticovestibular centers in the chick embryo in the absence of the afferent root fibers and of descending fiber tracts. J Comp Neurol 91:209–241.

Li L, Parkins CW, Webster DB (1999) Does electrical stimulation of deaf cochleas prevent spiral ganglion degeneration? Hear Res 13:27–39.

Liberman MC, Kiang NYS (1978) Acoustic trauma in cats: cochlear pathology and auditory nerve activity. Acta Otolaryngol Suppl 358:1–63.

Linthicum FH, Fayad J, Otto SR (1991) Cochlear implant histopathology. Am J Otol 12:245–311.

Loeb GE, Byers CL, Rebscher SJ, Casey DE, et al. (1983) Design and fabrication of an experimental cochlear prosthesis. Med Biol Eng Comput 21:241–254.

Lousteau RJ (1987) Increased spiral ganglion cell survival in electrically stimulated deafened guinea pig cochlea. Laryngoscope 97:836–842.

Lustig LR, Leake PA, Snyder RL, Rebscher SJ (1994) Changes in the cat cochlear nucleus following neonatal deafening and chronic intracochlear electrical stimulation. Hear Res 74:29–37.

Marsh MA, Coker NJ, Jenkins HA (1992) Temporal bone histopathology of a patient with a Nucleus 22-channel cochlear implant. Am J Otol 13:241–248.

Matsushima J-I, Shepherd RK, Seldon HL, Xu S-A, Clark GM (1991) Electrical stimulation of the auditory nerve in deaf kittens: effects on cochlear nucleus morphology. Hear Res 56:133–142.

Miller JM, Altschuler RA (1995) Effectiveness of different electrical stimulation conditions in preservation of spiral ganglion cells following deafness. Ann Otol Rhinol Laryngol 166:57–60.

Miller JM, Altschuler RA, Niparko JK, Hartshorn DO, Helfert RH, Moore JK (1991) Deafness induced changes in the central nervous system: reversibility and prevention. In: Dancer A, Henderson D, Salvi RJ, Hamernik RP, eds. Noise Induced Hearing Loss. St. Louis: Mosby Year Book, pp. 130–145.

Miller JM, Chi D, O'Keefe, Kruska P, Raphael Y, Altschuler RA (1997) Neurotrophins can enhance spiral ganglion cell survival after inner hair cell loss. Int J Dev Neurobiol 15:631–643.

Moore CM, Vollmer M, Leake PA, Snyder RL, Rebscher SJ (2002) Effects of chronic intracochlear electrical stimulation on inferior colliculus spatial representation in adult deafened cats. Hear Res 164:82–96.

Moore DR, Kitzes LM (1985) Projections from the cochlear nucleus to the inferior colliculus in normal and neonatally cochlea-ablated gerbils. J Comp Neurol 240:180–195.

Moore DR, Kowlachuk NE (1988) Auditory brainstem of the ferret: effects of unilateral cochlear lesions on cochlear nucleus volume and projections to the inferior colliculus. J Comp Neurol 272:503–515.

Moore JK, Niparko JK, Miller MR, Linthicum FH (1994) Effect of profound hearing loss on a central auditory nucleus. Am J Otol 15:588–595.

Moore JK, Niparko JK, Perazzo LM, Miller MR, Linthicum FH (1997) Effect of adult-onset deafness on the human central auditory system. An Otol Rhinol Laryngol 106:385–390.

Mostafapour SP, Cochran SL, del Puerto NM, Rubel EW (2000) Patterns of cell death in mouse AVCN neurons after unilateral cochlea removal. J Comp Neurol 426:561–571.

Nadol JB (1997) Patterns of neural degeneration in the human cochlea and auditory nerve: implications for cochlear implantation. Otolaryngol Head Neck Surg 117:220–228.

Nadol JB, Young Y-S, Glynn RJ (1989) Survival of spiral ganglion cells in profound sensorineural hearing loss: implications for cochlear implantation. Ann Otol Rhinol Laryngol 98:41–416.

Nadol JB, Ketten DR, Burgess BJ (1994) Otopathology in a case of multichannel cochlear implantation. Laryngoscope 104:299–303.

Niparko JK, Finger PA (1997) Cochlear nucleus cell size changes in the dalmatian: model of congenital deafness. Otolaryngol Head Neck Surg 117:229–235.

Nordeen KW, Killackey HP, Kitzes LM (1983) Ascending projections to the inferior colliculus following unilateral cochlear ablation in the neonatal gerbil, Meriones unguiculatus. J Comp Neurol 214:144–153.

O'Leary MJ, Fayad J, House WF, Linthicum FH Jr (1991) Electrode insertion trauma in cochlear implantation. Ann Otol Rhinol Laryngol 100:695–699.

O'Reilly BF (1981) Probability of trauma and reliability of placement of a 20 mm long model human scala tympani multielectrode array. Ann Otol Rhinol Laryngol Suppl 90:11–12.

Osofsky MR, Leake PA, Moore CM (2001) Does exogenous GM1 ganglioside enhance the effects of electrical stimulation in ameliorating degeneration after neonatal deafness. Hear Res 159:23–35.

Otte J, Schuknecht HF, Kerr AG (1978) Ganglion populations in normal and pathological human cochlea: implications for cochlear implantation. Laryngoscope 88:1231–1246.

Parks TN (1979) Afferent influences on the development of the brainstem auditory nuclei of the chicken: otocyst ablation. J Comp Neurol 183:665–677.

Ponton CW, Don M, Eggermont JJ, Waring MD, Kwong B, Masuda A (1996) Neuroreport 8:61–65.

Powell TPS, Erulkar SD (1962) Transneuronal cell degeneration in the auditory relay nuclei of the cat. J Anat 96:249–268.

Raggio MW, Schreiner CE (1994) Neuronal responses in cat primary auditory cortex to electrical cochlear stimulation. I. Intensity dependence of firing rate and response latency. J Neurophysiol 72:2334–2359.

Raggio MW, Schreiner CE (1999) Neuronal responses in cat primary auditory cortex to electrical cochlear stimulation. III. Activation patterns in short- and long-term deafness. J Neurophysiol 82:3506–3526.

Raphael Y, Frisancho JC, Roessler BJ (1996) Adenoviral-mediated gene transfer into guinea pig cochlear cells in vivo. Neurosci Lett 207:137–141.

Rebscher SJ, Heilmann M, Bruszweski W, Talbot N, Snyder R, Merzenich MM (1999) Strategies to improve electrode positioning and safety in cochler implants. IEEE Trans Biomed Eng 46:340–352.

Rebscher SJ, Snyder RL, Leake PA (2001) The effect of electrode configuration on threshold and selectivity of responses to intracochlear electrical stimulation. J Acoust Soc Am 109:2035–2048.

Roland JT, Fishman A, Alexiades G (2000) Electrode to modiolus proximity: a fluoroscopic and histologic analysis. Am J Otol 21:218–225.

Rubel EW, Fritzsch B (2002) Auditory system development: primary auditory neurons and their targets. Annu Rev Neurosci 25:51–101.

Russell RA, Moore DR (1995) Afferent reorganization within the superior olivary complex of the gerbil: development and induction by neonatal unilateral cochlear removal. J Comp Neurol 352:607–625.

Schindler RA, Gladstone HB, Scott N, Hradek GT, Williams H, Shah SB (1995) Enhanced preservation of the auditory nerve following cochlear perfusion with nerve growth factor. Am J Otol 16:304–309.

Schriener CE, Raggio MW (1996) Neuronal responses in cat primary auditory cortex to electrical cochlear stimulation: II. Repetition rate coding. J Neurophysiol 75:1283–1300.

Shah SB, Gladstone HB, Williams H, Hradek GT, Schindler RA (1995) An extended study: protective effects of nerve growth factor in neomycin-induced auditory neural degeneration. Am J Otol 16:310–314.

Shepherd RK, Clark GM, Pyman BC, Webb RL (1985) Banded intracochlear electrode array: evaluation of insertion trauma in human temporal bones. Ann Otol Rhinol Laryngol 94:55–59.

Shepherd RK, Matsushima J, Martin RL, Clark GM (1994) Cochlear pathology following chronic electrical stimulation of the auditory nerve: II. Deafened kittens. Hear Res 8:150–166.

Skinner MW, Ketten DR, Vannier MW, Gates GA, Yoffie RL, Kalender WA (1994) Determination of the position of Nucleus cochlear implant electrodes in the inner ear. Am J Otol 15:644–651.

Skinner MW, Ketten DR, Holden LK, Harding GW et al. (2002) CT-derived estimation of cochlear morphology and electrode array position in relation to word recognition in Nucleus-22 recipients. J Assoc Res Otolaryngol 3: 332–350.

Snyder RL, Rebscher SJ, Cao K, Leake PA, Kelly K (1990) Effects of chronic intracochlear electrical stimulation in the neonatally deafened cat. I: expansion of central spatial representation. Hear Res 50:7–33.

Snyder RL, Rebscher SJ, Leake PA, Kelly K, Cao K (1991) Chronic electrical stimulation in the neonatally deafened cat. II: temporal properties of neurons in the inferior colliculus. Hear Res 56:246–264.

Snyder RL, Leake PA, Rebscher SJ, Beitel RE (1995) Temporal resolution of neurons in cat inferior colliculus: effects of neonatal deafening and chronic intracochlear electrical stimulation on J Neurophysiol 72:449–467.

Snyder RL, Vollmer M, Moore CM, Rebscher SJ, Leake PA, Beitel RE (2000) Responses of inferior colliculus neurons to amplitude modulated intracochlear electrical pulses in deaf cats. J Neurophysiol 84:166–183.

Spoendlin HH (1975) Retrograde degeneration of the cochlear nerve. Acta Otolaryngol 79:266–275.

Staecker H, Kopke R, Malgrange B, Lefebvre P, Van De Water TR (1996) NT3 and BDNF therapy prevents loss of auditory neurons following loss of hair cells. NeuroReport 7:889–894.

Staecker H, Gabaizadeh R, Rederoff H, Van De Water TR (1998) Brain-derived neurotrophic factor gene therapy prevents spiral ganglion degeneration after hair cell loss. Otolaryngol Head Neck Surg 119:7–13.

Stebbins WC, Miller JM, Johnsson LG, Hawkins JE Jr (1969) Ototoxic hearing loss and cochlear pathology in the monkey. Ann Otol Rhinol Laryngol 78:1007–1025.

Svirsky MA, Robbine AM, Kirk KI, Pisone DB, Miyamoto RT (2000) Language development in profoundly deaf children with cochlear implants. Psychol Sci 11:153–158.

Tierney TS, Russell FA, Moore DR (1997) Susceptibility of developing cochlear nucleus neurons to deafferentation-induced death abruptly ends just before the onset of hearing. J Comp Neurol 378:295–306.

Trune DR (1982a) Influence of neonatal cochlear removal on the development of mouse cochlear nucleus: number, size and density of it's neurons. J Comp Neurol 209:409–424.

Trune DR (1982b) Influence of neonatal cochlear removal on the development of mouse cochlear nucleus: II. Dendritic morphometry of its neurons. J Comp Neurol 209:425–434.

Tycocinski M, Saunders E, Cohen LT (2001) The contour electrode array: safety study and initial patient trials of a new perimodiolar design. Otol Neurotol 22:23–41.

Vollmer M, Snyder RL, Leake PA, Beitel RE, Moore CM, Rebscher SJ (1999) Temporal properties of chronic electrical stimulation determine temporal resolution of neurons in cat inferior colliculus. J Neurophysiol 82:2883–2902.

Walsh EJ, Romand R (1992) Functional development of the cochlea and the cochlear nerve. In: Romand R, ed. Development of Auditory and Vestibular Systems 2. New York: Elsevier, pp. 161–219.

Webb RL, Clark GM, Shepherd RK, Franz BK-H, Pyman BC (1988) The biological safety of the cochlear corporation multiple-electrode intracochlear implant. Am J Otol 9:8–13.

Webster DB (1983) Late onset of auditory deprivation does not affect brainstem auditory neuron soma size. Hear Res 12:145–147.

Webster DB, Webster M (1977) Neonatal sound deprivation affects brainstem auditory nuclei. Arch Otolaryngol 103:392–396.

Webster DB, Webster M (1979) Effects of neonatal conductive hearing loss on brainstem auditory nuclei. Ann Otol Rhinol Laryngol 88:684–688.

Webster DB, Webster M (1981) Spiral ganglion neuron loss following organ of Corti loss: a quantitative study. Ann Otol Rhinol Laryngol 90(suppl 82):17–30.

Weiss MA, Frisancho C, Roessler JB, Raphael Y (1977) Viral-mediated gene transfer in the cochlea. Int J Dev Neurosci 15;577–583.

West BA, Brummett RE, Hines, DL (1973) Interaction of kanamycin and ethacrynic acid. Arch Otolaryngol 98:32–37.

Xu S-A, Shepherd RK, Chen Y, Clark GM (1993) Profound hearing loss in the cat following the single co-administration of kanamycin and ethacrynic acid. Hear Res 71:205–215.

Yagi M, Kanzaki S, Kawamoto K, Shin B, et al. (2000) Spiral ganglion neurons are protected from degeneration by GDNF gene therapy. JARO 1:315–325.

Ylikoski J (1974) Correlative studies of the cochlear pathology and hearing loss in guinea pigs after intoxication with ototoxic antibiotics. Acta Otolaryngol 79:266–275.

Ylikoski J, Privola U, Virkkala J, Suvanto P, et al. (1998) Guinea pig auditory neurons are protected by glial cell line-derived growth factor from degeneration after noise trauma. hear Res 124:17–26.

Zappia JJ, Niparko JK, Oviatt DL, Kemink JL, et al. (1991) Evaluation of the temporal bones of a multichannel cochlear implant patient. Ann Otol Rhinol Laryngol 100:914–921.

Zheng JL, Stewart RR, Gao W-Q (1995) Neurotrophin-4/5 enhances survival of cultured spiral ganglion neurons and protects them from cisplatin neurotoxitity. J Neurosci 15:5079–5087.

5
Biophysics and Physiology

PAUL J. ABBAS and CHARLES A. MILLER

1. Introduction

Electrical stimulation of the auditory nerve has been used as a treatment for sensory hearing loss for approximately 30 years. The typical cochlear implant candidate has sensorineural hearing loss, suggesting a lack of functional hair cells in the cochlea. The electrical fields produced by a cochlear implant directly excite auditory nerve fibers to elicit a hearing sensation. While the average performance with prosthetic devices has continued to improve over the years, there remains significant variation in the performance of individuals using the same implant design. Understanding the biophysical mechanisms of electrical stimulation and the response properties of the auditory nerve therefore can have considerable practical importance. Such information could lead to better designs of auditory prostheses, development of stimuli that could improve information transfer with current devices, and improved clinical procedures for fitting the device and assessing neurophysiological status.

1.1 Comparison of Responses to Acoustic and Electrical Stimuli

Important aspects of electrical excitation are illustrated by the physiologic response differences observed between direct electrical stimulation of the deafened ear and acoustic stimulation of the normal ear. The normal auditory nerve fiber is spontaneously active, stochastically discharging at rates of 0 to 80 spike/s (Liberman 1978). In an ear without functional hair cells, there is little or no spontaneous activity, presumably due to a lack of neurotransmitter release at the synapse (Hartmann and Klinke 1990a). A prominent feature of the normal mammalian auditory system is the frequency dependence of the cochlea's mechanical response—the traveling wave—and the resultant tuning properties of auditory nerve fibers. In contrast to sharp acoustic tuning, a fiber's response to sinusoidal electrical

149

stimuli exhibits broad tuning across frequency, with no apparent change in each fiber's tuning properties across the place of innervation within the cochlea (Kiang and Moxon 1972; Hartmann et al. 1984). In acoustically sensitive fibers, the growth of response with stimulus level is highly nonlinear, reflecting the mechanical response of the basilar membrane (Rhode 1971; Sachs and Abbas 1974; Yates et al. 1992). Fiber rate-level functions typically have a dynamic range on the order of 30 dB. With electrical stimulation, each auditory nerve fiber has a dynamic range of only 1.5 to 6 dB and the maximum discharge rate is typically much greater than that attained with acoustic stimulation (Kiang and Moxon 1972; Hartmann et al. 1984; Javel et al. 1987; Miller et al. 1999a). Responses of normal fibers to acoustic click (i.e., impulsive) stimuli reflect the filtering properties typical of resonant systems, resulting in temporal spread or *ringing* of the response. With electrical stimulation, fibers without functional hair cells typically produce post-stimulus time (PST) histograms with relatively short latencies and less temporal dispersion than is the case with acoustic stimulation in normal ears (Kiang and Moxon 1972). These marked differences between electrical and acoustic excitation of the auditory nerve have clear implications for the design of effective auditory prostheses. The acoustic signal being transduced for the cochlear implant must undergo significant processing and the resulting electrical stimuli must be delivered in a controlled fashion to overcome the loss of the normal signal transduction mechanisms (see Wilson, Chapter 2).

1.2 The Importance of Studying the Electrical Excitation of the Auditory Nerve

Long-term deafness results in changes to both the peripheral and central auditory system. There is evidence of degeneration and reorganization in the brainstem and cortex (see Leake and Rebscher, Chapter 4; Hartmann and Kral, Chapter 6), but the changes occurring at the periphery, within the cochlea, and in the auditory nerve are arguably more profound. The loss of a functional cochlea not only promotes degeneration of the auditory nerve, but also necessitates the use of a different physical stimulus and signal encoding. Significant aspects of cochlear signal processing—including frequency analysis, amplitude compression, and spontaneous activity—are lost and replaced by a biophysical system in which the electrical properties of the cochlea, nerve, and surrounding tissue are of paramount importance. As a result, an adequate understanding of the electrophysiological responses of the auditory nerve to electric stimulation is critical to both the design of cochlear prostheses and solid interpretations of central physiological and psychophysical responses (see Hartmann and Kral, Chapter 6; McKay, Chapter 7).

1.3 Outline of This Chapter

The focus of this chapter 15 is on the mechanisms by which the auditory nerve is stimulated, as well as the nerve's response properties. Section 2 summarizes the research methods used in the studies discussed throughout the chapter. The topic of electrical stimulation of the auditory nerve is somewhat complex due to the unique anatomy of the mammalian cochlea. For this reason, in section 3, we review basic properties of electrical stimulation of neural tissue independent of the specific characteristics of the auditory nerve and cochlea. Section 4 addresses the unique response characteristics produced by stimulating the auditory nerve through intracochlear electrodes. We focus on properties that include the growth of response with stimulus level, refractoriness, stochastic properties, and other characteristics, such as adaptation, that may affect the responses to the repetitive stimulation typical of cochlear implants. In addition, electrical stimulation of the nerve through cochlear implants requires an understanding of the population response. Issues such as fiber threshold distribution and fiber selectivity with different electrode placements and configurations are discussed in section 5. Most of the findings discussed in sections 4 and 5 were obtained in acutely deafened animal models, so that the data reflect the responses from a relatively intact neural population uninfluenced by functional hair cells. As the auditory nerves of typical cochlear implant users have some degree of neural degeneration—population loss and pathological changes in surviving fibers—the physiological changes that take place with neural degeneration are of interest. Such effects are discussed in section 6. In recent years, more individuals with significant levels of residual hearing have been implanted (von Ilberg et al. 1999; Gantz et al. 2000). The effects of functional hair cells on the response to electrical stimulation are discussed in section 7. In section 8, we speculate on how some of the aforementioned findings could be applied to future developments and applications of cochlear implants. These include the role of physiological measurements in determining specific parameters of stimulation in clinical use of cochlear implants, development of a better electrode–neural interface, use of conditioning stimuli to improve information transfer, and application of electrical stimulation in conjunction with acoustic stimulation in ears with significant hair-cell survival.

2. Major Research Approaches

A complete description of intracochlear electrical stimulation of the auditory nerve is a complex undertaking, considering variables such as the spatial arrangement of fibers, the diverse membrane properties along their peripheral-to-central dimension, extraneural tissue properties, and the influence of electrode position and configuration. The peripheral terminations of auditory nerve fibers are distributed along a helical path, reflecting the shape of the sensory epithelium. As they course centrally, the fibers converge within the modiolus (the central core of the cochlea) to form the trunk of the auditory nerve. Within the trunk, the fibers continue along spiral courses and do not maintain constant spatial interrelationships (Sando 1965). The morphology of single nerve fibers also varies along this peripheral-to-central dimension. Each auditory nerve fiber is bipolar, with a small-diameter peripheral process that is only partially myelinated and a larger, myelinated, central process. As a bipolar fiber it is unique in that action potentials generated at the peripheral process travel through the cell body. Furthermore, the tissue environment surrounding the auditory nerve fibers is diverse, ranging from fluid-filled space at their most peripheral aspect, to the thin bony plates of the osseous spiral lamina, and the thicker bone of the modiolus and internal meatus (see Fig. 4.1 in Leake and Rebscher, Chapter 4).

An ultimate research goal could be the characterization of the responses of all 30,000 afferent fibers of the human auditory nerve to a wide range of electrical stimuli. As this is not feasible, the pursuit of this goal by the research community has been divided into multiple, tractable, research domains that focus on particular aspects of neural excitation. Thus, a more complete understanding is afforded by combining the results from animal studies with results from computer simulations. No single research domain, however, provides the complete picture. In this section, we first discuss the use of animal models to characterize properties of electrical stimulation in humans. This is followed by a presentation of three major research approaches. The first focuses on the electrophysiology of single fibers; the second relies on the gross electrical response of an ensemble of responsive fibers; the third approach involves the use of computational models to simulate the electrical properties of an auditory nerve fiber. These various approaches can be used in a synergistic fashion to promote our general understanding of electrical excitation of the auditory nerve.

2.1 Animal Models

Research animals are the primary means of modeling the electrically stimulated auditory nerve of the human. Several species have been used, with the choice dictated, in part, by the research questions under study. For example, efforts that focus on fundamental properties of neural membranes can justifiably chose among a wider range of species than those seeking to understand the spread of current through cochlear tissues. With a few exceptions (e.g., Morse and Evans 1996), researchers use mammalian models, which afford intracochlear stimulation. Mice, gerbils, rats, and chinchillas have been used, but cats and guinea pigs are most widely used. Unlike those of small rodents, the larger dimensions of the cat cochlea (Igarashi et al. 1968)—and the guinea pig to a somewhat lesser degree—

allow for the insertion of multichannel electrode arrays similar to those in clinical devices, providing a wide range of experimental flexibility. While gross auditory nerve potentials can be recorded from several species, the surgical accessibility of its auditory nerve trunk has made the cat the nearly exclusive choice for single-fiber studies.

While many intracochlear anatomical features and relationships are preserved across mammalian species, the gross anatomical dimensions of cochleas vary significantly. These differences may result in various patterns of spatial spread of electrical excitation throughout the cochlea. For example, computer modeling of such properties indicates differences exist between guinea pig and human intracochlear excitation (Frijns et al. 2001). Gross dimensions of the cat cochlea better approximate those of the human, although there are interspecies differences in the cross-sectional areas of intrascalar regions (Fernandez 1952; Igarashi et al. 1976; Hatsushika et al. 1990). Another species-specific property is the degree of myelination of the spiral ganglion. While the cell bodies of guinea pigs and cats are myelinated (Thomsen 1966; Spoendlin 1972), surveys of human tissue show infrequent or relatively loose myelination of cell bodies (Ota and Kimura 1980; Arnold 1987; Spoendlin and Schrott 1989). The impact of these differences is unclear, and while they do not render the animal models inappropriate, they do require an awareness of them when extrapolating trends to humans.

It should be noted that, with some exceptions, electrophysiological studies have used acutely deafened animals, typically through the application of ototoxic drugs. This model is appropriate for approximating the condition of the deaf ear, as hair-cell function is obviated and yet the nerve fiber population is left relatively intact. It is widely used to understand the process of directly exciting the membranes of a population of fibers. In contrast, the typical ear of the implant candidate has suffered pathology to hair cells as well as degeneration of cochlear tissues and the auditory nerve. Thus, acutely deafened animal models typically provide a best case for electrical excitation of the auditory nerve. Other animal models, discussed in sections 6 and 7, take into account the effects of neural degeneration and the effects of residual hearing on the response properties to electrical stimulation.

2.2 Single-Fiber Studies

Over the last 30 years, concerted study of the electrically stimulated mammalian auditory nerve has paralleled the growing promise of cochlear implantation and the maturation of single-fiber auditory physiology. Much of this effort approached the auditory nerve and neuron as *systems* and sought to define input–output relationships, with the stimulus provided by one or more electrodes positioned within the fluid-filled space of the scala tympani. Single-fiber recordings require surgical access to the auditory nerve, which is impaled by a glass micropipette fashioned with a tip dimension suitable for isolating the extracellular potentials from a single fiber (i.e., ~1 μm diameter). Although recordings have been obtained from several mammalian species (e.g., monkey, cat, chinchilla, guinea pig, rat, gerbil), feline anatomy provides distinct advantages for recording electrically evoked action potentials. The large cranial space allows for adequate surgical exposure of the nerve trunk to record from central axons, providing two important advantages. First, the cochlear dimensions allow for the insertion of relatively large intracochlear electrodes arrays, comparable to those used in humans. Second, the intracranial position of the recording electrode (Kiang et al. 1965) greatly reduces the contamination of the evoked responses by the electrical stimulus artifact with intracochlear stimulation.

Much single-fiber research has attempted to describe the electrophysiology of the archetypal nerve fiber, assuming that most fibers of the nerve possess similar response properties. Major findings include descriptions of the temporal pattern of action potentials in response to pulsatile and continuous sinusoidal stimuli and frequency response characteristics. A limitation of single-fiber studies is that they provide only a statistical description of how the typical fiber responds to electrical stimulation. Although a very productive animal experiment can yield threshold measures from as many as 150 fibers (e.g., van den Honert and Stypulkowski 1987a), more extensive measures (beyond threshold estimation) will reduce that yield. Even with high-yield experiments, less than 1% of afferent fibers are assessed with standard micropipette techniques, rendering adequate sampling of the population problematic. This issue must be considered when interpreting single-fiber studies that attempt to characterize population characteristics, particularly as a function of cochleotopic place.

2.3 Gross Potential Studies

A second approach to characterizing electrophysiological responses relies on measurement of the electrically evoked compound action potential (ECAP). This potential represents the spatial summation of action potentials produced by a population of fibers and requires the synchronous firing of fibers to achieve a measurable potential. The ECAP is typically recorded using a relatively large ball (0.2–0.5 mm diameter) or needle electrode positioned in close proximity to the nerve or cochlea. Unlike the case of single-fiber recording, acquisition of the ECAP does not require surgical access of the auditory nerve or direct contact with it. An electrode positioned on the promontory or round window is sufficiently close to record this potential, as are intracochlear electrodes of a cochlear implant electrode array.

Under most recording conditions, time-domain signal averaging is needed to acquire an ECAP with a favorable signal-to-noise ratio. The use of signal averaging to successive stimuli in order to extract a response requires the assumption that the neural response to each individual stimulus is identical, i.e., no adaptation-like phenomena occur over the course of the repeated stimuli. Such assumptions are highly dependent on stimulus parameters such as level and rate of stimulus presentation and can also potentially be affected by the properties of stimulated neurons and their neuropathological state. Animal models that permit sufficient surgical assess (e.g., guinea pigs and cats) make it possible to position a recording electrode directly on or near the surface of the exposed nerve. Such preparations offer a technical advantage in that the resultant ECAP is large and more resistant to contamination by stimulus artifacts. It is possible, in some cases, to record potentials without resorting to signal averaging, which is a distinct advantage for studies of adaptation and ephemeral response properties.

Goldstein and Kiang (1958) developed a mathematical relationship between the compound action potential waveform, $A(t)$, and the underlying single-fiber activity of the N constituent fibers comprising the nerve:

$$A(t) = \sum_{n=1}^{N} \int_{-\infty}^{t} P(n,\tau) U(n, t-\tau) d\tau \qquad (1)$$

where $P(n,\tau)$ is the poststimulus time histogram of the nth fiber and $U(n,t)$ is the unit potential function. The latter function is the potential waveform contributed by the nth fiber as seen by the gross recording electrode. From the physiologist's perspective, the functions P and U represent two challenges. First, all N fibers must be described by their respective poststimulus time histograms in response to a specific stimulus. Likewise, the unit potential contribution of each fiber must be known. Note that while the histogram function $P(n,\tau)$ is specific to a particular stimulus, it is assumed that the unit potential is constant across stimulus conditions. However, the unit potential waveforms can be highly dependent on the location of the recording electrode. It is plausible to assume that, for relatively distant electrode-to-nerve positions, the unit potentials contributed by the constituent fibers will be comparable to one another. Based on measures from acoustically sensitive cats, Kiang et al. (1976) concluded that the unit potentials recorded by an electrode at the round window were comparable across the fiber population. For recording electrodes placed directly on the surface of the nerve, the unit potentials may be somewhat different. Doucet and Relkin (1997) suggested that they may be uniform only over a limited range of characteristic frequency (CF) or fiber position. The complex anatomy surrounding the nerve and different distances from fiber to recording electrode possibly distort the relative electrical contributions of each fiber. Nonetheless, modeling efforts suggest that the amplitude of the ECAP is proportional to the number of actively responding fibers (Miller et al. 1999b).

The ECAP recordings offer both advantages and disadvantages over single-fiber measures. A primary advantage is that they can be recorded from human cochlear implant patients whose internal device allows for direct electrical access to the intracochlear array (Brown et al. 1990) or is equipped with a telemetry system appropriate for neural recordings (Abbas et al. 1999a). As ECAP recordings are also readily obtained from animal subjects, parametric evaluations can be made and compared in animals and humans alike. Also, as described above, ECAP measures can be related to the underlying single-fiber responses with a relatively straightforward model. Consequently, they provide a link between single-fiber measures from animal studies and the ECAP studies that can be conducted with human participants. The primary disadvantage of the ECAP is that it is a population response, and as such, cannot directly assess the underlying single-fiber properties. As the ECAP requires highly synchronized neural activity, responses from nerve fibers that exhibit atypical latency or appreciable jitter will not be proportionally represented in this gross potential (Miller et al. 1999b).

The electrically evoked auditory brainstem response (EABR) is similar to the ECAP in that it is a population response that requires synchronous neural discharge. Interpretation of the auditory brainstem response is relatively complex, as several neural structures contribute to it and it is composed of several sequential response peaks (e.g., Jewett et al. 1970; Achor and Starr 1980; Melcher et al. 1996), with the first peak believed to correspond primarily to a far-field manifestation of the ECAP. Nevertheless, many studies of human cochlear implant users have used the brainstem response to assess both threshold and suprathreshold responses to stimulation through a cochlear implant (Starr and Brackmann 1979; van den Honert and Stypulkowski 1986; Abbas and Brown 1991; Aubert and Clarke 1994). Such measures are readily obtained from human implant users as only external electrodes on the surface of the scalp are necessary. Furthermore, the issues of stimulus artifact rejection are less severe, as the response latencies from brainstem structures are longer than that of the auditory nerve response and are more easily separable from the stimulus artifact. Measures of EABR and ECAP in the same subjects have demonstrated similar dependence on stimulus parameters (Brown et al. 2000), suggesting that, at least under some conditions, they depend on similar underlying neural generators.

2.4 Computational Models

While experimental animals provide the best means for modeling human auditory nerve excitation, computational models also play important roles in furthering our understanding of the physiology of the electrically stimulated system. The term *computer model* covers a wide range of investigative approaches; however, all use computers to rapidly solve equations designed to describe neural responses to electrical stimuli. Currently, no single computational approach adequately mimics the inherent sophistication of animal models, as complete mathematical descriptions of the electrical features of the cochlea, nerve fibers, and surrounding tissues are still prohibitively complex. Thus, computer-model approaches tend to focus on these different components to varying degrees of sophistication. While they have this disadvantage, computer models have a significant advantage: the tissue and neuronal elements described by the experimenter's equations can be readily manipulated with precise control, thereby isolating potentially critical variables in ways often difficult to achieve with in vivo studies. There is also a synergy between computer and animal models. While the former can be used to explore isolated physiological mechanisms with exquisite control, animal experiments are required for validation.

Computational models of the electrically stimulated periphery can be classified in multiple ways depending on the particular mechanism under investigation. Models that describe actual ionic currents and membrane parameters (e.g., Na^+ currents, internodal capacitance) can be referred to as *biophysical models* (e.g., Columbo and Parkins 1987; Finley et al. 1990; Frijns et al. 1995; Rubinstein 1995; Cartee 2000; Rattay et al. 2001). Such models offer a great deal of versatility in that arbitrary stimulus waveforms can be applied to them to produce unique solutions to novel conditions. A more limited approach employs a *phenomenological model* (e.g., Bruce et al. 1999; Miller et al. 1999b), which incorporates experimentally obtained functional relationships that describe neuronal responses (e.g., strength–duration, threshold–distance functions). Such models are somewhat less versatile in that they depend on known functional relationships to produce a model output.

Computational models can also be classified along two dimensions that describe neural behavior and the influence of the extraneural environment. *Field models* focus on predicting the distribution of potential and current gradients is the tissue surrounding nerve fibers. These models are indebted to work predating cochlear implant research that studied the conductive properties of the tissues of the inner ear (e.g., Strelioff 1973). *Neuronal models* predict the response of a nerve fiber when its membrane is subjected to a known electrical field. As we shall see, a complete computational description of intracochlear electrical stimulation of the auditory nerve must incorporate both of these model components, although many modeling efforts have primarily focused on only one of these two areas.

Field models can be realized using different techniques to specify potentials within a volume of tissue encompassing the cochlea and auditory nerve. Some are based on simple, analytic descriptions of an electrical field that typically require an assumption of an electrically uniform extracellular medium (e.g. Rubinstein 1993; O'Leary et al. 1995; Bruce et al. 1999; Cartee 2000). Others employ either lumped-parameter (e.g., Black and Clark 1980; Suessermann and Spelman 1993; Kral et al. 1998) or finite-element (Finley et al. 1990; Frijns et al. 1995; Rattay et al. 2001) techniques. These latter approaches are typically used to construct a three-dimensional model of the cochlea and nervous tissue that can account for the heterogeneous conductive properties of bone, fluid, and soft tissues of the inner ear. Finite-element model results have demonstrated that extracochlear tissues distort electric fields, rendering inadequate the assumption of a uniform extracellular medium. Finite-element models are particularly valuable in predicting excitation patterns with different electrode configurations, preliminary evaluation of new electrode designs, and comparisons of across-species differences. Such models require adequate descriptions of tissue properties and are, relative to other approaches, computationally intensive.

The second component of modeling focuses not on the external electrical field, but on realistic descriptions of the *neuronal elements* to account for the biophysical aspects of the fiber membrane. One of the simplest fiber models is that of a voltage-comparator circuit with a noisy reference (comparison) voltage. Such a model may include relatively simple circuit elements to approximate neural integration or refractoriness (Wilson et al. 1995) or use adjustable parameters to account for changes in stimulus parameters (Bruce et al. 1999). Such models find application in describing the presence or absence of a neural event under specific stimulus conditions, but cannot be generalized to an arbitrary stimulus. A more sophisticated approach is that of the biophysical model, which accounts for the spatially distributed membrane properties of axons. Such models rely on the empirical description of voltages generated at the node of Ranvier provided by the equations of Hodgkin and Huxley (1952). Following the approach of Reilly et al. (1985), Colombo and Parkins (1987) were the first to adapt such a model to the anatomy of a mammalian auditory nerve fiber. Models similar to theirs have been incorporated into field models (Finley et al. 1990; Frijns et al. 1995; Rattay et al. 2001), rendering a relatively complete description of intracochlear excitation of the nerve. Those biophysical models are deterministic in that threshold current is defined by a single level. An exception is a fiber model (Rubinstein 1995; Matsuoka et al. 2001), that incorporates stochastic ion channel kinetics at each node. This model accounts for the probabilistic behavior of fibers that has been observed experimentally and is an important consideration in the representation of electrical stimuli in cases where individual fibers are stimulated at levels within their dynamic range. Evidence from both human and animal experiments suggest that, in the case of the human nerve excited by a cochlear prosthesis, a significant subpopulation of excited fibers indeed responds within this range.

3. Basic Mechanisms of Neural Excitation

The electrophysiology of excitable membranes has been investigated for over a century. The classic work, which predates investigations of the auditory system, employed a wide variety of techniques—from dissected gastrocnemius muscles to patch clamps—to describe electrical properties and ionic mechanisms of the neural membrane that support the generation and propagation of an action potential. Early studies typically used relatively simple neural systems (e.g., frog motor neurons, squid giant axons) that could be isolated from surrounding tissue and excited by relatively well-defined electrical fields. For example, an isolated nerve bundle or fiber would be suspended along a straight line within an isotropic conductive medium (Rushton 1927). By using simple electrode configurations such as monopoles, dipoles, and parallel plates, experimenters assumed that the fibers had a negligible effect on the external electrical field, and that the basic field laws—inverse, inverse square, and constant—were applicable. Similarly, patch-clamp techniques (Neher and Sakmann 1976) were designed to measure membrane properties under conditions that controlled the external electrical state of the membrane. While the anatomical dimensions of motor neurons and the electrical medium of fibers measured in vitro vary considerably from that of the mammalian auditory nerve, the classic studies have provided many of the biophysical principles of excitable membranes that are routinely applied to studies of the auditory system. Basic biophysical properties such as the strength–duration relationship, effects of stimulus polarity, and electrode orientation, as well as the role of voltage-controlled ionic channels were well described in the first half of the twentieth century.

3.1 Membrane Biophysics

The properties of the neural membrane are important for understanding neural excitation by electrical stimulation. The term *neural membrane* typically refers to that of the axon, as it is the neural element that generates and propagates action potentials. The myelinated axon consists of a series of nodes of Ranvier separated by internodal regions. The nodes contain the voltage-sensitive ionic-specific channels responsible for action potential generation. The internodes are considered to be electrically passive elements whose voltages are driven by nodal activity, although the electrical properties of both regions determine the firing properties. In the case of auditory nerve fibers, consideration must also be given to the cell body and unmyelinated terminal of the peripheral process, as the electrical properties of those elements may also influence the generation and propagation of action potentials.

Although action potentials can be supported by a single type of voltage-gated channel, both Na^+ and K^+ specific channels have been isolated and described using patch-clamp techniques. Hodgkin and Huxley (1952) used voltage-clamp techniques to characterize the properties of ionic currents in the squid axon and described how action potentials are generated by these channels. From the experimental results, they developed empirical equations using three variables (m, h, and n) to describe the temporal proper-

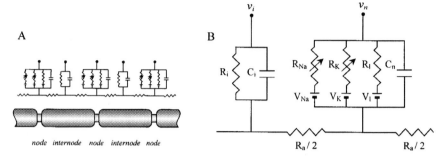

FIGURE 5.1. Lumped-parameter circuit model of a myelinated axon. A: Node and internodes are represented by repeated circuit elements joined by resistances (R_a) arising from the internodal segments of the axon. R_i and C_i represent internodal leakage resistance and capacitance. B: A single internodal segment and nodal segment. V_n is the voltage at the external surface of the node. Three batteries represent the Nernst potentials for Na$^+$, K$^+$, and nonspecific ionic carriers. Variable resistors represent voltage-controlled gates governed by Hogkin-Huxley descriptions. R_l and C_n represent passive electrical properties of the node.

ties of the Na$^+$ (m^3h) and K$^+$ (n^4) channels. A classic approach to understanding membrane physiology is through the use of electrical circuit models of the axon. As it provides an imperfect (leaky) separation of intra- and extracellular charges, the internodal membranes can be represented as a parallel combination of capacitance and resistance elements. The nodal regions include voltage-controlled resistances (to account for the voltage-gated channels) as well as passive capacitance and resistance terms. The Nernst potentials, arising from metabolically maintained ionic concentrations, are modeled as DC voltage sources. Figure 5.1 illustrates a relatively simple, lumped-parameter version of such a model, in which single circuit elements represent internodal and nodal capacitance and resistance. Axonal models have been developed with increasing sophistication that reflect advances in computer speed and memory. The model of McNeal (1976) was used to investigate the neural effects of externally generated stimulus currents. His model employed the active Hodgkin-Huxley equations at the node where current was injected and passive components at all other nodes. Bostock (1983), however, demonstrated that the near-threshold behavior of the membrane depended on both the active and passive membrane properties. Reilly et al. (1985) improved upon the McNeal model by using all active nodes, accounting for Bostock's observations and allowing for the investigation of more realistic extraneural fields that could influence more than one node.

While the model of Figure 5.1 collapses the distributed membrane properties of each internode into single, lumped, elements, more sophisticated models use multiple segments to represent their spatially distributed nature (e.g., Finley et al. 1990; Rubinstein 1995). In vitro measures of Tasaki (1955) showed that the specific internodal capacitance and conductance (per unit length) were orders of magnitude smaller than the nodal values. Some auditory nerve fiber models (e.g., Frijns et al. 1995) assume the internodal membranes to be perfect insulators. Even simple models, such as that of Figure 5.1, are useful for demonstrating the influence of neuron anatomy on electrical properties. For example, myelination greatly reduces internodal capacitance and conductance, reducing excitation threshold and improving conduction velocity. Also, increases in fiber diameter result in increased internodal capacitance and conductance. The deleterious effect of increased capacitance on the internodal time constant, however, is offset by larger reductions in axonal resistance (R_a).

Most of the aforementioned models are deterministic in that the response to a specific threshold-level stimulus would be identical for each presentation. Verveen (1961), however, described fluctuations in excitability of individual neurons in response to repeated electrical stimuli. He assumed that these fluctuations were due to properties of the neural membrane. More recent experimental work has used patch-clamp techniques to measure currents within single ionic channels. A key property of these channels is their stochastic nature. The pattern of membrane current typically varies in response to each stimulus presentation as a result of variations in the opening and closing of individual channels within the patch (Aidley 1998). Models that incorporate such detailed stochastic activity of ion channels (Clay and DeFelice 1983, Rubinstein 1995) can simulate stochastic properties of neurons in response to electrical stimulation and produce realistic, finite-sloped input–output functions that are observed experimentally.

3.2 Defining the Adequate Stimulus

Early physiological experiments sought to define the adequate electrical stimulus for initiating action potentials. In the case of an electrode positioned over an axon in a conductive medium (Fig. 5.2A), depolarization occurs at the node nearest the negatively charged (cathodic) electrode (Ranck 1975). A simple equivalent circuit is shown (Fig. 5.2B); which includes a nodal resistance (R_n) and two axonal resistances (R_a). Voltage source V_s puts a negative charge on the electrode, attracting positive charge carriers toward the electrode that results in source current I_s. For this circuit, we now assume the electrode is positioned so close to the node that all of its current flows through the node ($I_s = I_m$ where I_m is the current across the neural membrane at the node). Since positive current flows from the node (i.e., from point a to point e), membrane voltage (v_n) is positive by Ohm's law and the membrane is depolarized at that node. Furthermore, the converging axonal currents (i.e., those through R_a) must originate from inward currents at nodes flanking the depolarized node, resulting in hyperpolarization of those nodes. Using a frog gastrocnemius nerve suspended in a

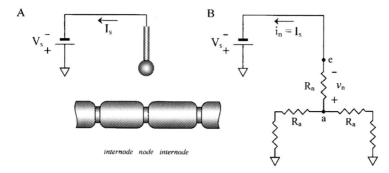

FIGURE 5.2. Schematic diagram of cathodal depolarization of a node. A: It is assumed a stimulating electrode is positioned so close to the node that all positive current (provided by the source voltage) flows out through the node ($I_n = I_s$). B: For simplicity, we only consider the passive resistance of the node (R_n). Within the axon, local circuit current flows toward the node through axonal resistances (R_a) and non-specific (unlabeled) resistances that represent more distance axonal segments. By Ohm's law, positive current drawn from the axon out of the node results in point a being at a higher potential than point e (i.e., depolarization).

presumably uniform electrical field, Rushton (1927) found that its excitability was proportional to the cosine of the angle between the current lines and the nerve. That is, most efficient excitation occurred with the nerve positioned parallel to the field gradient. Kato (1934) described how, by moving a monopolar electrode along the axis of isolated fibers, a strong threshold minimum occurred above each node. These findings established the need for a potential difference across nodes for neural excitation. For a linear axon, it is the component of the external stimulus current vector in parallel with the axon that is effective in depolarizing it. These findings underlie an oft-cited construct for locating the site of action potential initiation in the presence of an external electrical field. Reilly et al. (1985) suggested that the magnitude of membrane depolarization as a function of position along the axon could be used to predict the site of action potential initiation. Rattay (1986) formalized this notion, defining the *activating function*, f, as the second partial spatial derivative of $V_{em}(x)$,

$$f = \frac{d^2 V_{em}(x)}{dx^2} \qquad (2)$$

where $V_{em}(x)$ is the applied electric potential measured along the length, x, of the axon. The activating function plots the regions of the axon that are depolarized ($f > 0$) and hyperpolarized ($f < 0$).

We can gain an intuitive appreciation of f by considering a passive axon in a homogeneous conductive medium with particular applied electric fields. It is important to note that $V_{em}(x)$ is the voltage along the axis of the

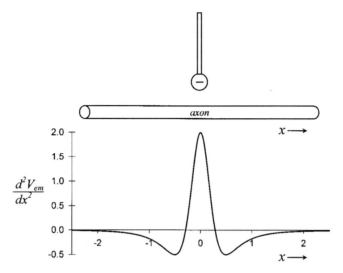

FIGURE 5.3. The activating function produced by a cathode positioned over a uniform axon within a homogeneous conductive medium. Distance along the axon, x, is measured from the point closest to the cathodic stimulating electrode. The positive portion of the curve represents the segment of the axon that would experience an outward flow of positive current (i.e., depolarization). For anodic stimulation, the curve is simply inverted such that the flanking regions of the curve represent depolarized regions.

fiber (but just external to it) and that the second derivative is taken along this x axis. Let us first assume that the first spatial derivative of V_{em} has a *constant* nonzero value, i.e., voltage changes along x at a constant spatial rate. Considering Ohm's law, the current along x therefore must be constant, with no current available to enter or leave any of the nodes. By definition, activating function f is zero at all x values, reflecting this lack of depolarizing or hyperpolarizing membrane currents. If, however, the first derivative changes along the axon, that implies some nodal currents are nonzero and the resting membrane voltage is altered. In that case, the activating function is also nonzero at those nodes. Figure 5.3 depicts the activating function for the simple case of a monopolar electrode positioned over a long axon with uniform membrane properties within a homogeneous medium. The region immediately below the electrode is shown by the function to be depolarized. Activating functions can be computed for curvilinear axons (Finley et al. 1990; Rattay et al. 2001) or axon terminations (Rubinstein 1993) and can have markedly different profiles from that shown in Figure 5.3.

The validity of the activating function concept depends on several simplifying assumptions. As it does not consider the equivalent circuit of the membrane, it assumes that the potentials created by the axon itself do not influence the pattern of membrane depolarization. This is, at best, a rough approximation. In his model axon, McNeal (1976) demonstrated that, in response to a current pulse, membrane potentials and currents vary significantly over the course of a few tens of microseconds. Warman et al. (1992) and Zierhofer (2001) demonstrated that the activating function is most accurate as a predictor of membrane potential at the onset of a stimulus; however, spatial filtering by the membrane renders it inaccurate for all but very short (i.e., a few microseconds) stimulus pulses. Importantly, the activating function cannot accurately predict threshold or compare thresholds for axons with a variety of electrode-to-axon distances. Nonetheless, given its simplicity, the activating function is a useful construct for gaining an appreciation of the field component effective in depolarizing the axon and locating likely points of greatest depolarization and hyperpolarization. If the stimulus polarity is reversed, the activating function is inverted, thus predicting the change in action potential initiation sites with polarity reversal. Additionally, this function can be used to demonstrate how different electrode configurations (e.g., bipolar, tripolar) can be used to alter the membrane depolarization profile (Rattay 1989; Spelman et al. 1995; Kral et al. 1998), a process sometimes referred to as *current focusing*.

3.3 Current Integration: The Strength–Duration Function

As excitable membranes integrate stimulus current, the threshold of excitation depends on the duration of the stimulus pulse. This relationship is described by the strength–duration function, which plots excitation threshold versus stimulus duration. The standard method of collecting these data employs a monophasic rectangular current pulse delivered to a monopolar electrode (Loeb et al. 1983). When plotted on linear axes, the function, as well as its first derivative, decreases monotonically and approaches a horizontal asymptote (i.e., *rheobase*). This curve is often characterized by a measure called *chronaxie*, i.e., the pulse duration at which threshold current is twice the rheobase value. These features of a typical strength–duration curve are showed in Figure 5.4.

Strength–duration data are often fit to a decaying exponential, following the model of Lapicque (1907):

$$I = I_{rh} / (1 - e^{(-t/\tau_{sd})}) \tag{3}$$

where I_{rh} is the rheobasic current and τ_{sd} is the strength–duration time constant. Chronaxie is linearly related to τ_{sd} and is a convenient way of characterizing the integrative property of membranes. Strength–duration data have also been fit to hyperbolic functions (Bostock 1983; Columbo and Parkins 1987) shown to provide better fits than Lapicque's function (Bostock 1983). However, both functional forms are used. Model work by Bean (1974) showed that the time constant of the passive membrane

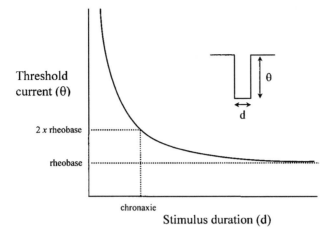

FIGURE 5.4. Hypothetical strength–duration curve. Threshold of activation is plotted as a function of the duration of a rectangular cathodic stimulus pulse (inset). "Rheobase" is the value of the horizontal asymptote. Chronaxie, derived from the rheobase measure, is proportional to the time constant of the curve.

depends on axial resistance as well as membrane capacitance and resistance. Bostock (1983) demonstrated that, when the active (i.e., voltage-sensitive) node is considered, the effective membrane time constant is greater than that of the passive model. A comparison of a strength–duration data with that of a perfect integrator (a hyperbola) demonstrates that neural membranes are leaky integrators that integrate current more efficiently at relatively short pulse durations. Strength–duration functions essentially describe a low-pass filtering of the electrical field by the membrane. Estimates of auditory nerve fiber chronaxie typically fall in the range of a few hundred microseconds. A mean value of 264 μs has been reported from cats (van den Honert and Stypulkowski 1984) while Parkins and Columbo (1987) reported a mean of 350 μs from squirrel monkeys.

Chronaxie has been shown to depend on the shape of the external field and the membrane depolarization profile (Davis 1923; Grundfest 1932; Bostock 1983). The strength–duration time constant may decrease with decreasing electrode–axon distance, an important consideration for interpreting data obtained under different stimulus conditions and from fibers of differing thresholds (Bostock 1983; van den Honert and Stypulkowski 1984). Chronaxie is also dependent on the stimulus waveform, with biphasic pulses producing more steeply sloped strength–duration functions than those obtained with monophasic pulses (Reilly et al. 1985; Miller et al. 1995a). The influence of the second phase of a biphasic pulse on the strength–duration characteristic will vary with pulse duration, as the membrane has a finite integration time window. The effect of the second phase would be expected to be large at short pulse durations and nil for durations that exceed the integration time window.

3.4 Threshold–Distance Relationship

For a monopolar electrode within an isotropic medium, field strength varies inversely with the distance from the electrode. Available single-fiber data (BeMent and Ranck 1969a; Ranck 1975) and model simulations of an axon (BeMent and Ranck 1969b; Rattay 1989) indicate that threshold current increases approximately with the square of this distance. As can be appreciated from the activating function concept, the region of axonal depolarization varies with electrode–axon distance; with increasing distance, more nodes fall into the depolarized region of the function (Rattay 1989). As noted above, however, the activating function does not predict threshold current (Warman et al. 1992; Zierhofer 2001) and cannot be used to estimate threshold as a function of distance. At any rate, the anatomical complexity of the cochlea and nerve requires more sophisticated approaches to the issue of threshold–distance relationships.

Given the location of intracochlear stimulation electrodes in cochlear implants and the spatial distribution of the target neurons (see Leake and Rebscher, Chapter 4), it is likely that electrode-to-axon distance has a significant effect on the distribution of fiber thresholds. Changing the intrascalar location of the stimulating electrode can result in large threshold changes and growth rate of the nerve's electrically evoked gross potential (Miller et al. 1993; Shepherd et al. 1993). In their finite-element model of the human cochlea, Frijns et al. (2001) demonstrated that fiber recruitment with level was more gradual when the modeled electrode was moved to a position closer to the medial wall of the scala tympani. They observed the greatest reduction in fiber threshold for those fibers located closest to the electrode, a result consistent with a distance effect.

The integrative property of axons results in a potentially useful interaction between stimulus pulse duration and the distribution of thresholds across the fiber population. As pulse duration is decreased, the range of fiber thresholds (in linear units of current) increases. This relationship has been investigated by Grill and Mortimer (1996) and can be readily appreciated by considering the relationship between the strength–duration functions of two fibers with identical time constants but differing rheobases. As threshold is a decreasing exponential function of pulse duration, the difference in the two fiber thresholds is larger at short durations. Thus, the use of relatively short-duration stimulus pulses offers not only greater efficiency, but also a greater range of usable current values.

3.5 Threshold vs. Fiber Diameter

The well-known observation that threshold decreases with increasing fiber diameter has typically been inferred through measurements of threshold and conduction velocity or across-study comparisons (Ranck 1975; Fang and Mortimer 1991). Computational models indicate that these threshold reductions become smaller as fiber diameter increases (McNeal 1976; Rattay 1986). Although estimates of fiber diameter in cats and humans vary somewhat across studies, they are comparable. Most measures have been made of myelinated fibers at the level of the central axons within the nerve trunk. Mean diameters of cat fibers have been estimated between 2.5 and 4 μm (Arneson and Osen 1978; Romand et al. 1981; Liberman and Oliver 1984). Measures from humans indicate mean values ranging from about 3 to 5 μm (Nadol 1990; Spoendlin and Schrott 1989; Felix et al. 1992). The distribution of diameters around the mean are relatively narrow in both species, with reported standard deviations typically ranging from about 0.7 to 1.0 μm. The diameters of the peripheral processes are reported to be about half that of the central axons (Liberman and Oliver 1984; Spoendlin and Schrott 1989). While variations in fiber diameter tend to be small, they have been shown to correlate with spontaneous activity, i.e., high-sponta-neous-rate fibers tend to have larger diameters than those with low spon-taneous activity. In addition, the smaller diameter, low-spontaneous-rate nerve endings are segregated from the larger diameter, high-spontaneous-rate nerve endings on the inner hair cells (Liberman 1982; Liberman and Oliver 1984). To the extent that peripheral processes may survive in implanted ears, this segregation of nerve endings with different fiber diam-eter could influence the fiber threshold distribution.

In his axonal model, McNeal (1976) showed that thresholds for small fibers (<5 μm), approach an inverse-square relationship with diameter, i.e., a twofold range in fiber diameter can account for a fourfold (12 dB) range in threshold. Using model simulations, van dan Honert and Stypulkowski (1987a) demonstrated larger effects of diameter at greater electrode–fiber distances. Taking into account these observations, it is plausible that the combined effects of a range of distance and fiber diameter can account for the 15 to 20-dB range of fiber thresholds observed in cats (van den Honert and Stypulkowski 1987a; Miller et al. 1999b) and in realistic spatial models of the cochlea (Frijns et al. 2001).

3.6 Site of Action Potential Initiation

An area of considerable conjecture concerns the identification of mem-brane sites at which action potentials are initiated by intracochlear stimulation. This is complicated by the spiral course of the fibers, the distal-to-proximal changes in diameter and myelination, and the distribu-tion of electrode-to-fiber distances. Action potentials could theoretically be generated at nodes either peripheral or central to the cell body. Some research has attempted to infer the site of spike initiation through measures that presumably reflect membrane properties at that site. van den Honert and Stypulkowski (1984) measured single-fiber strength–duration functions both in cats with intact fibers and in cats in which the peripheral processes were mechanically ablated. Strength–duration time constants were significantly greater for the intact fibers. As peripheral processes likely have different integrative properties than those of the central processes (the presumed initiation site in the ablated preparations), that observation suggests that the peripheral processes were excited at threshold stimulus levels in the intact system. Additional evidence of peripheral excitation was provided by ECAP measures. Stypulkowski and van den Honert (1984) observed two negative peaks at slightly different latencies (labeled N0 and N1) in their feline recordings. As the relative amplitudes of these two peaks varied systematically with stimulus level, they suggested that such data provided evidence of a shift in excitation site across the cell body.

Inferences based on gross potentials are complicated by the possibility that different neural subpopulations could contribute to a response waveform. Single-fiber recordings offer the advantage that changes in place of excitation will be reflected in a change in mean spike latency. If, at low levels of stimulation, action potentials are initiated peripheral to the cell body and, at higher levels, central to the cell body, one may expect a discontinuity in the latency functions, resulting in a bimodal distribution. Single-fiber recordings, however, have not consistently demonstrated such discontinuities. Using monophasic pulses, van den Honert and Stypulkowski (1984) observed none and Miller et al. (1999a) observed them in only 2% percent of sampled fibers. Latency discontinuities are more common with biphasic stimulation (Javel et al. 1987; Javel 1990), although they are, in part, attributable to the biphasic stimulus (van den Honert and Stypulkowski 1987b; Miller et al. 1999b). Additional insight is provided by finite-element models of the cochlea (Frijns et al. 1995, 2001), which can estimate spatial fiber excitation profiles. The plots of Figure 5.5 show both fiber threshold and site-of-excitation data as a function of cochlear place for a modeled guinea pig cochlea stimulated by monopolar electrodes. They demonstrate that, in general, only a small proportion of fibers near the stimulating electrode are excited at the peripheral processes, while most fibers are excited within the modiolus. Comparisons of feline ECAPs produced by monopolar and bipolar intracochlear stimulation suggest that the latter mode more readily excites peripheral processes, while monopolar stimulation generally results in modiolar excitation (Miller et al. 2003).

4. Response Properties of Auditory Nerve Stimulation

Most of the studies discussed above focused on membrane properties to determine how fibers respond under various stimulus conditions. Although they used several approaches, including in vitro models, invertebrate or muscle tissue, and simplified mathematical models, they have provided many of the biophysical principles of excitable membranes routinely

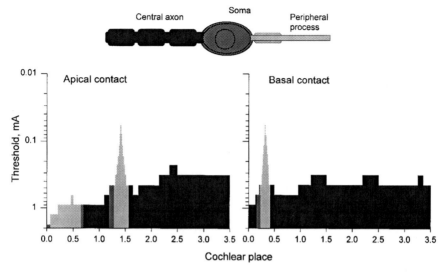

FIGURE 5.5. Spatial excitation patterns of the guinea pig auditory nerve produced by monopolar stimulation, as estimated by a finite-element model of Frijns et al. (2001). Cochleotopic fiber location and stimulus current are shown on the axes, while the distal-to-proximal site of excitation of each modeled fiber is depicted by three different shadings. Excitation patterns produced by an electrode in the base (A) and at a more apical location (B) are shown. (Modified from Frijns et al. 2001.)

applied to auditory nerve studies. It is only within the last 25 years that extensive studies specific to the auditory system have been performed, spurred on by the success of cochlear implantation. This section reviews properties of the auditory nerve response, including growth of response with stimulus level, refractory recovery, and characteristics unique to sustained electrical stimulation. Assessments of spatial spread of the population response are described in section 5.

4.1 Input–Output Functions

4.1.1 Stimulus Paradigm

Electrical stimulation of the nerve can be delivered by a variety of electrode configurations and stimulus waveforms (see Wilson, Chapter 2). Current is passed between a minimum of two electrodes. With cochlear stimulation, the electrode configuration is considered to be bipolar when the two electrodes reside within the cochlea. When the return electrode is remote (i.e., extracochlear) the configuration is considered to be monopolar. A typical stimulus used in cochlear implants is the biphasic pulse, consisting of a cathodic (negative) phase and an anodic (positive) phase that deliver equal, but opposite, charge. Neural excitation can result from either the anodic or cathodic phase, which can complicate the interpretation of electrophysiological responses. Furthermore, bipolar stimulation presents the neural tissue with both cathodic and anodic fields. Therefore, much of the experimental work described in this chapter has used monophasic, monopolar stimulation to simplify the excitation pattern and data interpretation.

Several aspects of single-fiber responses are illustrated in Figure 5.6. Unless otherwise noted, the data in this and subsequent figures were measured in response to monopolar stimulation. In Figure 5.6A, responses to 40-μs monophasic cathodic pulses are plotted for 100 repeated stimulus presentations. They exhibit a large stimulus artifact followed by an action potential in approximately 80% of the traces. These waveforms illustrate a common problem in studies of electrical stimulation, i.e., separation of the neural response—in this case action potentials—from stimulus artifacts. There is also evidence of the ECAP in each trace. Investigators have devised various ways to extract the desired response from the waveforms. Here, we illustrate a method in which we construct a *template* waveform by averaging all traces with no action potential present. The resulting waveforms after template subtraction, shown in the lower traces, reveal the action potentials without the two aforementioned contaminants. They illustrate the stochastic nature of the neural discharge, in both the probabilistic nature of the response and the spike-to-spike variations in latency (i.e., jitter). Finally, we note that these and other data discussed in this section represent responses measured in an ear without functional hair cells. The response latencies are short (<1 ms), indicating that the responses arise from direct depolarization of the nerve. There is generally no evidence of spontaneous neural activity, but as noted there is still considerable stochastic variation in the response to individual stimuli.

4.1.2 Basic Single-Fiber Response Properties

The characteristics of latency and firing efficiency (FE) as a function of stimulus current level are demonstrated in Figure 5.6B. Verveen (1961) measured similar FE-vs.-level functions in nonauditory fibers and found that they were fit well by a cumulative (integrated) gaussian function. Based on that curve, he defined threshold as the current level evoking 50% FE, and relative spread (RS) as the standard deviation of the gaussian function divided by the threshold level. The input–output functions illustrate the probabilistic nature of the responses as well as the limited dynamic range of fibers, and RS can be used to characterize these properties. Reported values of RS for auditory nerve fibers range from 5% to 10%, corresponding to a dynamic range (using 10% to 90% FE response criteria) of 2 dB or less (Dynes 1996; Javel 1990; Shepherd and Javel 1997; Miller et al. 1999a).

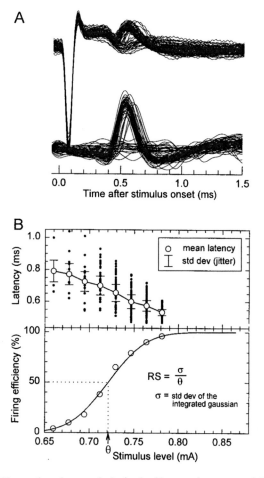

FIGURE 5.6. A: Example of recorded single-fiber action potentials evoked by a 40-μs cathodic pulse delivered by a monopolar electrode. Responses to 100 repeated stimuli are shown. The top traces include large stimulus artifacts (at about 0.1 ms), which are removed by a template-subtraction procedure in the lower traces. B: Example input–output functions of a fiber obtained by 100 repeated stimuli presented at various levels. In the upper graph, small dots denote timing of individual action potentials, circles indicate mean spike latency, and bars indicate jitter, or standard deviation of the spike times. The lower graph depicts firing efficiency (probability) as a function of level, with the data fit to an integrated gaussian curve. (From Miller et al. 1999a.)

As seen in Figure 5.6B, mean latency decreases with stimulus current, typically reaching an asymptote at high current levels (van den Honert and Stypulkowski 1984). It is also apparent that jitter—the standard deviation of the pool of action-potential latencies—decreases with increasing level and firing efficiency. Poststimulus time histograms to single-pulse stimuli typically have the form of a Poisson functions with a delay (Miller et al. 1999b). Although level-dependent reductions in mean latency are caused by reduced jitter, other factors are also present. Reduced latency at higher levels may be caused by a shorter time-to-charge delay and a spread in the region of depolarized nodes to more central regions of the fibers. Miller et al. (1999a) showed that, in a limited number of cases, discrete jumps to shorter latencies can be observed, suggesting the possibility of a shift of excitation site across the cell body. They also demonstrated that threshold, latency, jitter, and RS are somewhat interdependent.

4.1.3 Electrically Evoked Compound Action Potential (ECAP)

The ECAP is typically measured using a recording electrode positioned on the nerve trunk, in the scala tympani, or in the middle ear near the round window. Averaged response waveforms for both anodic and cathodic monophasic pulses, in this case recorded directly from the auditory nerve in a cat, are illustrated in Figure 5.7A. The ECAP is usually quantified as an amplitude (negative to positive peak) and a latency (negative or positive peak). Plots of both as a function of current level are shown in Figure 5.6B and C. These functions are qualitatively similar to those for single-fiber FE and latency, but differ significantly in dynamic range. The wider range of the gross potential reflects, to a great degree, changes in the population of responding fibers. Based on the model of Goldstein and Kiang (1958) (see section 2.3), if one assumes the same unit potential and latency of response for all neurons in the auditory nerve, the amplitude of the ECAP would be simply proportional to the number of active fibers in the population. While such assumptions are not exactly correct, the ECAP growth can be reasonably approximated from the fiber threshold distribution for the particular configuration of stimulating electrode (Miller et al. 1999b). Such measures therefore can be useful in estimating patterns of fiber recruitment (Merzenich and White 1977; Hall 1990; see section 6).

4.1.4 Effects of Stimulus Polarity

As is demonstrated in Figure 5.7, stimulation with either cathodic or anodic current pulses is effective in eliciting a response from the auditory nerve. Sensitivity of individual nerve fibers in cats tends to be better for cathodic pulses, and the average latency at a particular FE in response to cathodic pulses is typically longer (Miller et al. 1999a; Shepherd and Javel 1999). The difference in latency across the range of sensitivity (Fig. 5.7B) also suggests different sites of spike initiation for anodic and cathodic stimuli, consistent with well-established trends with electrically stimulated tissue (Ranck 1975). Data collected using biphasic stimuli also demonstrate a separate peak in the response to each polarity, suggesting two sites of excitation. With biphasic pulses, van den Honert and Stypulkowski (1984) observed two peaks in the response. Data with continuous sinusoidal or other periodic

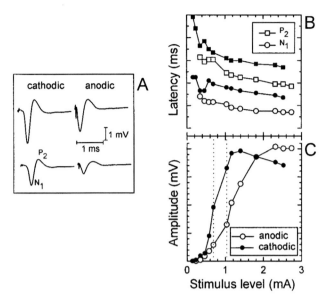

FIGURE 5.7. Example of electrically evoked compound action potential (ECAP) waveforms, latency-level functions, and amplitude levels functions recorded from a cat. Stimuli were anodic and cathodic monophasic pulses (39 µs in duration) delivered by a monopolar electrode placed in the scala tympani. A: The recorded response at the two levels indicated by the dashed lines in panel C. The relatively large stimulus artifact at the beginning of each trace has been suppressed using a template subtraction procedure described by Miller et al. (1998). B: The latency of the N_1 and P_2 peaks as indicated measured as indicated on the response waveforms. C: The ECAP amplitude, N_1 minus P_2 peaks, as a function of stimulus level. (Modified from Miller et al. 1998.)

stimuli have demonstrated responses to both stimulus polarities, with apparent preference to cathodic stimulation (Hartmann et al. 1984; van den Honert and Stypulkowski 1987b).

The greater sensitivity to cathodic stimuli, however, is dependent on specific anatomy and electrode configuration. Although with monopolar scala tympani electrodes the cat ECAP is generally more sensitive to cathodic stimulation, a small percentage of neurons demonstrate better sensitivity to anodic pulses (Miller et al. 1999a). In contrast, guinea pigs exhibit ECAPs at lower levels with monopolar anodic stimulation (Miller et al. 1998). Responses to longitudinal bipolar stimulation in both cats and monkeys show better sensitivity to cathodic stimulation on the more apical electrode (van den Honert and Stypulkowski 1987b; Parkins 1989). Consequently, while models using monophasic pulses with simplified cochlear anatomy (Ranck 1975; Rattay 1989) predict lower cathodic threshold, the anatomy of the cochlea and the specific electrode configuration can have a significant effect on fiber sensitivity.

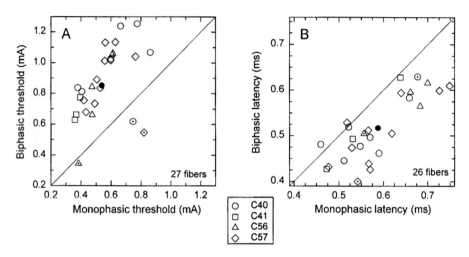

FIGURE 5.8. Comparison of monophasic and biphasic threshold and latency in data obtained from 28 auditory nerve fibers from one cat (shown by open symbols). Stimuli are 40 μs/phase delivered through a monopolar electrode placed in the scala tympani. A: Biphasic threshold [level at 50% firing efficiency (FE)] is plotted vs. monophasic threshold. B: Mean biphasic latency at 50% FE is plotted vs. the corresponding value for monophasic stimulus. Mean values are shown by the filled circles. Three fibers where the biphasic threshold is less that the monophasic threshold are plotted in both panels with dotted symbols. (Modified from Miller et al. 2001c.)

The difference in latency of the ECAP in response to cathodic and anodic stimulation is relevant to a very practical issue regarding artifact reduction in recording the ECAP or EABR. One option for artifact reduction is to average the responses to both stimulus polarities. As the artifacts to cathodic and anodic stimulation tend to cancel, the stimulus artifact is reduced in the resulting averaged response. However, the differences in sensitivity, and particularly the difference in response latency, could clearly affect the morphology and the amplitude of the ECAP derived by summing response to both polarities.

4.1.5 Monophasic vs. Biphasic Pulsatile Stimulation

Pulsatile stimulation through cochlear implants is delivered by charge-balanced biphasic pulses to avoid accumulation of charge that could produce toxic tissue reactions. Many physiological studies have also used biphasic stimuli. As neural membranes integrate current and anodic and cathodic pulses evoke different within- and across-fiber responses, the responses evoked by the two phases of biphasic stimuli require some attention.

Results of single-fiber comparisons of biphasic and monophasic stimulation are shown in Figure 5.8. The plots summarize how threshold and mean latency differ in response to monophasic (40 µs) and biphasic (40 µs/phase) stimuli. Individual single-fiber data are shown by open symbols and mean values are shown by filled circles. Monophasic thresholds tend to be lower than biphasic thresholds, and mean spike latency (assessed at 50% FE) tends to be greater for monophasic stimuli. ECAP comparisons using the same stimuli demonstrated threshold and latency differences consistent with the single-fiber measures and a simple model of the nerve fiber as a leaky integrator coupled to a delay element. The resistive and capacitive properties for the nerve membrane serve to integrate stimulus current and the voltage-gated ionic channels introduce a time delay (van den Honert and Mortimer 1979). Rubinstein et al. (2001) demonstrated that a relatively simple linear integrative membrane model can predict the threshold differences between monophasic and biphasic stimulation over a range of pulse durations.

Lower monophasic thresholds can result in more spatially restricted patterns of neural stimulation (Frijns et al. 1996; McIntyre and Grill 2000). Reduced spatial spread could decrease channel interaction in a multichannel implant (see section 5), and the lower thresholds could reduce device power consumption. Some of the advantages of monophasic stimuli could be realized with charge-balanced stimuli in the form of asymmetric pulses (second phase longer that the first), biphasic stimuli with an interphase delay (van den Honert and Mortimer 1979; Shepherd and Javel 1999), or triphasic stimuli (Coste and Pfingst 1996; Shepherd and Javel 1999). Figure 5.9 plots single-fiber thresholds vs. second-phase duration for asymmetric, charge-balanced stimuli (i.e., *pseudomonophasic* pulses). Significant threshold reductions are consistently observed for second-phase durations as short as 500 µs with the initial (excitation) phase at 40 µs. However, prolonging total stimulus duration may complicate or limit the nonsimultaneous delivery of pulses across multiple channels at high rates [e.g., continuous interleaved sampling (CIS) processing, Wilson, Chapter 2].

4.1.6 Applications to Clinical Practice

Physiological measures of peripheral function may find significant application to the clinical use of cochlear implants. One possible use is with young children and other patients who may be unable to provide accurate behavioral responses to electrical stimulation. Programming the speech processor of a cochlear implant typically requires establishing threshold and an upper level of stimulation on as many as 22 electrodes. For very young children, this can be difficult and time-consuming. To a great extent, the utility of physiological measures depends on how well they correlate with behavioral thresholds. In general, when the stimuli used for both physiological and behavioral measures are similar, the correlation between the two measures of threshold has been shown to be strong (Abbas and Brown 1991; Smith et al. 1994; Miller et al. 1995a; Brown et al. 1996). Typically, however,

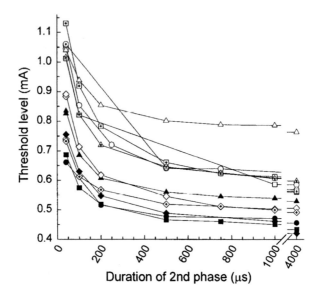

FIGURE 5.9. Reponses to pulses that are charge-balanced but asymmetric (second phase longer that the first). Single-fiber threshold is plotted as a function of the second phase duration. Data are shown from 12 fibers of two cats. (Modified from Miller et al. 2001c.)

the stimuli used to measure ECAP or EABR thresholds differ significantly from those used to program the speech processor. Pulse trains at rates of 40 to 80 pulses per second (pps) are generally used to measure ECAP and EABR thresholds, whereas trains at 250 pps or greater are generally for behavioral fitting procedures. Behavioral thresholds are influenced by central temporal integration effects unrelated to peripheral physiological measures. Considering that central integration varies across subjects (Brown et al. 1999), it is not surprising that correlations between physiological and behavioral thresholds are only moderate (Shallop et al. 1990, 1991; Brown et al. 1999). Nevertheless, Brown et al. (2000) outlined a method of assessing across-subject variations in temporal integration to predict programming levels in children using low-rate speech processors.

Although ECAP threshold measures are being applied in clinical settings, other aspects of the ECAP input–output functions could be useful. While the growth of ECAP amplitude with stimulus level measured in implant users is similar to measures from experimental animals, the human growth functions do not typically demonstrate a saturation in amplitude at high levels (Brown et al. 1990; Brown and Abbas 1990). This is likely due to uncomfortable loudness percepts that limit stimulus levels to values below ECAP amplitude saturation in awake humans but not in anesthetized animals. As a result, comparison of the dynamic range of ECAP growth in human implant users and animals is not possible.

Although physiological dynamic ranges may not be obtainable, there are significant variations in the slope of ECAP and EABR growth across users and also across electrodes within users (Abbas and Brown 1991; Brown et al. 1996). In as much as the growth of the evoked potential reflects the fiber threshold distribution for a particular electrode (Miller et al. 1999b), that information may be useful in programming the implant as well. Generally speaking, there is significant flexibility in the selection of active electrodes, the electrical configuration of those electrodes, and the programmed range of stimulus levels. Information regarding the sensitivity and location of excitable neural elements could be useful in choosing the best combination of stimulating electrodes and levels for individual implant users.

4.2 Two-Pulse Stimuli: Summation and Refractory Characteristics

4.2.1 Stimulus Nomenclature

The stimulation protocols of typical cochlear prostheses employ either trains of pulses or continuous (analog) stimulation. As a result, both refractory and integrative properties of the neural membrane are relevant. These properties can be assessed using a two-pulse stimulus paradigm. In cases where a response is elicited by the first pulse, subsequent effects are primarily refractory, and we refer to the first pulse as a *masker*. Where there is no response to the first pulse, the effect on the response to the second pulse is primarily due to integration, in which case we refer to the first pulse as a *conditioner*. In either case, we refer to the second pulse as a *probe*.

4.2.2 Time Course of Recovery

Single-fiber refractory recovery functions have been assessed using two-pulse stimuli by a number of investigators (Hartmann et al. 1984; Dynes 1996; Cartee et al. 2000; Miller et al. 2001b). Figure 5.10 plots data from Miller et al. (2001b), where the masker level was set to elicit a firing efficiency of 100%. Figure 5.10A plots firing efficiency (FE) in response to the probe as a function of masker–probe interval (MPI) for several probe levels. Based on those data, threshold current level is plotted as a function of MPI using three different threshold criteria (20%, 50%, and 80% FE) in Figure 5.10B. The trends of Figure 5.10 are generally consistent with observations made from single fibers as well as ECAP recordings from both animals and humans (Brown and Abbas 1990; Brown et al. 1990). Estimates of the absolute refractory period have recently been reported to be as short as 330 μs (Miller et al. 2001b), somewhat less than previous estimates of 500 to 700 μs (Dynes 1996; Bruce et al. 1999). Following the absolute refractory

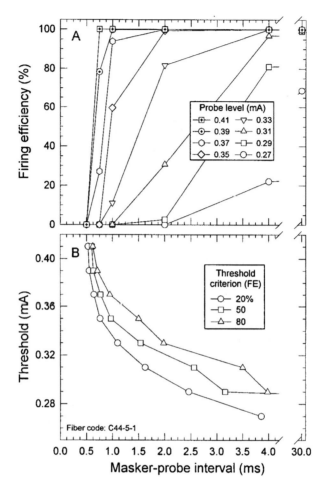

FIGURE 5.10. Example of refractory recovery functions from a single auditory-nerve fiber in a cat. A: Firing efficiency (FE) is plotted as a function of masker–probe interval (MPI) from several levels of the probe as indicated in the legend. Masker level was set in all cases to a level that evoked an FE of 100%. B: Interpolated values obtained from the plot in A are replotted as a function of MPI using three different criteria FE values to define threshold. (From Miller et al. 2001b.)

period is a period of reduced excitability, or relative refractory period, that persists for about 5 ms after masker presentation (Stypulkowski and van den Honert 1984; Dynes 1996; Miller et al. 2001b). Recovery time constants have been estimated at 400 to 1300 μs (Bruce et al. 1999; Cartee et al. 2000; Miller et al. 2001b). Feline ECAP data have been used to infer single-fiber refractory properties. They suggest that the relative refractory period is about 5 ms long and the absolute refractory period is between 300 (Stypulkowski and van den Honert 1984) and 500 μs (Brown and Abbas 1990).

4.2.3 Effects of Stimulus Level

The integrative and recovery properties are highly dependent on stimulus level. Probe level has a significant effect on the time course of recovery, as illustrated in Figure 5.10A, with higher levels producing faster recovery (Miller et al. 2001b). Not surprisingly, the higher probe levels can overcome reduced sensitivity at the small MPI values. Similar data have been reported in ECAP measures from both animals and humans, where recovery rate increases with increasing probe level (Brown and Abbas 1990; Finley et al. 1997a; Hong et al. 1998).

Masker level affects recovery in a very different way. In ECAP measures, varying masker level above that of the probe results in little change in recovery characteristics (Abbas et al. 1998). This is due to the fact that, with high-level masking, fibers responding to the probe likely will have responded to the masker. Thus, the response of the masker is effectively saturated, and further increases in masker level have no additional effect on the probe response. The situation is quite different when masker and probe levels are equal. At low stimulus levels, effects attributable to neural integration can occur. Stypulkowski and van den Honert (1984) observed a nonmonotonic recovery of the ECAP. As MPI was reduced to intervals less than 0.5 ms, they observed an *increase* in ECAP amplitude, suggesting integrative effects. Finley et al. (1997a) reported a similar effect in human implant users and demonstrated a particularly large effect at low current levels. Dynes (1996) and Cartee et al. (2000) specifically investigated the effect of a subthreshold conditioning stimulus, demonstrating facilitation presumably due to favorable biasing of residual charge on the neural membranes. These findings are consistent with a model of both integrative and refractory effects influencing the response to the second pulse, with integrative effects dominating at low stimulus levels. At high levels, refractory effects will likely dominate the response characteristics.

Psychophysical forward-masking using pulse-train stimuli can have relatively long recovery times (Shannon 1990). When a two-pulse stimulus paradigm is used, the time course of recovery has a relatively fast component, similar to ECAP recovery, and a slower component that can last several hundred milliseconds and may be due to central masking (Brown et al. 1996; Nelson and Donaldson 2001). Nelson and Donaldson (2001) also observed nonmonotonic behavior or facilitation at short masker-probe intervals, consistent with integrative properties evident in neural responses as noted above.

4.2.4 Refractory Effects on Suprathreshold Response Properties

The effects of a masker pulse on the probe response are not limited to changes in threshold or FE. Miller et al. (2001b) observed that, in addition to decreased FE during the refractory period, there was also a diminution in action-potential amplitude. This is consistent with whole-cell patch-clamp data in cultured spiral ganglion cells (Lin 1997) and model results (Matsuoka et al. 2001), and is relevant to the interpretation of gross evoked potentials such as the ECAP. A decrease in ECAP amplitude may be due not only to fewer action potentials, but also to decreased amplitudes. Diminished amplitudes may also have implications for central processing, as they could result in weaker excitation of the cochlear nucleus and could result in a higher probability of conduction failure. Thus, the characterization of refractory effects based only on spike-count data (or FE) may overestimate the output of the auditory nerve. If that is the case, gross-potential based estimates of neural recovery may yield recovery times that are faster than functional rates.

Model data also suggest that stochastic properties may be altered during the refractory period (Matsuoka et al. 2001). Both Dynes (1996) and Miller et al. (2001b) observed increased variability at the smallest MPI values. Miller et al. (2001b) also observed that, at the shortest MPI tested (500 μs), mean RS increased almost twofold over the mean value obtained at a 4000-μs MPI. This degree of RS increase is much smaller than that predicted by Matsuoka et al. (2001) with a computational model. Nevertheless, the observations are consistent with the idea that the degree of stochasticity in single-fiber responses increases during the relative refractory period. Such increases in RS could provide improved representation of signals by the nervous system due to higher levels of stochastic activity (see section 4.3.7).

4.2.5 Applications to Clinical Devices

Integrative and recovery characteristics can affect the encoding of stimuli provided by cochlear implants. Data from the previous section imply that implants using pulse trains at rates of 250 pps and above will be affected. Refractory effects can distort the neural representation of such pulse trains, as has been observed in ECAP measures from humans and animals (Wilson et al. 1997a,b; Vischer et al. 1997; Abbas et al. 1999b). These effects may be particularly important with very high rates of stimulation (>1000 pps) that are used in CIS processing schemes with many state-of-the-art devices.

Significant variations in ECAP recovery have been observed with stimulus level, across electrodes within a subject, and across subjects (Brown et al. 1990, 1996; Abbas et al. 1999a). Such variations are consistent with a hypothesis that refractory properties are dependent on the neural membrane properties and also that those properties may change with neural degeneration that takes place with hearing loss (see section 6). If these differences in refractory recovery affect signal processing in the auditory nerve, choosing stimulating electrodes or levels on the basis of refractory measurements could be useful.

Finally, specific observations of refractory measures in experimental animals may also suggest novel stimulation approaches in a cochlear implant. For instance, differences in ECAP recovery have been observed between cathodic and anodic stimuli (Matsuoka et al. 2000a). One possible reason for these differences may relate to the nerve membrane properties at different sites of spike initiation. The notion that recovery rate is related to site of excitation suggests the possibility of actively controlling the mode of stimulation in order to produce favorable refractory characteristics.

4.3 Responses to Sustained Stimulation

4.3.1 Stimulus Paradigm

One- and two-pulse paradigms are useful to evaluate neural growth, refractoriness, and integration. However, stimulation through a typical cochlear implant involves more complex stimuli—either continuous analog stimulation or amplitude-modulated pulse trains. Responses to such stimuli are affected by properties such as growth, refractoriness, and integration as well as more long-term cumulative effects of electrical stimulation. Such effects have been evaluated by simulating conditions of normal implant stimulation. Typically, continuous sinusoidal stimulation, constant-amplitude pulse trains, or amplitude-modulated trains have been the stimuli of choice.

4.3.2 Sinusoidal Stimuli

Continuous analog stimulation was used in early multichannel implants (Eddington 1980) and, more recently, in the signal processing selected by many users of the Clarion device (Osberger and Fisher 1999; Battmer et al. 1999b). With sinusoidal electrical stimulation, threshold varies little as stimulus frequency is changed, and for all fibers the threshold is lowest at about 100 Hz (Kiang and Moxon 1972; Hartmann et al. 1984).

The timing of action potentials relative to the phase of the sinusoidal electrical stimulus can be assessed using a period histogram synchronized to the phase of the input stimulus. Histograms show a narrow peak compared to acoustic stimulation, indicating greater synchrony (Kiang and Moxon 1972; Hartmann et al. 1984). In addition, the response to electric stimulation can be evident to both phases of the sinusoid compared to a single excitatory phase typical with acoustic stimulation. The histograms in response to square waves or pulse trains also demonstrate a similar high degree of synchrony (van den Honert and Stypulkowski 1987b; Parkins 1989; Hartmann and Klinke 1990b; Dynes and Delgutte 1992), suggesting that fine details of the stimulus wave shape may be poorly represented in the response. As a result of this high degree of synchrony, an ECAP can be obtained in response to continuous sinusoidal stimulation by averaging the response waveforms triggered by a constant stimulus phase (Hartmann et al. 1994; Runge-Samuelson et al. 2001). That potential has a more complex morphology than that evoked with pulses, although its amplitude and latency variations with stimulus level are comparable (Fig. 5.7).

4.3.3 Responses to Pulse Trains: Alternation

As cochlear implants often use amplitude-modulated pulse trains, insight is gained by examining the physiological response to constant-amplitude pulse trains. In animal studies, Javel (1990) and Shepherd and Javel (1997) reported single-fiber PST histograms for biphasic pulse trains; the histogram produced by a 800-pps train is shown in Figure 5.11. Two stereotypic response patterns to such stimuli are evident. First, there is a general decrease in response probability over time relative to the response to the first pulse (i.e., an adaptation-like phenomena discussed in section 4.3.4). There is also an alternation between high and low probability of response to successive pulses. Such data demonstrate that the prior history of the response is important in determining the response to each subsequent stimulus pulse. Wilson et al. (1994, 1997a,b) made ECAP measures from humans and reported similar alternation in the response amplitude for successive pulses of a train. Alternation patterns have also been reported in ECAP responses from animals (Vischer et al. 1997; Haenggeli et al. 1998; Matsuoka et al. 2000a). Such phenomena are limited to a critical range of pulse rates and presumably result from a combination of refractory and stochastic properties (Wilson et al. 1994).

Although both animal and human responses exhibit ECAP amplitude alternation, there are differences between the two groups. Similar to the single-fiber data of Figure 5.11, ECAP alternation in animals tends to be restricted to a short period (<50 ms) after onset of the stimulus train (Vischer et al. 1997; Haenggeli et al. 1998; Matsuoka et al. 2000a). In contrast, the human ECAP measures can demonstrate large alternation patterns that do not decay appreciably, persisting for several hundred milliseconds (Wilson et al. 1994, 1997a,b). The differences between human and animal data may arise from the presumably greater degree of neuropathology present in the former group. While most human implantees have suffered deafness (and subsequent degeneration) over extended periods, most animal preparations are deafened immediately prior to data collection. One could speculate that a fiber population with greater degeneration (e.g., loss of peripheral processes) may exhibit less membrane noise and a more uniform response across fibers. In that case, the response pattern could be defined more by refractory properties than by stochastic properties.

4.3.4 Responses to Pulse Trains: Adaptation

For acoustic tone-burst stimuli, PST histograms show an initially high response probability that decreases within several hundred milliseconds to

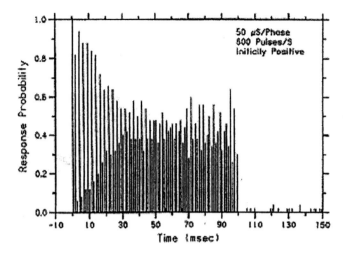

FIGURE 5.11. Example of the response of a single auditory nerve fiber to a biphasic pulse train with bipolar intracochlear stimulation. Response probability is plotted in response to each pulse in the train as a function of time after stimulus onset. (From Javel 1990.)

a constant probability (Kiang et al. 1965). Such adaptation is at least partly attributable to the hair-cell synapse (Smith and Brachman 1982). Electrical stimulation, as demonstrated in Figure 5.11, can produce adaptation even though it bypasses this synapse. In addition to the data of Javel and Shepherd, Parkins (1989) used relatively short (20 ms) tone burst stimuli and noted a decrease in excitability over that stimulus duration. Litvak et al. (2001a) obtained single-fiber responses with high-rate stimulation (up to 24,000 pps) and observed significant decreases during the first 100 ms. Such changes are generally consistent with the course of adaptation observed with ECAP stimulation (Killian et al. 1994; Vischer et al. 1997; Haenggeli et al. 1998; Matsuoka et al. 2000a).

In addition, van den Honert and Stypulkowski (1987b) observed adaptation over the course of several seconds during continuous sinusoidal stimulation, particularly at high frequencies. Miller et al. (1999a) noted that single-fiber response probability could be reduced at relatively low stimulus rates (30-ms interpulse interval) and that such changes depended on stimulus polarity. More recent data obtained with electrical pulse trains suggest that cumulative effects of electrical stimulation may be evident over even longer periods. Abbas et al. (2001b) reported changes in ECAP amplitude in response to pulse trains over the course of 50 to 100 seconds. These changes were evident for interpulse intervals as long as 16 ms. Finally, using very high rates of stimulation (5000 pps), Litvak (2001b) noted changes in responses of single fibers over several hundred seconds, in some fibers the response was reduced to zero after a period of time.

It is evident that adaptation or forward-masking phenomena can exhibit a wide range of recovery periods. As noted in section 4.2, short-term refractory effects (<10ms) are evident. With sustained stimulation, additional decreases in response over time (on the order of 100ms) are observed. With more intense stimulation or longer periods of exposure, even longer time constants (>1s) are evident. Perceptual measures also have demonstrated a similar range of temporal effects. Chatterjee (1999) described psychophysical forward-masking recovery functions by two time constants, one on the order of 2 to 5ms (similar to refractory recovery), and a second on the order of 50 to 200ms. There have also been reports of loudness adaptation in response to continuous electrical stimulation (Shannon 1983; Brimacombe and Eisenberg 1984). Shannon noted loudness changes over the course of 20 to 40 seconds, a time scale similar to that observed in ECAP measures. Recent observations (Rubinstein et al. 2002) with high-rate conditioning stimuli, in implant users as well as with extracochlear stimulation, have noted consistent decay of the percept of such stimuli. The extent to which these perceptual characteristics are the result of peripheral changes has not been determined.

4.3.5 Stochastic Responses at High Stimulus Rates

Adaptation in auditory nerve fibers can have a significant effect on the responses to normal cochlear implant stimulation. Figure 5.12 illustrates input–output functions for single neurons in response to constant-amplitude pulse trains at different pulse rates. On the left, the overall discharge rate is plotted as a function of current level. Maximum discharge rate for each stimulus saturates at or below each stimulus pulse rate. As a result, the dynamic range for the high-rate pulse trains (800 pulses/s) is much greater than the dynamic range for the low-rate (100 pulses/s) pulse train. The graph on the right in Figure 5.12 shows spike probability (i.e., FE) in response to each pulse in the train for the same data. These plots emphasize the similarity in threshold among the stimuli at various rates and the correspondingly large increase in dynamic range. It can be seen how, at high pulse rates, a more accurate encoding of a stimulus waveform could be accomplished with amplitude-modulated pulse trains. With broader dynamic ranges, neurons will be capable of encoding more detailed change in stimulus amplitude. These data strongly support the notion of greater stochasticity at higher stimulus rates, presumably due to refractory effects.

Wilson et al. (1994, 1997a) reported human ECAP data relevant to this issue. Using constant-amplitude pulse trains, they noted decreased response amplitudes with increasing pulse rate, consistent with the spike probability data in Figure 5.12. Amplitude alternations (as discussed above) occurred for pulse rates near 1000pps, but above 3000 to 4000pps response amplitudes were smaller and no alternation was observed. The alternation in amplitude of the population response suggests consistency in the pattern of

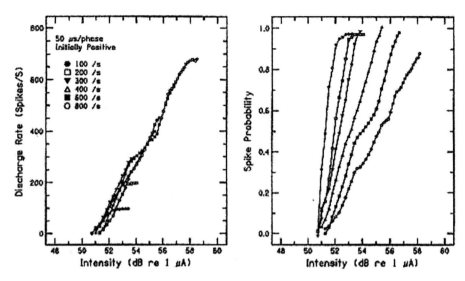

FIGURE 5.12. Response of an auditory nerve fiber to biphasic pulse train presented at different rates as indicated in the legend. (Left) Discharge rate is plotted as function of current level (in dB re 1μA). (Right) The same data are plotted as spike probability (or FE) as a function of current level. (From Javel 1990.)

responses across fibers. The loss of alternation and small amplitudes at high rates suggest that, while some individual fibers may persist in alternation, the temporal response pattern across the population is relatively diverse, with a high degree of stochasticity. Computational model simulations support this notion, demonstrating relatively greater stochastic response patterns at rates of 5000pps (Rubinstein et al. 1999). This pattern was termed *pseudospontaneous activity*, as the model fiber demonstrated stochastic properties similar to those observed in the spontaneous activity of acoustically sensitive fibers. Litvak et al. (2001a) measured cat single-fiber response to pulse-train stimulation and demonstrated some stochastic properties consistent with Rubinstein's model, although it should be noted that not all fibers demonstrated the predicted Poisson-like behavior of spontaneous activity. Furthermore, the stochastic responses occurred over limited ranges of stimulus levels and modulation depths. Nevertheless, these observations have led to proposals that attempt to exploit this property of nerve fibers to produce a stochastic resonance effect in responses (Rubinstein et al. 1999).

4.3.6 Amplitude-Modulated Pulse Trains

Signal encoding by cochlear implants generally involves filtering of the acoustic signal (for multichannel stimulation), compression, possibly envelope detection, and the delivery of one or more time-varying current wave-forms to the auditory nerve (see Wilson, Chapter 2). The ability of the nerve to transmit stimulus level differences (as characterized by input–output functions) and the temporal sequence of those differences (affected by inte-gration, refractoriness, and adaptation) are important for effective trans-mission of information to the central nervous system. Several studies have assessed the responses to amplitude-modulated pulse trains in efforts to simulate the time-varying stimuli typically delivered to the nerve through a cochlear implant. ECAP amplitude measured in response to sequences of amplitude-modulated pulses has been used to assess the nerve's ability to encode changes in the stimulus (Wilson et al. 1995; Abbas et al. 1999b). Those measures are generally consistent with the data on growth and tem-poral response properties discussed in the previous sections. Relatively little distortion of the signal is evident at low-modulation depths and low-modulation frequencies. At high-modulation depths where the dynamic range of the neural response is exceeded, and at high-modulation frequen-cies where refractory properties are evident, the ability of the response to follow the signal is compromised and distorted responses are evident.

4.3.7 Stochastic Resonance

The loss of spontaneous activity in deaf ears can be viewed as a deficit in signal processing, as a more deterministic system can reduce the transmis-sion of fine temporal detail. Several studies have proposed adding noise to stimuli to increase information transmission, essentially an attempt to produce a stochastic resonance effect. Morse and Evans (1996) added noise to speech stimuli and measured improved vowel representation in the tem-poral response of the frog sciatic nerve. Matsuoka et al. (2000b) demon-strated decreases in ECAP amplitude alternation in response to pulse trains in the presence of a background electrical noise. Experiments with human implant users have shown positive effects of background noise. Zeng et al. (2000) demonstrated increases in threshold sensitivity, and, more recently, Chatterjee and Robert (2001) demonstrated increased sensitivity to ampli-tude modulation. Although some reported effects are small, they are con-sistent with physiological observations.

An alternative to adding noise may be the presentation of a high-rate pulse train as a conditioning stimulus. The production of stochastic responses at high stimulus rates could result in a more accurate represen-tation of stimuli (Wilson et al. 1994; Rubinstein et al. 1997, 1999). Obser-vations in both human and animal experiments are consistent with the hypothesis that the random activity produced by a high-rate conditioning stimulus will have a positive effect on stimulus encoding. In both human and animal subjects, a decrease in response alternation to a low-rate pulse train is observed when presented in a background of a high-rate pulse train (Rubinstein et al. 1999). More recently, Runge-Samuelson et al. (2001)

observed a decrease in the growth of the ECAP response to sinusoidal stimuli when presented in a background of a high-rate pulse train. Using similar stimuli in Clarion implant users, Rubinstein et al. (2002) observed significant decreases in threshold as well as increased behavioral dynamic range. Such data suggest that a novel stimulus encoding strategy (using a high-rate pulse train as a conditioning stimulus) may produce more stochastic activity in the auditory nerve, thus simulating properties of the normal cochlea with acoustic stimulation and facilitating better speech perception.

5. Spatial Patterns of Neural Excitation

Multichannel implant users demonstrate better performance than users of single-channel devices (Tyler and Tye-Murray 1991; see also Zeng, Chapter 1). An inherent assumption is that multichannel devices provide a degree of independent excitation of multiple neural channels. An important limiting factor for performance with such devices may be the degree of interaction that occurs when overlapping subsets of nerve fibers are stimulated by the various electrodes. Such channel or spatial interactions may impose significant limitations with present implant designs (White 1984; Fu et al. 1998). The ability to assess the spatial patterns of the response, or which neurons of the population are being stimulated by each electrode, therefore is of considerable interest.

5.1 Spatial Spread of Neural Response

We distinguish the assessment of spatial response patterns from what is termed *channel interaction*. The spatial pattern of neural excitation is a direct assessment of the responding neurons to a particular stimulus, essentially a fiber recruitment pattern. Channel interaction assesses the effect of stimulation of one intracochlear electrode site on the response to a second electrode site. Spatial spread information may be useful in assessing the degree of channel independence in a multichannel implant, but also has implications for the stimulus encoding by a particular implant, electrode design, or electrode placement. For example, while the single-fiber surveys of van den Honert and Stypulkowski (1987a) showed that a basal monopolar electrode can produce very broad patterns of spatial excitation, other single-fiber work suggests that spatial specificity is possible with monopolar stimulation (Kral et al. 1998; Liang et al. 1999). Monopolar stimulation can also produce ordered pitch percepts corresponding to the general tonotopic organization of fibers of the cochlea (Eddington 1980). In this section, we describe various approaches to the assessment of spatial spread of neural activity. Assessment of channel interaction is discussed in section 5.2.

5.1.1 Computational Model Data

Computational models are a useful means of assessing the spatial spread of electrical fields and neural excitation under a variety of stimulus conditions. Numerous studies have established that monopolar stimulation produces broader excitation than does bipolar stimulation. *Tripolar* or *quadrupolar* stimulation has been shown to be even more spatially specific (Spelman et al. 1995; Jolly et al. 1996; Kral et al. 1998). Most studies of electrode configuration have employed electrode arrays oriented along the longitudinal dimension of the cochlea. However, both finite-element (Finley et al. 1990; Frijns et al. 1996) and analytical (Rubinstein 1988) models have also demonstrated the superior spatial selectivity of a bipolar configuration oriented along the radial dimension of the cochlea. The finite-element work of Frijns has also demonstrated that, across the excited fiber population, the radial site-of-initiation may vary with cochlear place, as is evident in Figure 5.5.

In addition to electrode configuration, electrode position (within the cross-sectional space of the scala tympani) has strong effects on fiber recruitment (Frijns et al. 1996, 2001). Issues of intrascalar position of electrodes are particularly relevant to new electrode arrays. For example, there has been some interest in perimodiolar electrode designs (Saunders et al. 1998; Battmer et al. 1999a; Lenarz 2001; Pelizzone et al. 2001). Such designs place electrodes closer to the neural elements, resulting in lower thresholds and provide the possibility for greater spatial selectivity in stimulation, as was demonstrated in a spatial computational model of the human cochlea (Frijns et al. 2001). It should be noted, however, that not all spatial effects reported with computational approaches have been observed in animal studies, underscoring the importance of model validation.

5.1.2 ECAP and EABR Measures

As noted above (section 4.1.3), the amplitude of the ECAP can approximate the number of neurons responsive to a particular stimulus. The growth of ECAP or EABR with stimulus level can consequently be used to assess nerve fiber recruitment patterns in both experimental animals and human subjects. Such studies have provided insight into how electrode configuration and position influence fiber threshold, recruitment, and selectivity. For instance, recordings of the EABR demonstrate markedly different growth functions as the position of an intrascalar electrode is moved along a radial dimension (Miller et al. 1993; Shepherd et al. 1993). Threshold was lower and growth of amplitude with level was more gradual when electrode placements were closer to neural elements. Electrode configuration can also have a significant effect on evoked potential growth and, presumably, fiber recruitment patterns. Relative to bipolar stimulation, monopolar stimuli produce lower thresholds and steeper growth functions (Merzenich and White 1977; Marsh et al. 1981; Smith et al. 1994; Miller et al. 1995b; Brown et al. 1996). Similar effects on threshold and growth of response have also been seen as spacing between bipolar stimulating electrodes is increased (Abbas and Brown 1991; Shepherd et al. 1993). In general, growth functions were more shallow with wider electrode spacing and closer placement to the neurons, suggesting poorer spatial resolution. However, as discussed below (section 5.3), a more stochastic pattern of response may be produced by such broad excitation.

Another method of assessing spatial spread is the use of different intra-cochlear electrodes to record the ECAP and to measure response amplitude as a function of longitudinal position in the cochlea (Finley et al. 1997b, 2001; Abbas et al. 2000a). This method assumes that the recording electrode primarily measures the response from neurons near that electrode; however, the degree of spatial filtering has not yet been determined. Consequently, the response amplitude as a function of recording electrode position for a fixed level and position of stimulating electrode depends on two primary factors. The response amplitude profile reflects both the *spread of excitation* across fibers and the *spread of the response fields* from each active neuron to the recording electrode. The interpretation of such measures therefore should be tempered in that they do not directly represent only the spread of excitation.

5.1.3 Single-Fiber Recordings

Accurate assessment of the spatial spread of excitation with single-fiber techniques requires some method of assessing the cochleotopic origin of each recorded neuron. The tonotopic organization of the cochlea provides such a means if the hair cells remain functional so that acoustic tuning curves can be obtained (van den Honert and Stypulkowski 1987a). This approach is problematic for all but the least invasive intracochlear electrode arrays. An alternate technique uses microstimulation of neurons in the deafened cochlea and inferior colliculus recordings to assess fiber location (van den Honert et al. 1997). Results from animal experiments are generally consistent with model predictions and evoked potential data in that radial-bipolar stimulation provides more spatially specific excitation of fibers than do either longitudinal-bipolar or monopolar stimulation (van den Honert and Stypulkowski 1987a; Hartmann and Klinke 1990a). Kral et al. (1998) also demonstrated that tripolar stimulation results in significantly improved spatial selectivity. Liang et al. (1999) demonstrated considerable variation in the sensitivity of an individual fiber to stimulation at different longitudinal electrode positions.

Given the orderly tonotopic arrangement of its lamellae, the central nucleus of the inferior colliculus (IC) is a useful site for assessing regions of neural activity produced by acoustic and electrical cochlear stimulation (Merzenich and White 1977; Snyder et al. 1990, 2000). Several single-cell IC studies have estimated regions of excitation produced by various intracochlear electrode configurations. Black et al. (1983) showed that closely spaced bipolar electrodes produced a much sharper IC spatial tuning curve than did monopolar intracochlear stimulation. Differences in the width of spatial tuning curves were reported by Rebscher et al. (2001), who specifically examined issues of spatial tuning in the IC produced by different monopolar and bipolar intracochlear electrode configurations. Bierer and Middlebrooks (2002) have recently reported comparisons of monopolar and bipolar response patternsi in the auditory cortex of guinea pigs. These central measures of spatial tuning have demonstrated changes in tuning associated with the different configurations; however, the degree of differences were rather modest when compared to the results of Black et al. (1983) and, in particular, the auditory nerve maps of van den Honert and Stypulkowski (1987a). All studies, however, consistently reported the lowest thresholds for monopolar stimulation. It is not clear what underlies the different degrees of spatial tuning observed in the peripherally based study of van den Honert and Stypulkowski and the IC-based study of Rebscher et al. One possibility, suggested by Rebscher et al., is that the different electrode-carrier designs used in the two studies likely produced different intracochlear electric field patterns. It is also possible that neural adaptation, which can be a significant effect with repetitive electric stimulation, may differentially affect the auditory nerve and IC. Further comparisons of central and peripheral assessments of spatial selectivity are warranted.

5.2 ECAP and EABR Channel-Interaction Measures

Instead of measuring the spatial pattern of neural excitation, one may stimulate two channels of a multichannel implant to more directly assess the degree to which stimulation of one channel affects the response to the second. Interactions can occur from the superposition of the electrical fields. Either field summation or cancellation may result, depending on the field orientations. Such effects would be evident only with simultaneous stimulation on the two channels. Processing schemes such as CIS provide nonsimultaneous stimulation to avoid such interaction. Even with nonsimultaneous stimulation, however, interactions may result from overlapping regions of neural excitation provided by two channels. That interaction may be the result of nonlinear growth so that effects of stimulation from two electrodes may not linearly sum (e.g., due to saturation as is Fig. 5.6). Alternatively, responses to stimulation of the second channel may be affected by refractory properties (Fig. 5.10) and/or adaptation resulting from stimulation of the first.

Channel interaction can be assessed with gross potentials (e.g., EABR) using two intracochlear sites stimulated simultaneously, as well as with a nonsimultaneous, forward-masking procedure. The simultaneous method (Gardi 1985; Abbas and Brown 1988) uses stimulus pulses on two electrodes presented both in phase and out of phase. If the excitation patterns from the two electrodes are non-overlapping, the relative phase of stimulation should have little effect. If there are overlapping stimulus fields, summation and cancellation effects will change the response according to the relative phase, and the difference between the in-phase and out-of-phase responses will reflect the degree of interaction. This method is clearly dependent on the interaction of the stimulus fields in the tissue surrounding the neurons.

The nonsimultaneous method uses a two-pulse forward-masking paradigm to assess the effect of stimulation of one electrode on the response to stimulation of another electrode. It has been implemented using both EABR (Abbas and Purdy 1990) and ECAP (Finley et al. 1997b; Abbas et al. 1998). If masker and probe pulses are presented to two different electrodes, the probe response depends on the extent of overlap in populations stimulated by each electrode. Measures of probe response as a function of probe position would then reflect the degree of overlap between the population of neurons responding to the masker and the population that is stimulated by the probe. This method exploits the refractory property of the neuron and is not affected by the overlap or interaction in electrical fields inherent in the simultaneous method. Recent results suggest that limited spatial interactions are possible even with monopolar stimulation (Cohen et al. 2001; Miller et al. 2001a; Abbas et al. 2003).

5.3 Application to Clinical Devices

Several physiologic methods have been used to assess spatial spread and channel interaction with multielectrode cochlear implant arrays. Due to the advent of telemetry systems capable of recording intracochlear potentials, the ECAP measures are likely to be the most useful in clinical applications. The ability to determine specific electrode combinations that result in little channel interaction could provide important information in choosing parameters for stimulation. This is an area of active clinical research.

Some studies have shown that monopolar stimulation can result in better speech perception and sound quality than is produced with bipolar electrodes (Pfingst et al. 1997, 2001). This may be due to a more favorable pattern of ensemble fiber activity by evoking lower within-fiber firing rates, as suggested by the modeling work of White (1984). Low firing probabilities correlate with greater spike jitter, another dimension of neural *stochasticity* (Miller et al. 1999a) that could be perceptually relevant. Recent single-fiber and ECAP analyses suggest that a greater level of stochastic response can indeed occur with monopolar stimulation at the level of the auditory nerve (Miller et al. 2003).

6. Changes with Neural Degeneration

The effectiveness of a prosthetic device depends, in part, on the status of the auditory nerve and central auditory pathways, as well as on the relationship between the surviving neural elements and the stimulating electrodes. Loss of hair-cell function can result in degeneration of both the auditory nerve and brainstem nuclei. It is therefore important to understand pathological changes resulting from hair-cell losses, how those changes affect the temporal and spatial patterns of stimulation, and the extent to which the assessment of those changes physiologically may be useful to clinical applications of cochlear implants.

In general, the pattern of auditory nerve degeneration is closely associated with the loss of inner hair cells innervating those neurons (Spoendlin 1975, 1984; Terayuma et al. 1979). Changes in the peripheral processes occur soon after hair-cell loss and include swelling, vacuolization, and disrupted lamellae of the myelin. The complete degeneration of these processes can occur with little reduction in the population of cell bodies. These peripheral changes then result in the following sequence of degeneration: demyelination and shrinkage of the cell soma, demyelination of the central axon, and, finally, disappearance of the spiral ganglion cell (Leake and Hradek 1988). Each of these changes could affect the neural responses to electrical stimulation.

Under the assumption that spiral ganglion cell survival is an important parameter in predicting performance with a cochlear implant, several studies have attempted to correlate evoked-potential measures, typically the EABR, with cell survival. The underlying assumption of these studies is that spiral ganglion cell loss will have a significant effect on the amplitude of the evoked response. The EABR has been used to monitor changes in animals with chronically implanted electrodes and is sensitive to both neural degeneration as well as bony growth (Walsh and Leake-Jones 1982; Miller et al. 1983). Smith and Simmons (1983) and Lusted et al. (1988) demonstrated that slope of the EABR growth was correlated with spiral ganglion cell survival in chronically deafened cats. Hall (1990) found the strongest correlations using wave I of the EABR, which is the auditory nerve component of the response (Melcher et al. 1996). Finally, Miller et al. (1994) observed the highest correlation with EABR threshold measures rather than slope, but also noted that the data were best fit by a nonmonotonic function of spiral ganglion cell count. They noted slight decreases in threshold with extensive loss, suggesting a possible change in current pathways with that degree of cochlear pathology. Despite these encouraging results, other studies have shown poor correlations (Shepherd et al. 1983; Steel and Bock 1984; Stypulkowski et al., 1986) and human studies have yielded poor correlations with performance (van den Honert and Stypulkowski 1986; Gantz et al. 1988; Abbas and Brown 1991). Patient performance is likely affected by factors not assessed by such basic electrophysiological measures. Thus, while positive correlations exist between EABR and cell counts, evoked-potential measures for the purpose of assessing neural survival in humans have not proved to be clinically useful.

Although neural loss can affect the amplitude and/or sensitivity of evoked potentials, the changes that take place in the surviving neurons may also significantly affect their responses. For instance, changes in the condition of the peripheral processes of auditory neurons and their degree of myelination may affect threshold and refractory properties. Stypulkowski and van den Honert (1984) noted differences in refractory characteristics between neomycin-deafened animals and those in which the cochleas had been mechanically destroyed. Zhou et al. (1995) measured EABR refractory properties in transgenic mice engineered with a myelin deficiency and observed slower recovery.

Other studies have demonstrated pathology-induced differences in integrative properties, as measured through the strength–duration function. Bostock's (1983) model showed that the slope of the strength–duration function decreases with membrane time constant, and Bostock et al. (1983) observed that experimentally induced demyelination decreased the strength–duration slope. Columbo and Parkins (1987) observed significant changes in the strength–duration function produced by their model with changes in length of the unmyelinated peripheral dendrite. These observations are consistent with findings of Miller et al. (1995b), who described decreases in the slope of the strength–duration function over time in chronically implanted animals. They attributed the reduction in efficiency of current integration to effects of neural degeneration. Shepherd et al. (2001) measured strength–duration functions from single fibers of the colliculus in acutely deafened and chronically deafened cats. They noted steeper slopes from the chronic animals, opposite to the results of Miller et al. in guinea pigs. While the reason for the difference is not completely clear, one may hypothesize that it is due to differing degrees of neural degeneration in the two studies. Perhaps the changes observed in Miller et al. (1995b) were due to peripheral changes in the neurons (such as demyelination), whereas the trends in Shepherd et al. (2001) cases, where there was profound degeneration of the nerve, arose from stimulation of the central axons, which may produce steeper strength–duration slopes.

Other changes in neural responses likely occur with degeneration. Computer modeling work suggests that the stochastic properties of jitter and relative spread may be affected by degenerative changes in peripheral processes (Rubinstein 1995). Effects on refractory properties, integration, and stochastic properties could result in significant change in the response to sustained stimulation with different degrees of neural degeneration. Finally, membrane degeneration may affect action potential amplitude, conduction velocity, and incidence of spike failures.

The successful use of cochlear implants by individuals with auditory neuropathy suggests that increased neural synchrony through electric stimulation may actually be beneficial in specific neural disorders. The term *auditory neuropathy* designates a disorder characterized by normal outer hair-cell function (evidenced by normal otoacoustic emissions) and poor auditory nerve function [evidenced by lack of or distorted auditory brainstem response (ABR)] (Starr et al. 1996). Although the specific site of lesion is not known or necessarily the same in patients with auditory neuropathy, many have generalized peripheral nerve disease. Such subjects also demonstrate relatively poor temporal processing abilities in contrast to other hearing-impaired individuals (Zeng et al. 1999). Nevertheless, as a group, they have been successful with cochlear implantation, showing performance similar to other implanted groups (Buss et al. 2002; Madden et al. 2002). Physiological measures with electrical stimulation also demonstrate response patterns similar to those of other implant users whose deafness was presumed primarily sensory. Where no ABRs were recorded before implantation, electrically evoked ABR and ECAP with normal morphology have been recorded from subjects after implantation (Shallop et al. 2001). More detailed psychophysical measures of temporal responses suggest that while processing is not the same as in normal listeners, it can clearly be improved with electric stimulation (Zeng et al. 2002).

7. Response in Cochlea with Viable Hair Cells

Over the past several years, there has been an increase in the number of individuals with significant hearing that have received cochlear implants. As implant candidate selection criteria are relaxed, the most obvious expansion of the patient cohort would involve more individuals with high-frequency hearing loss and significant low-frequency hearing. Histological studies in implanted animals suggest that hair cells apical to the implanted electrode array can survive over chronic periods of implantation (Ni et al. 1992; Xu et al. 1997). Similar patterns of apical hair-cell survival may also occur in the cochleas of implanted humans. Indeed, clinical data suggest that patients with residual acoustic sensitivity can retain that sensitivity after implantation (Kiefer et al. 1998). Perception of speech and nonspeech sounds can be enhanced through combined acoustic and electrical stimulation (Kiefer et al. 2001; Turner and Gantz 2001). The implantation of individuals with significant hearing raises the possibility that the presence of functional hair cells may affect the electrical stimulation of auditory nerve fibers (Risberg et al. 1990). A physiological evaluation of the acoustic–electrical interactions on the neural response in those subjects may prove useful in understanding the variations in performance among individuals as well as to develop appropriate stimulus coding strategies that may either avoid interference between the two modes of stimulation or take advantage of positive interactive effects.

Up to this point we have discussed the responses of the auditory nerve to electrical stimulation without the presence of hair cells. This is usually referred to as the *direct neural response* to indicate that the applied electrical field directly depolarizes the neural membrane. The latency of the direct response, termed the α response by Moxon (1971), is less than 0.5 ms. This mode is distinguished from a hair-cell–mediated response in which the response to an electrical stimulus requires the presence and activity of a functional hair cell. Due to synaptic delays, the latency of hair-cell–mediated responses is longer.

One type of hair-cell–mediated activation is manifested as the electrophonic (Jones et al. 1940) or β (Moxon 1971) response. It is assumed that the electrical stimulus induces a mechanical vibration of the basilar membrane that results in changes in hair-cell polarization and release of neurotransmitter. The latency of this response (1–2 ms) is likely due to both traveling wave delays as well as synaptic mechanisms. Also, importantly, the tuning properties of the basilar membrane affect the electrophonic response. Using sinusoidal electric stimuli, Moxon obtained electrical tuning curves and demonstrated that the characteristic (most sensitive) frequencies of single fibers were the same for both electric and acoustic stimulation. Within the V-shaped portion of the electrical tuning curve, the dynamic range of the β response was comparable (essentially linear) to the fiber's acoustic dynamic range. However, in the less sensitive tail regions, the α response growth was precipitous. Moxon suggested that the highly tuned β responses arose from mechanical transduction and the broadly tuned α responses arose from neural membrane depolarization. Moxon also found that β responses could be most effectively masked by an acoustic sinusoid tuned to the same frequency, and that masking was sensitive to the relative phases of the masker and probe. Through stapes fixation and round window vibration measures, Moxon demonstrated that electrical stimulation produced vibration of the cochlear partition. Finally, he noted that the β response was vulnerable to aminoglycoside administration as well as excitation of the olivo-cochlear bundle.

Another hair-cell–mediated response is believed to be due to depolarization of inner hair-cell membranes by the electrical stimulus. van den Honert and Stypulkowski (1984) reported so-called δ single-fiber responses, characterized by a mean latency (0.9–1.3 ms) intermediate to those of the α and β responses. A similar response pattern was reported by Javel et al. (1987). Examples of PST histograms revealing these three response modes are shown in Figure 5.13. These histograms, constructed from thousands of spike events, clearly illustrate the unique temporal epochs of the α, δ, and β responses. van den Honert and Stypulkowski (1984) noted that thresholds for the β response were lowest, followed by the δ, and then the α response. Because of neural refractoriness, the occurrence of the earlier α response after a stimulus pulse would usually obviate a δ response, particularly at higher stimulus levels. In contrast, the later β responses would occur even if an α response was elicited. Thus, following a single stimulus, α and δ responses are typically mutually exclusive, while α and β responses are not. In contrast to the β response, the δ response did not vary across fibers. Along with its short latency (relative to β) and vulnerability to

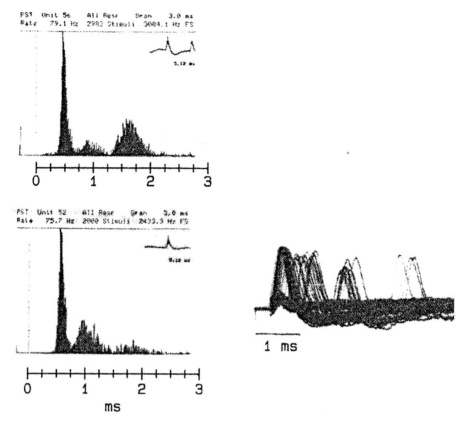

FIGURE 5.13. Examples of α, δ, and β responses evoked from auditory nerve fibers by electrical stimulation. Stimuli were 100 μs current pulses. (Left) Poststimulus time histograms from two fibers that responded in all three modes at a single stimulus level, a relatively rare event, as the three modes typically have unique dynamic ranges. The histograms were created using 2983 stimuli (top) and 2000 stimuli (bottom). (Right) Action potential waveforms illustrating the three clusters of spike times. (From van den Honert and Stypulkowski, 1984.)

cochlear insult, this suggests that the δ response arises from direct depolarization of inner hair cells. Javel and Shepherd (2000) have suggested further differentiation among different sites of excitation based on response latency: outer hair cells (β), inner hair cells (δ), unmyelinated peripheral processes, myelinated peripheral processes, and myelinated central axons. Response characteristics were observed to differ with latency, presumably due to characteristics of different spike initiation sites.

A number of investigators have demonstrated the electrophonic effect in the EABR of animals with relatively intact acoustic sensitivity. Yamane et al. (1981) obtained amplitude-level functions from guinea pigs that had a slow-growing, low-amplitude response that appeared at low stimulus levels. This response was abolished in neomycin-treated animals and had a lower threshold than that presumably mediated by direct neural depolarization. The same low-level *tail* of the growth function was observed in cats by Black et al. (1983). Low levels of wide-band noise could mask this tail, while a derived-band technique indicated that the electrophonic response arose from a broad region of the cochlea. Measures from chronically implanted cats indicated that this electrophonic response could be preserved in cochleas that did not suffer damage from electrode insertion or tissue infection (Shepherd et al. 1983). Furthermore, the expression of electrophonic EABR was correlated with hair-cell survival (Shepherd et al. 1983). Measurement of electrophonic activity with the EABR is complicated by the fact that the response epoch to direct depolarization of nerve fibers overlaps with that of the electrophonic response. In some cases, it is difficult to unambiguously separate the two response components (Black et al. 1983, van den Honert and Stypulkowski 1986).

McAnally and colleagues have investigated hair-cell–mediated responses using the ECAP recorded at or near the round window. Using acoustic probe tones and sinusoidal electric maskers, they noted that the greatest ECAP masking occurred when the frequency of the masker equaled that of the probe (McAnally et al. 1993). They also noted that the growth of ECAP masking was gradual for maskers and probes at the same frequency and steep for off-frequency maskers (McAnally et al. 1997). Using a comparable acoustic/electric masking paradigm with guinea pigs, Kirk and Yates (1994) also found a high degree of frequency selectivity with electrical stimuli. All of these results are consistent with those of Moxon (1971) for α and β response characteristics of single fibers. Using pulsatile electric maskers, McAnally and Clark (1994) and McAnally et al. (1997) noted that the spectral extent of masking was dependent on the spectrum of the masking waveform. This observation may have implications in *tuning* electrical stimuli to optimize their effectiveness in exciting specific regions of the cochlea electrophonically.

Although hair cells are required for β and δ responses, the mere presence of viable hair cells may also influence the α (direct) neural response. As we discussed above, increasing stochastic activity has been suggested for improving the neural transmission of information (Morse and Evans 1996; Moss et al. 1996; Rubinstein et al. 1999). It is possible that hair cells may affect the neural representation of electrical stimuli through a stochastic-resonance effect, thereby increasing sensitivity, dynamic range, and stochastic properties of the responses. The spontaneous activity due to the presence of hair cells has been shown to simulate a Poisson stochastic process with dead time (Johnson 1996). The random nature of this process, resulting from neurotransmitter release at the hair cell–neuron synapse provides an important level of background activity for improving the temporal response characteristics as illustrated by the comparison of responses to acoustic and electrical stimulation (Kiang and Moxon 1972). While neural membranes have characteristics that produce stochastic response properties (Verveen 1961), the degree of noise is clearly less than that of normal auditory nerve fibers. Both model and ECAP measures (Abbas et al. 2000b, 2001a) are generally consistent with hypotheses of decreased threshold, more gradual growth with stimulus level, and increased temporal jitter in response to electrical stimuli with spontaneous activity resulting from functional hair cells.

8. Summary

The established success of cochlear prostheses has brought about higher clinical expectations for better perception of speech in noise, and representation of complex stimuli such as music (Stainsby et al. 1997). Several factors have been identified as limiting clinical performance, and the design of the implant itself is certainly among them. In that respect, ongoing and future efforts to improve the electrode–neural interface may be fruitful, and physiological measurements will likely assist in such efforts. Better understanding of the mechanisms of electrical stimulation—and of how new designs interact with those mechanisms—has contributed to and will continue to advance the development of novel, safe, and effective electrode arrays, surgical techniques, and stimulation strategies. Thus, nearly all aspects of cochlear implant design are aided by physiological research. The design cycle is both informed and completed by detailed physiological measures that are only possible through animal experimentation. Through their exquisite control of underlying biophysical processes and parameters, computational models can be used to extend the power of animal experiments and contribute to hypothesis and design development.

The continued improvement of the clinical cohort has resulted in expansion of clinical indications for implantation (see Niparko, Chapter 3). Such expansion is occurring in young children (Gantz et al. 2000), where the clinical utility of physiological measures may be most apparent. Furthermore, there is considerable across-subject variation in performance with well-fitted prostheses. Such variations may be due, in part, to differences occurring in the peripheral system, and we have seen that some physiological measures are sensitive to difference across subjects and across the electrode array of single subjects. There are active research efforts to improve these measures and examine correlations with pathology. We have described how ECAP thresholds are used in young children to assist with programming the implant's stimulation levels. Additional measures, such as growth of response, temporal interactions, and channel interactions, could expand the role of the ECAP. For instance, more accurate predictions of programming levels may be made by combining ECAP threshold measures with suprathreshold measures or responses to pulse-train stimuli, in order to better estimate behavioral responses to the high-rate stimuli used in prostheses. Furthermore, as ECAP responses have been observed to vary in some subjects across their electrodes, ECAP measures that assess neural temporal and spatial interactions may help choose the most effective electrodes for each subject or tailor stimulus parameters for each electrode to maximize performance.

Implant users with residual hearing (Kiefer et al. 2001; Turner and Gantz 2001) are an additional population for with physiological measures may be particularly helpful. These individuals open the possibility of simultaneous acoustic and electrical stimulation of the same ear. To adequately explore the feasibility of combining acoustic and electrical stimulation, however, it is important to understand how the two types of stimuli may interact. For instance, it is possible that the integration of this information occurs centrally, i.e., different populations of neurons (apical vs. basal) carry separate *channels* of information provided by the two modalities. If acoustic and electrical information is transmitted via different neural populations, independent manipulations of the processing of each are relatively straightforward. The presence of functional hair cells along with simultaneous acoustic stimulation, however, can significantly affect the response of neurons to electrical stimulation. Physiological measures such as the ECAP are particularly suited to test these potential interactions in both experimental animals and in cochlear implant users who have significant residual hearing.

A final area of research concerns relationships among the processing capabilities of neurons at different levels of the auditory system. There is a wide range of parameters that may affect an individual's ability to perceive speech (see Shannon et al., Chapter 8). These include amplitude mapping (Zeng and Galvin 1999), stimulation rate (Brill et al. 1997), number of processing channels (Fishman et al. 1997; Dorman et al. 1998; Fu et al. 1998), choice of stimulus channels (Lawson et al. 1996), electrode spacing and position (Fu and Shannon 1999) and electrode discrimination (Donaldson and Nelson 2000; Henry et al. 2000). We have discussed auditory nerve response properties relevant to each to these parameters. However, there are differences among the response capabilities of the nerve, the central nuclei (see Hartmann and Kral, Chapter 6) and the organism, as measured psychophysically (see McKay, Chapter 7). Although we have stressed the importance of understanding peripheral response properties (section 1.2), central neural processing clearly modifies neural coding and influences the perceptual ability to discriminate among certain stimulus features. Clearly, the relationships among the physiological response capabilities of the ascending neural stations must be better understood for a comprehensive understanding of prosthetic excitation through a cochlear implant.

Acknowledgments. The authors wish to thank their colleagues Barbara Robinson, Jay Rubinstein, Carolyn Brown, Bruce Tomblin, Kirill Nourski, and John Nichols for their critical comments relative to the manuscript. The research of the authors in the area of electrically evoked responses of the auditory nerve was sponsored by National Institutes of Health (NIH) contracts N01-DC-9-2106 and N01-DC-9-2107. Research into clinical applications of electrically evoked potential was supported by NIH grant DC00242.

References

Abbas PJ, Brown CJ (1988) Electrically evoked brainstem potentials in cochlear implant patients with multi-electrode stimulation. Hear Res 36:153–162.

Abbas PJ, Brown CJ (1991) Electrically evoked auditory brainstem response: growth of response with current level. Hear Res 51:123–138.

Abbas PJ, Purdy SJ (1990) Use of forward masking of the EABR to evaluate channel interaction in cochlear implant users. (Abstract). Association for Research in Otolaryngology Midwinter Research Meeting, St. Petersburg Beach, FL.

Abbas PJ, Miller CA, Matsuoka AJ, Rubinstein JT (1998) The neurophysiological effects of simulated auditory prosthesis stimulation. Fourth quarterly progress report, NIH contract N01-DC-6-2111.

Abbas PJ, Brown CJ, Shallop JK, Firszt JB, et al. (1999a) Summary of results using the Nucleus CI24M implant to record the electrically evoked compound action potential. Ear Hear 20:45–59.

Abbas PJ, Rubinstein JT, Miller CA, Matsuoka AJ, Robinson BK (1999b) The neurophysiological effects of auditory prosthesis stimulation. Contract N01-DC-6-2111, final report.

Abbas PJ, Brown CJ, Hughes ML, Gantz BJ, et al. (2000a) Electrically evoked compound action potentials (EAP) recorded from subjects who use the Nucleus CI24M device. Ann Otol Rhinol Laryngol 185:6–9.

Abbas PJ, Miller CA, Rubinstein JT, Abkes BA, Runge-Samuelson C, Robinson BK (2000b) Effects of remaining hair cells on cochlear implant function. Third quarterly progress report, NIH contract N01-DC-9-2106.

Abbas PJ, Miller CA, Rubinstein JT, Robinson BK (2001a) Effects of remaining hair cells on cochlear implant function. Seventh quarterly progress report, NIH contract N01-DC-9-2106.

Abbas PJ, Miller CA, Rubinstein JT, Robinson BK, Hu N (2001b) The neurophysiological effects of simulated auditory prosthesis stimulation. Eighth quarterly progress report, NIH contract N01-DC-9-2107.

Abbas PJ, Hughes ML, Brown CJ, Miller CA (2003) Physiological assessment of spatial tuning and channel interaction in cochlear implants. (Abstract) Association for Research in Otolaryngology Midwinter Research Meeting, Daytona Beach, FL.

Achor LJ, Starr A (1980) Auditory brainstem responses in the cat. II. Effects of lesions. Elecroencephalogr Clin Neurophysiol 48:174–190.

Aidley DJ (1998) Physiology of Excitable Cells, 4th ed. Cambridge: Cambridge University Press.

Arnesen AR, Osen KK (1978) The cochlear nerve in the cat: topography, cochleotopy, and fiber spectrum. J Comp Neurol 178:661–678.

Arnold W (1987) Myelination of the human spiral ganglion. Acta Otolaryngol Suppl (Stockh). 436:76–84.

Aubert LR, Clarke GP (1994) Reliability and predictive value of the electrically evoked auditory brainstem response. Br J Audiol 28:121–124.

Battmer RD, Kuzma J, Frohne C (1999a) Better modiolus-hugging electrode placement: electrophyiological and clinical results of new Clarion electrode positioner. 7th symposium on cochlear implantation in children, Iowa City, IA.

Battmer RD, Zilberman Y, Haake P, Lenarz T (1999b) Simultaneous analog stimulation (SAS)–continuous interleaved sampler (CIS) pilot comparison study in Europe (clinical trial). Ann Otol Rhinol Laryngol 177:69–73.

Bean CP (1974) A theory of microstimulation of myelinated fibers. Appendix to Abzug C, Maeda M, Peterson BW, Wilson VJ, eds. Cervical branching of lumbar vestibulospinal axons. J Physiol (Lond) 243:499–522.

BeMent SL, Ranck JB Jr (1969a) A quantitative study of electrical stimulation of central myelinated fibers. Exp Neurol 24:147–170.

BeMent SL, Ranck JB Jr (1969b) A model for electrical stimulation of central myelinated fibers with monopolar electrodes. Exp Neurol 24:171–186.

Bierer JA, Middlebrooks JC (2002) Auditory cortical images of cochlear-implant stimuli: dependence on electrode configuration. J Neurophysiol 87:478–492.

Black RC, Clark GM (1980) Differential electrical excitation of the auditory nerve. J Acoust Soc Am 67:868–874.

Black RC, Clark GM, O'Leary SJ, Walters C (1983) Intracochlear electrical stimulation of normal and deaf cats investigated using brainstem response audiometry. Acta Otolargyngol (Stockh) Suppl 399:5–17.

Bostock H (1983) The strength-duration relationship for excitation of myelinated nerve: computed dependency on membrane parameters. J Physiol (Lond) 341:59–74.

Bostock H, Sears TA, Sherratt RM (1983) The spatial distribution of excitability and membrane current in normal and demyelinated mammalian nerve fibers. J Physiol (Lond) 341:41–58.

Brill SM, Gstottner W, Helms J, von Ilberg C, et al. (1997) Optimization of channel number and stimulation rates for the fast continuous interleaved sampling strategy in the Combi40+. Am J Otol 18:S104–S106.

Brimacombe JA, Eisenberg LS (1984) Tone decay in subjects with the single-channel cochlear implant. Audiology 23:321–332.

Brown CJ, Abbas PJ (1990) Electrically evoked whole-nerve action potentials. II. Parametric data from the cat. J Acoust Soc Am 88:2205–2210.

Brown CJ, Abbas PJ, Gantz B (1990) Electrically evoked whole-nerve action potentials. I. Data from symbion cochlear implant users. J Acoust Soc Am 88:1385–1391.

Brown CJ, Abbas PJ, Borland J, Bertschy MR (1996) Electrically evoked whole nerve action potentials in Ineraid cochlear implant users: responses to different stimulating electrode configurations and comparison to psychophysical responses. J Speech Hear Res 39:453–467.

Brown CJ, Hughes ML, Lopez SM, Abbas PJ (1999). The relationship between EABR thresholds and levels used to program the Clarion speech processor. Ann Otol Rhinol Laryngol Suppl 108(177):150–157.

Brown CJ, Hughes ML, Luk B, Abbas PJ, Wolaver A, Gervais J (2000) The relationship between EAP and EABR thresholds and levels used to program the Nucleus CI24M speech processor: data from adults. Ear Hear 21:151–163.

Bruce IC, White MW, Irlicht LS, O'Leary SJ, Clark GM (1999). The effects of stochastic neural activity in a model predicting intensity perception with cochlear implants: low-rate stimulation. IEEE Trans Biomed Eng 46:1393–1404.

Buss E, Labadie RF, Brown CJ, Gross AJ, Grose JH, Pillsbury HC (2002) Outcome of cochlear implantation in pediatric auditory neuropathy Otol Neurotol 23:328–332.

Cartee LA (2000) Evaluation of a model of the cochlear neural membrane. II. Comparison of model and physiological measures of membrane properties measured in response to intrameatal electrical stimulation. Hear Res 146:153–166.

Cartee LA, van den Honert C, Finley CC, Miller RL (2000) Evaluation of a model of the cochlear neural membrane. I. Physiological measurement of membrane characteristics in response to intrameatal electrical stimulation. Hear Res 146:143–152.

Chatterjee M (1999) Temporal mechanisms underlying recovery from forward masking in multielectrode-implant listeners. J Acoust Soc Am 105:1853–1863.

Chatterjee M, Robert ME (2001) Noise enhances modulation sensitivity in cochlear implant listeners: stochastic resonance in a prosthetic sensory system? J Assoc Res Otolaryngol 2:159–171.

Clay JR, DeFelice LJ (1983) Relationship between membrane excitability and single channel open-close kinetics. Biophysics J 42:151–157.

Cohen LT, Knight MR, Saunders E, Cowan RSC (2001) Characteristics of NRT measurement in CI24 Nucleus contour electrode and straight arrays. (Abstract) Conference on Implantable Auditory Prostheses, Pacific Grove, CA.

Columbo J, Parkins CW (1987) A model of electrical excitation of the mammalian auditory-nerve neuron. Hear Res 31:287–312.

Coste RL, Pfingst BE (1996) Stimulus features affecting psychophysical detection thresholds for electrical stimulation of the cochlea. III. Pulse polarity. J Acoust Soc Am 99:3099–3108.

Davis H (1923) The relationship of the "chronaxie" of muscle to the size of the stimulating electrode. J Physiol 57:81–82.

Donaldson GS, Nelson DA (2000) Place-pitch sensitivity and its relation to consonant recognition by cochlear implant listeners using the MPEAK and SPEAK speech processing strategies. J Acoust Soc Am 107:1645–1658.

Dorman MF, Loizou PC, Fitzke J (1998) The identification of speech in noise by cochlear implant patients and normal-hearing listeners using 6-channel signal processors. Ear Hear 19:481–484.

Doucet JR, Relkin EM (1997) Neural contributions to the prestimulus compound action potential: implications for measuring the growth of the auditory nerve spike count as a function of stimulus intensity. J Acoust Soc Am 101:2720–2734.

Dynes SB (1996) Discharge characteristics of auditory nerve fibers for pulsatile electrical stimuli. Ph.D. thesis, MIT, Cambridge, MA.

Dynes SB, Delgutte B (1992) Phase-locking of auditory-nerve discharges to sinusoidal electric stimulation of the cochlea. Hear Res 58:79–90.

Eddington DK (1980) Speech discrimination in deaf subjects with cochlear implants. J Acoust Soc Am 68:885–891.

Fang ZP, Mortimer JT (1991) Selective activation of small motor axons by quasitrapezoidal current pulses. IEEE Trans Biomed Eng 38:168–174.

Felix H, Johnsson LG, Gleeson MJ, deFraissinette A, Conen V (1992) Morphometric analysis of the cochlear nerve in man. Acta Otolaryngol 112:284–287.

Fernandez C (1952) Dimensions of the cochlea (guinea pig). J Acoust Soc Am 24:529–532.

Finley CC, Wilson BS, White MW (1990) Models of neural responsiveness to electrical stimulation. In: Miller JM, Spelman FA, eds. Cochlear Implants: Models of the Electrically Stimulated Ear. New York: Springer-Verlag, pp. 55–96.

Finley CC, Wilson BS, van den Honert C (1997a) Fields and EP responses for electrical stimuli: spatial distributors, channel interactions and regional differences along the tonotopic axis. Abstracts of Conference on Implantable Auditory Prostheses. New York: Springer.

Finley CC, Wilson BS, van den Honert C, Lawson D (1997b) Speech processors for auditory prostheses. Sixth quarterly progress report, NIH contract N01-DC-5-2103.

Finley CC, Segel P, Boyle P, Faltys M (2001) Are variable spatial distributions of intracochlear evoked responses across subjects measures of variable nerve survival? (Abstract) Conference on Implantable Auditory Prostheses, Pacific Grove, CA.

Fishman K, Shannon RV, Slattery WH (1997) Speech recognition as function of the number of electrodes used in the SPEAK cochlear implant processor. J Speech Hear Res 40:1201–1215.

Frijns JHM, de Snoo SL, Schoonhoven R (1995) Potential distributions and neural excitation patterns in a rotationally symmetric model of the electrically stimulated cochlea. Hear Res 87:170–186.

Frijns JHM, de Snoo SL, ten Kate JH (1996) Spatial selectivity in a rotationally symmetric model of the electrically stimulated cochlea. Hear Res 95:170–186.

Frijns JHM, Briaire JJ, Grote JJ (2001) The importance of human cochlear anatomy for the results of modiolus-hugging multichannel cochlear implants. Otol Neurotol 22:340–349.

Fu Q, Shannon RV (1999) Effects of electrode location and spacing on phoneme recognition with the Nucleus-22 cochlear implant. Ear Hear 20:321–331.

Fu Q, Shannon RV, Wang X (1998) Effects of noise and spectral resolution on vowel and consonant recognition: acoustic and electric hearing. J Acoust Soc Am 104:1–11.

Gantz BJ, Tyler RS, Knutson JF, Woodworth G, et al. (1988) Evaluation of five different cochlear implant designs: audiologic assessment and predictors of performance. Laryngoscope 98:1100–1106.

Gantz B, Rubinstein J, Tyler R, Teagle H, et al. (2000) Long term results of cochlear implants in children with residual hearing. Ann Otol Rhinol Laryngol 12:33–36.

Gardi JN (1985) Human brainstem and middle latency responses to electrical stimulation: preliminary observations. In: Schindler RA, Merzenich MM, eds. Cochlear Implants. New York: Raven Press, pp. 351–363.

Goldstein MH, Kiang NYS (1958) Synchrony of neural activity in electric response evoked by transient acoustic stimuli. J Acoust Soc Am 30:107–114.

Grill WM Jr, Mortimer JT (1996) The effect of stimulus pulse duration on selectivity of neural stimulation. IEEE Trans Biomed Eng 43:161–166.

Grundfest J (1932) Excitation and accommodation in nerve. Proc R Soc B 119:305–355.

Haenggeli A, Zhang JS, Vischer MW, Pelizzone M, Rouiller EM (1998) Electrically evoked compound action potential (ECAP) of the cochlear nerve in response to pulsatile electrical stimulation of the cochlea in the rat: effects of stimulation at high rates. Audiology 37:353–371.

Hall RD (1990) Estimation of surviving spiral ganglion cells in the deaf rat using the electrically evoked auditory brainstem response. Hear Res 45:123–136.

Hartmann R, Klinke R (1990a) Response characteristics of nerve fibers to patterned electrical stimulation. In: Miller JM, Spelman FA, eds. Cochlear Implants: Models of the Electrically Stimulated Ear. New York: Springer-Verlag, pp. 135–160.

Hartmann R, Klinke R (1990b) Impulse patterns of auditory nerve fibres to extra- and intracochlear electrical stimulation. Acta Oto-Laryngol 469:128–134.

Hartmann R, Topp G, Klinke R (1984) Discharge patterns of cat primary auditory fibers with electrical stimulation of the cochlea. Hear Res 13:46–62.

Hartmann R, Pfennigdorff T, Klinke R (1994) Evoked potentials from the auditory nerve following sinusoidal electrical simulation of the cochlea: new possibilities for preoperative testing in cochlear-implant candidates? Acta Otolaryngol (Stockh) 114:495–500.

Hatsushika S-I, Shepherd RK, Tong YC, Clark GM, Funasaka S (1990) Dimensions of the scala tympani in the human and cat with reference to cochlear implants. Ann Otol Rhinol Laryngol 99:871–876.

Henry BA, McKay CM, McDermott HJ, Clark GM (2000) The relationship between speech perception and electrode discrimination in cochlear implantees. J Acoust Soc Am 108:1269–1280.

Hodgkin AL, Huxley AF (1952) A quantitative description of membrane current and its application to conduction and excitation in nerve. J Physiol (Lond) 117:500–544.

Hong SH, Brown CJ, Hughes ML, Abbas PJ (1998) Electrically evoked compound action potentials using neural response telemetry in CI24M: refractory recovery function of the auditory nerve. (Abstract) Association for Research in Otolaryngology Midwinter Research Meeting, St. Petersburg Beach, FL.

Igarashi M, Mahon RG Jr, Konishi S (1968) Comparative measurements of cochlear apparatus. J Speech Hear Res 11(2):229–335.

Igarashi M, Takahashi M, Alford BR (1976) Cross sectional area of scala tympani in human and cat. Arch Otolaryngol 102:428–429.

Javel E (1990) Acoustic and electrical encoding of temporal information. In: Miller JM, Spelman FA, eds. Models of the Electrically Stimulated Cochlea. New York: Springer-Verlag, pp. 247–292.

Javel E, Shepherd RK (2000) Electrical stimulation of the auditory nerve. III. Response initiation sites and temporal fine structure. Hear Res 140:45–76.

Javel E, Tong YC, Shepherd RK, Clark GM (1987) Responses of cat auditory nerve fibers to biphasic electrical current pulses. Ann Otol Rhinol Laryngol 96 (Suppl 128):26–30.

Jewett DL, Romano MN, Williston JS (1970) Human auditory evoked potentials: possible brainstem components detected on scalp. Science 167:1517–1518.

Johnson DH (1996) Point process models of single-neuron discharges. J Comput Neurosci 3:275–299.

Jolly CN, Spelman FA, Clopton BM (1996) Quadrupolar stimulation for cochlear prostheses: modeling and experimental data. IEEE Trans Biomed Eng 43:857–865.

Jones RC, Stevens SS, Lurie MH (1940) Three mechanisms of hearing by electrical stimulation. J Acoust Soc Am 12:281–290.

Kato G (1934) The Microphysiology of Nerve. Tokyo: Maruzen.

Kiang NYS, Moxon EC (1972) Physiological considerations in artificial stimulation of the inner ear. Ann Otol 81:714–730.

Kiang NYS, Watanabe T, Thomas EC, Clark LF (1965) Discharge Patterns of Single Auditory Nerve: Fibers in the Cat's Auditory Nerve. Cambridge: MIT Press.

Kiang NYS, Moxon EC, Kahn AR (1976) The relationship of gross potentials recorded from the cochlea to single-unit activity in the auditory nerve. In: Rubin RJ, Elberling C, Salomon G, eds. Electrocochleography. Baltimore: University Park Press, pp. 96–115.

Kiefer J, von Ilberg C, Reimer B, Knecht R, et al. (1998) Results of cochlear implantation in patients with severe to profound hearing loss—implications for patient selection. Audiology 37:382–395.

Kiefer J, Tillein J, Sturzebecher E, Pfennigdorff T, et al. (2001) Combined electric-acoustic stimulation of the auditory system. Results of an ongoing clinical study. (Abstract) Conference on Implantable Auditory Prostheses, Pacific Grove, CA.

Killian MJP, Klis SFL, Smoorenburg GF (1994) Adaptation in the compound action potential response of the guinea pig VIIIth nerve to electric stimulation. Hear Res 81:66–82.

Kirk DL, Yates GK (1994) Evidence for electrically evoked traveling waves in the guinea pig cochlea. Hear Res 74:38–50.

Kral A, Hartmann R, Mortazavi D, Klinke R (1998) Spatial resolution of cochlear implants: the electrical field and excitation of auditory afferents. Hear Res 121:11–28.

Lapicque L (1907) Recherches quantitatifs sur l'excitation electrique des nerfs traitee comme un polarisation. J Physiol (Paris) 9:622–635.

Lawson DT, Wilson BS, Zerbi M, Finley CC (1996) Speech processors for auditory prostheses. Third quarterly progress report, NIH contract N01-DC-5-2103.

Leake PA, Hradek GT (1988) Cochlear pathology of long term neomycin induced deafness in cats. Hear Res 33:11–34.

Lenarz T (2001) Channel interactions with different electrode arrays: results from animal studies and psychophysics and relation to possible damage. (Abstract) Conference on Implantable Auditory Prostheses, Pacific Grove, CA.

Liang DH, Lusted HS, White RL (1999) The nerve-electrode interface of the cochlear implant: current spread. IEEE Trans Biomed Eng 46:35–43.

Liberman MC (1978) Auditory-nerve responses from cats raised in a low-noise chamber. J Acoust Soc Am 63:442.

Liberman MC (1982) The cochlear frequency map for the cat: labeling auditory-nerve fibers of known characteristic frequency. J Acoust Soc Am 72:1441–1449.

Liberman MC, Oliver ME (1984) Morphometry of intracellularly labeled neurons of the auditory nerve: correlations with functional properties. J Comp Neurol 223:163–176.

Lin X (1997) Action potentials and underlying voltage-dependent currents studied in cultured spiral ganglion neurons of the postnatal gerbil. Hear Res 108:157–179.

Litvak L, Delgutte B, Eddington D (2001a) Auditory nerve fiber responses to electric stimulation: modulated and unmodulated pulse trains. J Acoust Soc Am 110:368–379.

Litvak L, Smith Z, Delgutte B, Eddington DK (2001b) Study of a potential stimulation strategy that utilizes a conditioning high-frequency pulse train: Single unit recordings. (Abstract) Conference on Implantable Auditory Prostheses, Pacific Grove, CA.

Loeb GE, White MW, Jenkins WM (1983) Biophysical considerations in electrical stimulation of the auditory nervous system. Ann NY Acad Sci 405:123–136.

Lusted HS, Shelton C, Simmons FB (1988) Comparison of electrode sites in electrical stimulation of the cochlea. Laryngoscope 94:878–882.

Madden C, Hilbert L, Rutter M, Greinwald J, Choo D (2002) Pediatric implantation in auditory neuropathy. Otol Neurotol 23:163–168.

Marsh RR, Yamane H, Potsic PP (1981) Effect of site of stimulation on the guinea pig's electrically evoked brainstem response. Otolaryngol Head Neck Surg 89:125–130.

Matsuoka AJ, Abbas PJ, Rubinstein JT, Miller CA (2000a) The neuronal response to electrical constant-amplitude pulse train stimulation: evoked compound action potential recordings. Hear Res 149:115–128.

Matsuoka AJ, Abbas PJ, Rubinstein JT, Miller CA (2000b) The neuronal response to electrical constant-amplitude pulse train stimulation: additive gaussian noise. Hear Res 149:129–137.

Matsuoka AJ, Rubinstein JT, Abbas PJ, Miller CA (2001) The effects of interpulse interval on stochastic properties of electrical stimulation: modes and measurements. IEEE Trans Biomed Eng 48:416–424.

McAnally KI, Clark GM (1994) Stimulation of residual hearing in the cat by pulsatile electrical stimulation of the cochlea. Acta Otolaryngol (Stockh) 114:366–372.

McAnally KI, Clark GM, Syka J (1993) Hair cell mediated responses to the auditory nerve to sinusoidal electrical stimulation of the cochlea in cat. Hear Res 67:55–68.

McAnally KI, Brown M, Clark GM (1997) Acoustic and electric forward-masking of the auditory nerve compound action potential: evidence for linearity of electromechanical transduction. Hear Res 106:146–153.

McIntyre CC, Grill WM (2000) Selective microstimulation of central nervous system neurons. Ann Biomed Eng 28:219–233.

McNeal DR (1976) Analysis of a model for excitation of myelinated nerve. IEEE Trans Biomed Eng 23:329–337.

Melcher JR, Guinan JJ Jr, Knudson IM, Kiang NY (1996) Generators of the brainstem auditory evoked potential in cat. II. Correlating lesion sites with waveform changes. Hear Res 93:28–51.

Merzenich MM, White MW (1997) Cochlear implant: the interface problem. Biomed Eng Inst Funct Elect Stim 3:321–340.

Miller CA, Abbas PJ, Brown CJ (1993) Electrically evoked auditory brainstem response to stimulation of different sites in the cochlea. Hear Res 66:130–142.

Miller CA, Abbas PJ, Robinson BK (1994) The use of long-duration current pulses to assess nerve survival. Hear Res 78:11–26.

Miller CA, Woodruff KE, Pfingst BE (1995a) Functional responses from guinea pigs with cochlear implants. I. Electrophysiological and psychophysical measures. Hear Res 92:85–99.

Miller CA, Faulkner MJ, Pfingst BE (1995b) Functional responses from guinea pigs with cochlear implants. II. Changes in electrophysiological and psychophysical measures over time. Hear Res 92:100–111.

Miller CA, Abbas PJ, Rubinstein JT, Robinson BK, Matsuoka AJ, Woodworth G (1998) Electrically evoked compound action potentials from cat: responses to monopolar, monophasic stimulation. Hear Res 119:142–154.

Miller CA, Abbas PJ, Robinson BK, Rubinstein JT, Matsuoka AJ (1999a) Electrically evoked single-fiber action potentials from cat: responses to monopolar, monophasic stimulation. Hear Res 130:197–218.

Miller CA, Abbas PJ, Rubinstein JT (1999b) An empirically based model of the electrically evoked compound action potential. Hear Res 135:1–18.

Miller CA, Abbas PJ, Brown CJ (2001a) Physiological measurements of spatial excitation patterns produced by electrical stimulation. (Abstract) Conference on Implantable Auditory Prostheses, Pacific Grove, CA.

Miller CA, Abbas PJ, Robinson BK (2001b) Response properties of the refractory auditory nerve fiber. J Assoc Res Otolaryngol 2:216–232.

Miller CA, Robinson BK, Rubinstein JT, Abbas PJ, Runge-Samuelson C (2001c) Auditory nerve responses to monophasic and biphasic electric stimuli. Hear Res 151:79–94.

Miller CA, Abbas PJ, Nourski KV, Hu N, Robinson BK (2003) Electrode configuration influences action potential initiation site and ensemble stochastic response properties. Hear Res 175:200–214.

Miller JM, Duckert LG, Malone MA, Pfingst BE (1983) Cochlear prostheses: stimulation-induced damage. Ann Otol Rhinol Laryngol 92:599–609.

Morse RP, Evans EF (1996) Enhancement of vowel coding for cochlear implants by addition of noise. Nat Med 2:928–932.

Moss F, Chiou-Tan F, Klinke R (1996) Will there be noise in their ears? Nat Med 2:860–862.

Moxon EC (1971) Neural and mechanical responses to electrical stimulation of the cat's inner ear. Doctoral dissertation, MIT, Cambridge, MA.

Nadol JB Jr (1990) Degeneration of cochlear neurons as seen in the spiral ganglion of man. Hear Res 49:141–154.

Neher E, Sakmann B (1976) Single channel currents recorded from membrane of denervated frog muscle cells. Nature 260:799–802.

Nelson DA, Donaldson GS (2001) Psychophysical recovery from single-pulse forward masking in electric hearing. J Acoust Soc Am 109:2921–2933.

Ni D, Shepherd RK, Seldon HL, Xu S-A, Clark GM, Millard RE (1992) Cochlear pathology following chronic electrical stimulation of the auditory nerve. I. Normal hearing kittens. Hear Res 62:63–81.

O'Leary SJ, Clark GM, Tong YC (1995) Model of discharge rate from auditory nerve fibers responding to electrical stimulation of the cochlea: identification of cues for current and time-interval coding. Ann Otol Rhinol Laryngol 166:121–123.

Osberger MJ, Fisher L (1999) SAS-CIS preference study in postlingually deafened adults implanted with the CLARION cochlear implant (Clinical Trial). Ann Otol Rhinol Laryngol 177:74–79.

Ota CY, Kimura RS (1980) Ultrastructural study of the human spiral ganglion. Acta Otolaryngol (Stockh) 89(1–2):53–62.

Parkins CW (1989) Temporal response patterns of auditory nerve fibers to electrical stimulation in deafened squirrel monkeys. Hear Res 41:137–168.

Parkins CW, Colombo J (1987) Auditory-nerve single-neuron thresholds to electrical stimulation from scala tympani electrodes. Hear Res 31:267–286.

Pelizzone M, Boex C, de Balthasar C, Kos M-I (2001) Electrode interactions in Ineraid and Clarion subjects. (Abstract) Conference on Implantable Auditory Prostheses, Pacific Grove, CA.

Pfingst BE, Zwolan TA, Holloway LA (1997) Effects of stimulus configuration on psychophysical operating levels and on speech recognition with cochlear implants. Hear Res 112:247–260.

Pfingst BE, Franck KH, Xu L, Bauer EM, Zwolan TA (2001) Effects of electrode configuration and place of stimulation on speech perception with cochlear prostheses. J Assoc Res Otolaryngol 2:87–103.

Ranck JB Jr (1975) Which elements are excited in electrical stimulation of mammalian central nervous system: a review. Brain Res 98:417–440.

Rattay F (1986) Analysis of models for external stimulation of axons. IEEE Trans Biomed Eng 33:974–977.

Rattay F (1989) Analysis of models for extracellular fiber stimulation. IEEE Trans Biomed Eng 36:676–682.

Rattay F, Leao RN, Felix H (2001) A model of the electrically excited human cochlear neuron. II. Influence of the three-dimensional cochlear structure on neural excitability. Hear Res 153:64–79.

Rebscher SJ, Snyder RL, Leake PA (2001) The effect of electrode configuration and duration of deafness on threshold and selectivity of responses to intracochlear electrical stimulation. J Acoust Soc Am 109:2035–2048.

Reilly JP, Freeman VT, Larkin WD (1985) Sensory effects of transient electrical stimulation—evaluation with neuroelectric model. IEEE Trans Biomed Eng 23:1001–1011.

Rhode WS (1971) Observations of the vibration of the basilar membrane in squirrel monkeys using Mossbauer technique. J Acoust Soc Am 49:1218–1231.

Risberg A, Agelfors E, Lindstrom B, Bredberg G (1990) Electrophonic hearing and cochlear implant. Acta Otolaryngol 469:156–163.

Romand R, Romand MR, Marty R (1981) Regional differences in fiber size in the cochlear nerve. J Comp Neurol 198:1–5.

Rubinstein JT (1988) Quasi-static analytical model for electrical stimulation of the auditory nervous system. Doctoral dissertation, University of Washington, Seattle, WA.

Rubinstein JT (1993) Axon termination conditions for electrical stimulation. IEEE Trans Biomed Eng 40:654–663.

Rubinstein JT (1995) Threshold fluctuations in an N sodium channel model of the node of Ranvier. Biophysical J 68:779–785.

Rubinstein JT, Matsuoka AJ, Abbas PJ, Miller CA (1997) The neurophysiological effects of simulated auditory prosthesis stimulation. Second quarterly progress report, NIH contract N01-DC-6-2111.

Rubinstein JT, Wilson BS, Finley CC, Abbas PJ (1999) Pseudospontaneous activity: stochastic independence of auditory nerve fibers with electrical stimulation. Hear Res 127:108–118.

Rubinstein JT, Miller CA, Abbas PJ, Mino H (2001) Effects of remaining hair cells on cochlear implant function. Sixth quarterly progress report, NIH contract N01-DC-9-2106.

Rubinstein JT, Hong R, Wehner D, Chen G (2002) Stochastic resonance in CI patients. (Abstract) Association for Research in Otolaryngology Midwinter Research Meeting, St. Petersburg Beach, FL.

Runge-Samuelson CL, Rubinstein JT, Abbas PJ, Miller CA, et al. (2001) Sinusoidal electrical stimulation of the auditory nerve with and without high-rate pulses. (Abstract) Association for Research in Otolaryngology Midwinter Research Meeting, St. Petersburg Beach, FL.

Rushton WAH (1927) The effect upon the threshold for nervous excitation of the length of nerve exposed and the angle between current and nerve. J Physiol (Lond) 63:357–377.

Sachs MB, Abbas PJ (1974) Rate versus level functions for auditory-nerve fibers in cats: tone-burst stimuli. J Acoust Soc Am 56:1835–1847.

Sando I (1965) The anatomical interrelationships of the cochlear nerve fibers. Acta Otolaryngol (Stockh) 59:417–435.

Saunders E, Cohen LT, Treaba C (1998) A new precurved electrode array: benefits as measured by initial psychophysics. 7th symposium on cochlear implantation in children, Iowa City, IA.

Shallop JK, Beiter AL, Goin DW, Mischke RE (1990) Electrically evoked auditory brainstem responses (EABR) and middle latency responses (EMLR) obtained from patients with the Nucleus multichannel cochlear implant. Ear Hear 11:5–15.

Shallop JK, VanDyke L, Goin DW, Mischke RE (1991) Prediction of behavioral threshold and comfort values for Nucleus 22-channel implant patients from electrical auditory brainstem response test results. Ann Otol Rhinol Laryngol 100:896–898.

Shallop JK, Peterson A, Facer GW, Fabry LB, Driscoll CL (2001) Cochlear implants in five cases of auditory neuropathy: postoperative findings and progress. Laryngoscope 11:155–162.

Shannon RV (1983) Multichannel electrical stimulation of the auditory nerve in man. I. Basic psychophysics. Hear Res 11:157–189.

Shannon RV (1990) Forward masking in patients with cochlear implants. J Acoust Soc Am 88:741–744.

Shepherd RK, Javel E (1997) Electrical stimulation of the auditory nerve. I. Correlation of physiological responses with cochlear status. Hear Res 108:112–114.

Shepherd RK, Javel E (1999) Electrical stimulation of the auditory nerve. II. Effect of stimulus waveshape on single fibre response properties. Hear Res 130:171–188.

Shepherd RK, Clark GM, Black RC (1983) Chronic electrical stimulation of the auditory nerve in cats. Physiological and histopathological results. Acta Otolaryngol (Stockh) 399:19–31.

Shepherd RK, Hatsushika S, Clark GM (1993) Electrical stimulation of the auditory nerve: the effect of electrode position on neural excitation. Hear Res 66:108–120.

Shepherd RK, Hardie NA, Baxi JH (2001) Electrical stimulation of the auditory nerve: single neuron strength-duration functions in deafened animals. Ann Biomed Eng 29:195–201.

Smith DW, Finley CC, van den Honert C, Olszyk VB, Konrad KEM (1994) Behavioral and electrophysiological responses to electrical stimulation in the cat absolute threshold. Hear Res 81:1–10.

Smith L, Simmons FB (1983) Estimating eighth nerve survival by electrical stimulation. Ann Otol Rhinol Laryngol 92:19–23.

Smith RL, Brachman ML (1982) Adaptation in auditory-nerve fibers: a revised model. Biol Cybernet 44:107–120.

Snyder RL, Rebscher SJ, Cao KL, Leake PA, Kelly K (1990) Chronic intracochlear electrical stimulation in the neonatally deafened cat. I. Expansion of central representation. Hear Res 50:7–33.

Snyder RL, Sinex DG, McGee JD, Walsh EW (2000) Acute spiral ganglion lesions change the tuning and tonotopic organization of cat inferior colliculus neurons. Hear Res 147:200–220.

Spelman FA, Pfingst BE, Clopton BM, Jolly CN, Rodenhiser KL (1995) Effects of electrical current configuration on potential fields in the electrically stimulated cochlea: field models and measurements. Ann Otol Rhinol Laryngol 166:131–136.

Spoendlin H (1972) Innervation densities of the cochlea Acta Otolaryng 73:235–248.

Spoendlin H (1975) Retrograde degeneration of the cochlear nerve. Acta Otolaryngol 79:266–275.

Spoendlin H (1984) Factors inducing retrograde degeneration in the auditory nerve. Ann Otol Rhinol Laryngol Suppl 112:76–82.

Spoendlin H, Schrott A (1989) Analysis of the human auditory nerve. Hear Res 43:25–38.

Stainsby TH, McDermott HJ, McKay CM, Clark GM (1997) Preliminary results on spectral shape perception and discrimination of musical sounds by normal hearing subjects and cochlear implantees. Proceedings of the International Computer Music Conference, Thessaloniki, Hellas.

Starr A, Brackmann DE (1979) Brain stem potentials evoked by electrical stimulation of the cochlea in human subjects. Ann Otol Rhinol Laryngol 88:550–556.

Starr A, Picton TW, Sininger Y, Hood LJ, Berlin CI (1996) Auditory neuropathy. Brain 199:741–753.

Steel KP, Bock GR (1984) Electrically evoked responses in animals with progressive spiral ganglion cell degeneration. Hear Res 15:59–67.

Strelioff D (1973) A computer simulation of the generation and distribution of cochlear potentials. J Acoust Soc Am 54:620–629.

Stypulkowski PH, van den Honert C (1984) Physiological properties of the electrically stimulated auditory nerve. I. Compound action potential recordings. Hear Res 14:225–243.

Stypulkowski PH, van den Honert C, Kvistad SD (1986) Electrophysiologic evaluation of the cochlear implant patient. Otolaryngol Clin North Am 19(2):249–257.

Suesserman MF, Spelman FA (1993) Lumped-parameter model for in vivo cochlear stimulation. IEEE Trans Biomed Eng 40:237–245.

Tasaki I (1955) New measurements of the capacity and the resistance of the myelin sheath and the nodal membrane of the isolated frog nerve fiber. Am J Physiol 181:639–650.

Terayuma Y, Kaneko K, Tanaka K, Kawamoto K (1979) Ultrastructural changes of the nerve elements following disruption of the organ of Corti. II. Nerve elements outside the organ of Corti. Acta Otolaryngol 88:27–36.

Thomsen E (1966) The ultrastructure of the spiral ganglion in the guinea pig. Acta Otolaryngol (Stockh) Suppl 224:442–448.

Turner C, Gantz B (2001) Combining acoustic and electric hearing for patients with high frequency hearing loss. (Abstract) Conference on Implantable Auditory Prostheses, Pacific Grove, CA.

Tyler RS, Tye-Murray N (1991) Cochlear implant signal processing strategies and patient perception of speech and environmental sounds. In: Cooper H, ed. Cochlear Implants. London: Whurr Publishers, pp. 58–83.

van den Honert C, Mortimer JT (1979) The response of the myelinated nerve fiber to short duration biphasic stimulating currents. Ann Biomed Eng 7:117–125.

van den Honert C, Stypulkowski PH (1984) Physiological properties of the electrically stimulated auditory nerve. II. Single fiber recordings. Hear Res 14:225–243.

van den Honert C, Stypulkowski PH (1986) Characterization of the electrically evoked auditory brainstem response (ABR) in cats and humans. Hear Res 21:109–126.

van den Honert C, Stypulkowski PH (1987a) Single fiber mapping of spatial excitation patterns in the electrically stimulated auditory nerve. Hear Res 29:195–206.

van den Honert C, Stypulkowski PH (1987b) Temporal response patterns of single auditory nerve fibers elicited by periodic electrical stimuli. Hear Res 29:207–222.

van den Honert C, Finley CC, Xue S (1997) Microstimulation of auditory nerve for estimating cochlear place of single fibers in a deaf ear. Hear Res 113:140–154.

Verveen AA (1961) Fluctuation in Excitability. Amsterdam: Drukkerjj Holland NV.

Vischer M, Haenggeli A, Zhang J, Pelizzone M, Hausler R, Rouiller EM (1997) Effect of high-frequency electrical stimulation of the auditory nerve in an animal model of cochlear implants. Am J Otol 18:S27–S29.

von Ilberg C, Kiefer J, Tillein J, Pfenningdorff T, et al. (1999) Electric-acoustic stimulation of the auditory system. New technology for severe hearing loss. ORL J Otorhinolaryngol Relat Spec 61:334–340.

Walsh SM, Leake-Jones PA (1982) Chronic electrical stimulation of auditory nerve in cat: physiological and histological results. Hear Res 7:281–304.

Warman EN, Grill WM, Durang D (1992) Modeling the effects of electric fields on nerve fibers: determination of excitation thresholds. IEEE Trans Biomed Eng 39:1244–1254.

White MW (1984) Psychophysical and neurophysiological considerations in the design of a cochlear prosthesis. Audiol Ital 1:77–117.

Wilson BS, Finley CC, Zerbi M, Lawson DT (1994) Speech processors for auditory prostheses. Seventh quarterly progress report, NIH contract N01-DC-2-2401.

Wilson BS, Finley CC, Zerbi M, Lawson DT (1995) Speech processors for auditory prostheses. Eleventh quarterly progress report, NIH contract N01-2-2401, Center for Auditory Prosthesis Research.

Wilson BS, Finley CC, Lawson DT, Zerbi M (1997a) Temporal representations with cochlear implants. Am J Otol 18:S30–S34.

Wilson BS, Zerbi M, Finley CC, Lawson DT, van den Honert C (1997b) Speech processors for auditory prostheses. Seventh quarterly progress report, NIH contract N01-DC-5-2103.

Xu J, Shepherd RK, Millard RE, Clark GM (1997) Chronic electrical stimulation of the auditory nerve at high stimulus rates: a physiological and histological study. Hear Res 105:1–29.

Yamane H, Marsh RR, Potsic WP (1981) Brain stem response evoked by electrical stimulation of the round window of the guinea pig. Otolaryngol Head Neck Surg 89:117–124.

Yates GK, Johnstone BM, Patuzzi RB, Robertson D (1992) Mechanical preprocessing in the mammalian cochlea. Trends Neurosci 15:57–61.

Zeng FG, Galvin JJ 3rd (1999) Amplitude mapping and phoneme recognition in cochlear implant listeners. Ear Hear 20:60–74.

Zeng FG, Oba S, Garde S, Sininger Y, Starr A (1999) Temporal and speech processing deficits in auditory neuropathy. Neuroreport 10:3429–3435.

Zeng FG, Fu QJ, Morse R (2000) Human hearing enhanced by noise. Brain Res 869:251–255.

Zeng FG, Liu S, Kong YY, Starr A, Michalewski H, Shallop J (2002) Treatment of auditory neuropathy: cochlear implants or hearing aids? Paper presented at the International Hearing Aid Conference, Lake Tahoe, CA.

Zhou R, Abbas PJ, Assouline JG (1995) Electrically evoked auditory brainstem response in peripherally myelin-deficient mice. Hear Res 88:98–106.

Zierhofer CM (2001) Analysis of a linear model for electrical stimulation of axons—critical remarks on the "activating function concept." IEEE Trans Biomed Eng 48:173–184.

6
Central Responses to Electrical Stimulation

Rainer Hartmann and Andrej Kral

1. Introduction

The morphology and function of the central auditory system has been studied extensively over more than 100 years using a wide variety of approaches (Popper and Fay 1992; Webster et al. 1992). The study of central responses to *cochlear implant* (CI) stimulation, however, can provide additional answers to questions that cannot be addressed with acoustical stimulation. Moreover, questions of clinical interest can be addressed.

The mammalian organ of Corti does not regenerate lost cells after damage. Neuroprosthetic devices consequently represent the only possibility to investigate functional properties of the auditory system in complete deafness (*auditory deprivation*). Furthermore, perceptions in general, and those evoked by cochlear implants, arise in the central, not the peripheral, auditory system. Therefore, it is necessary to understand the central processing of the electrical stimuli to be able to understand the evoked perceptions.

This chapter addresses the following questions: Are any degenerative processes in the central auditory system setting in as a consequence of peripheral hearing loss? If yes, what is their impact on the functionality of the auditory system? Congenital hearing loss might affect normal auditory development. If it is so, what are the determinants of this process and what are the critical developmental steps that condition a successful cochlear implantation in congenitally deaf subjects? Can we outline an optimal age span for identification of congenital (perinatal) deafness and cochlear implantation in children?

First, we briefly review the anatomy and physiology of the auditory system (section 2) and describe responses of the adult, normally developed auditory system to electrical stimulation (section 3). Then we concentrate on the effect of electrical stimulation on the deprived auditory system (section 4).

2. The Central Auditory System

This section gives a general overview of the anatomy and of physiological auditory phenomena relevant for cochlear implants. It is not intended as a comprehensive review of the subject and the relevant original literature. For details the reader is referred to the reviews in Webster et al. (1992) and Popper and Fay (1992), and the literature referenced therein.

2.1 Anatomy

The main anatomic relations of the ascending auditory pathway are briefly summarized in Figure 6.1. Webster (1992) defines the parts of the auditory system as follows:

Inner and outer hair cells of the cochlea
Spiral ganglion and cochlear portion of the vestibulo-cochlear nerve
Cochlear nuclear (CN) complex
Medial nuclei of the trapezoid body (MNTB)
Superior olivary complex (SOC)

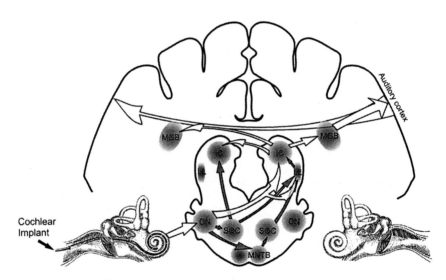

FIGURE 6.1. Simplified schematic illustration of the main projections in the afferent auditory pathway. The cochlear implant stimulates the left auditory nerve. Arrows indicate functional connections, gray arrows indicate connections thought to participate in early binaural processing. CN: cochlear nucleus; SOC: superior olivary complex; MNTB: medial nucleus of the trapezoid body; LL: lateral lemniscus; IC: inferior colliculus; MGB: medial geniculate body.

Lateral lemniscus (LL)
Inferior colliculus (IC)
Medial geniculate body (MGB)
Auditory portions of the cerebral cortex

2.1.1 Cochlea

The transduction of sound to neural activity takes place in the inner and outer hair cells. Ninety percent of the afferent auditory nerve fibers (type I) are myelinated and contact inner hair cells, and each type I nerve fiber contacts only a single inner hair cell. A small number of thin unmyelinated axons (type II) contact outer hair cells. Type I fibers convey the coded auditory information to the central auditory system; the function of type II afferents is not known yet. Efferent fibers innervating the cochlea come from the ipsilateral and contralateral olivary complex.

Type I fibers are axons of bipolar neurons lying in the bony part of the cochlea. Their neuronal bodies form the spiral ganglion, their central axons constitute the auditory part of the VIIIth nerve. The axons leave the petrous bone in the internal meatus and enter the brainstem at the level of the cochlear nucleus (for details see Leake and Rebscher, Chapter 4).

2.1.2 Cochlear Nucleus (CN)

The CN is divided into three parts: the dorsal cochlear nucleus (DCN), the anteroventral cochlear nucleus (AVCN), and the posteroventral cochlear nucleus (PVCN). Type I afferents bifurcate: one branch innervates the AVCN, the other the PVCN and DCN. Each of these nuclei receives information from all parts of the cochlea in an ordered manner (cochleotopically organized).

The CN contains a number of morphologically and functionally distinct cell types. A complete description of their functional properties is beyond the scope of this overview. Some units, mainly in the AVCN, show response properties similar to those of the auditory nerve fibers ("primary-like" units), units in the PVCN and especially in DCN have complex response properties resulting from complex excitatory-inhibitory interactions.

Neural activity in CN is relayed to more central nuclei:
The olivary complex and inferior colliculus of the ipsilateral side
The medial trapezoid body, olivary complex, and lateral lemniscus of the
    contralateral side
The inferior colliculus of the contralateral side

2.1.3 Medial Nuclei of the Trapezoid Body (MNTB)

The neurons of the MNTB are also cochleotopically organized. They are activated by cells from the contralateral AVCN. The MNTB principal cells respond similarly to primary afferents and send their information to the ipsilateral olivary complex.

2.1.4 Superior Olivary Complex (SOC)

The olivary complex is divided into several nuclei. The SOC with the lateral (LSO) and medial part (MSO) receives input from both ears. It is an important relay station for binaural signal processing and the first station involved in localization of a sound sources in space. The SOC is also a part of the stapedius reflex arch (spiral ganglion neurons, AVCN bushy cells, MSO/LSO principal cells, and facial nucleus neurons innervating the stapedius muscle).

2.1.5 Nuclei of the Lateral Lemniscus (LL)

The nuclei of LL receive afferent inputs from the SOC and CN and project bilaterally to the inferior colliculus. The LL, like all nuclei central to the CN, takes part in binaural signal processing. It provides robust inhibitory inputs mainly to the inferior colliculus. In addition, the LL plays a role in the short latency acoustic-startle response.

2.1.6 Inferior Colliculus

The IC is composed of a central nucleus (ICc) and two surrounding nuclei: the external-lateral nucleus (ICx) and the dorsal nucleus (ICd). The main excitatory input to ICc comes from the contralateral cochlear nucleus. The cells (primarily small multipolar fusiform cells with tufted dendrites) in the larger central part have a curved laminar arrangement. They are cochleotopically organized. The ICc sends strong ascending projections to the MGB and descending projections to the SOC and CN.

The ICx receives weak inputs both from the lower auditory brainstem nuclei and from the auditory cortex. Additionally, somatosensory projections influence the ICx. The main output runs to the medial geniculate body and acousticomotor systems. The ICd receives ascending weak input from the LL and descending input bilaterally from the auditory cortex.

The ascending outputs of the inferior colliculus are directed to the medial geniculate body predominantly at the ipsilateral side. Descending connections have been shown to target the ispilateral and contralateral superior colliculus, lateral lemniscus, cochlear nucleus, and the trapezoid body. Additionally, a projection to the contralateral medial geniculate body exists.

2.1.7 Medial Geniculate Body (MGB)

The MGB (the auditory thalamus) is composed of four nuclei (ventral, dorsal, medial, and lateral posterior nucleus) with a variety of large multipolar neurons. It receives ascending inputs from IC and the LL, but also descending inputs from the superior colliculus. It receives also nonauditory inputs, e.g., from vestibular nuclei. The axons of the cells in MGB project to the auditory cortex, to some nonauditory cortical regions, the putamen, and the amygdala. Descending inputs from the cortex to the nuclei of the MGB form corticothalamic feedback loops. The MGB also integrates multisensory information.

2.1.8 Auditory Cortex

The auditory cortex represents the highest center of the auditory system. The auditory areas of the cortex differ between species, even within mammals. There is abundant information on the cell types, connections, their neuronal activity, and histochemistry from cats, monkeys, guinea pigs, rats, rabbits, and gerbils. The auditory cortex is differentiated in primary and nonprimary areas. Primary areas are defined by short-latency responses and direct connections from cochleotopically organized thalamic nuclei (Reale and Imig 1980). In consequence, primary cortical areas are cochleotopically organized. The cat primary auditory cortex comprises fields A1 and the anterior auditory field (AAF). The surrounding areas are considered higher-order in the cat (posterior auditory; ventroposterior auditory and dorsal auditory fields; field AII, the secondary auditory cortex; auditory fields of the posterior ectosylvian cortex; temporal cortex; and insular cortex). The main input to higher-order auditory areas comes from the primary areas and from noncochleotopically organized auditory thalamic nuclei. The output of the auditory cortex is mainly directed to other cortical fields (see below).

In primates primary auditory fields lie in the superior temporal gyrus (core areas) and are surrounded by belt and parabelt areas comprising higher order auditory fields (Pandya and Sanides 1973; Rauschecker et al. 1997; Hackett et al. 1998; review in Kaas et al. 1999). In humans (see Fig. 6.16) the primary auditory cortex lies in Heschl's gyrus in the depth of the sylvian sulcus (area Tel.1, Tel.2, Tel.3, corresponding to Brodmann area 41, invisible from lateral view; Morosan et al. 2001). The secondary fields surround the primary areas and are located in part on the planum temporale, in part on the lateral side of the brain in superior temporal gyrus (areas Te2, Te3, TI, corresponding to Brodmann areas 42 and 22). These auditory areas have an important function also in speech recognition and are discussed in respect to cochlear implants in Section 4.5.

Cortical neurons are divided into pyramidal and nonpyramidal cells according to the form of the neuronal body and the organization of the dendrites. According to the presence of dendritic spines a further division into spiny and nonspiny neurons is possible. Cortical cells in the primary auditory cortex are organized into six layers parallel to the cortical surface:

Layer I, the most superficial cortical layer, is thin and contains few cell bodies, but many axons from thalamus and other cortical areas (secondary cortex, contralateral auditory cortex). These run parallel to the cortical surface and form "en-passant" synapses with dendrites of neurons whose cell body is located in deeper cortical layers.

Layer II cells are mostly nonpyramidal. They form intrinsic circuitry and receive input from neurons in layer III. Their outputs are directed to layers I and III and to secondary auditory fields.

Layer III is rich in pyramidal cells and receives input from the MGB. Additionally, layer III receives strong input from layer IV. It sends output to layers V, II, and I, and secondary auditory areas. It is the main source of commissural fibers.

Layer IV is the main recipient of thalamocortical fibers from MGB. The layer is devoid of pyramidal cells and contains mainly so-called stellate cells receiving input from MGB. The main output goes to layer III.

Layer V receives main inputs from layer II and III, and some input from the MGB. Its outputs are directed to layer III, ipsilateral secondary auditory fields, MGB, and IC.

Layer VI receives main input from layer III and some input from MGB, and sends output fibers back to MGB and to layer IV.

The present model of information processing in the primary auditory cortex assumes more or less parallel inputs from the thalamus (ventral nucleus of MGB) to layers IV (main input layer, granular layer), III, I, V, and VI. From layer IV, activity is relayed to *supragranular* layers III, II, and I. Cells in supragranular layers send projections to higher order auditory cortex and to *infragranular* layers V and VI. Infragranular layers are sources of further output back to the subcortical auditory nuclei and to the higher order cortex in both hemispheres (Mitani and Shimokouchi 1985; Mitani et al. 1985). Nonprimary auditory areas share a similar laminar organization. However, the thicknesses of the layers differ between primary and secondary areas, so that they can be differentiated also according to their cellular organization (cytoarchitectonics).

2.2 Physiology—Acoustical Stimulation

In the cochlea, sound is first mechanically decomposed in its spectral components by the physical properties of the basilar membrane. Different frequencies activate different parts of the cochlea, constituting an orderly tonotopic map of sound frequencies along the basilar membrane with high frequencies being represented at the basal end (*place code*). Functional properties of auditory nerve fibers are correlated with their cochlear innervation. Auditory nerve fibers spontaneously generate action potentials at irregular (Poisson-like) time intervals. They respond to tones of different frequencies and intensities with an increase in firing rate. The responses are a function of frequency and intensity (so-called response area of the neuron). The response thresholds are V-shaped on a frequency-intensity plot. The lower border of the response area represents the frequency threshold curve (FTC) or tuning curve. The frequency of the sound to which the fiber is most sensitive is called characteristic frequency (CF). Very sharp and narrow FTCs were determined in normal-hearing animals. The CF of a fiber is correlated with its innervation site inside the cochlea (for cats see Liberman 1982; for correlation of postmortem cochlear anatomy with audiograms in humans see Otte et al. 1978). Over a wide range along the basilar membrane there is a near-logarithmic relationship of CF to position of the recorded fiber in the cochlea both in cats and humans (cochlear place, Fig. 6.2A). The only exception is the most apical part of the cochlea (site of low CFs), where frequencies are more densely "packed."

When a neuron responds to pure tone stimulation of frequencies below ~5 kHz, the action potentials in the auditory nerve are generated at a certain phase of the stimulus ("phase locking"). Consequently, the interspike intervals can provide information on the stimulus period (*temporal code*) complementary to the place code. Stimulus intensity is most probably coded by the average firing rate in the auditory system (*rate code*).

2.2.1 Dynamic Range

With increasing sound level the mean firing rate of primary fibers increases up to a maximum level. With sound level increasing further, the firing rate does not increase any more (monotonic rate-intensity functions) or it even decreases (nonmonotonic functions). The range of stimulus intensities over which the firing rate of a neuron increases from spontaneous to maximum is called dynamic range. The dynamic range of individual auditory nerve fibers is mostly between 20 to 60 dB. The full dynamic range of hearing results from integration of activity from fibers with different thresholds and dynamic ranges.

Voiced speech signals like vowels are characterized by the fundamental frequency (F0) and higher dominant frequency components (formants F1–4). The discharge rate of a fiber with CF corresponding to one of formants increases with increasing stimulus intensity. The timing of the action potentials is synchronized ("locked") to the time structure of the signal components. Already at moderate speech intensities [60–80 dB sound pressure level (SPL)] the discharge rate of primary afferents saturates and the formants cannot be discriminated using the place code in the majority of auditory nerve fibers (Delgutte and Kiang 1984; Sachs 1984). Additionally, the time structure of the firing pattern also conveys the information on the formants to the central auditory system up to high levels. The importance of the temporal code for speech understanding is indicated by the observation that single-channel speech processors have allowed open-set speech understanding in some patients (McKay, Chapter 7; Shannon et al., Chapter 8).

A

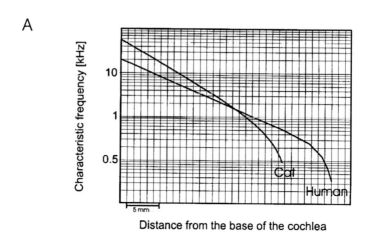

Distance from the base of the cochlea

B

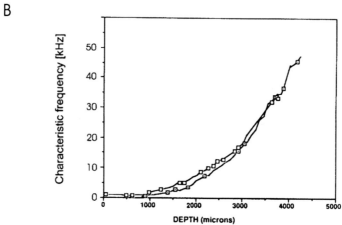

C

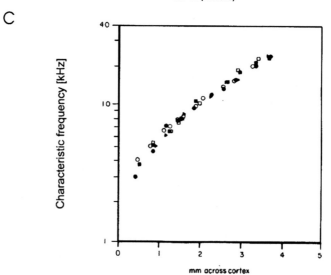

2.2.2 Nonlinear Cochlear Effects

Loud sounds can suppress the perception of following sounds for several hundred milliseconds ("forward masking"). Presentation of several pure tones simultaneously can additionally produce combination effects. Combination of two tones with frequencies f_1 and f_2 can also produce the perception of further (new) components called combination tones (e.g., of the frequency $2*f_1 - f_2$). The response of an auditory nerve fiber to a tone can be reduced by the presence of a second tone (two-tone rate suppression). Whether one tone causes suppression of responses to another depends strongly on the levels and frequencies of the two tones. Because most natural sounds, especially speech sounds, are a complex spectrotemporal pattern of sound pressure, all these effects influence the excitation pattern in the auditory nerve.

2.2.3 Neural Interactions in the Afferent Auditory Pathway

Neural activity from the auditory nerve is relayed through the auditory brainstem nuclei to the IC, thalamus, and the auditory cortex. Response areas of neurons are modified by neuronal (excitatory and inhibitory) interactions in the afferent auditory pathway. The response areas vary from simple ("primary like") response areas through response areas with inhibitory side bands at low and high frequencies to complex nonmonotonic spectrotemporal response areas. The frequency tuning curves to pure tone stimuli, however, preserve their sharp tuning. A *cochleotopic organization* with respect to CF can consequently be found in a part of the nuclei ("tonotopic system") at each level of the auditory pathway (Fig. 6.2B,C). There, anatomically neighboring neurons have similar CFs.

The temporal response properties, on the other hand, change significantly on the way from the auditory nerve to the cortex. Neurons in the central auditory system react to simple tone stimuli with an on-response, sustained response, off-response, or a complex (e.g., on-off) temporal response. Sustained responses can reliably transmit the temporal code in the central auditory system; however, a low-pass filtering effect for responses to simple tones is observed on the way up the afferent auditory system. Neurons in the central auditory system show tuning for modulation frequencies of

---

FIGURE 6.2. The afferent auditory pathway is cochleotopically organized and reveals a place code for sound frequency. Shown is the distribution of characteristic frequencies of neurons along anatomical axes in different auditory structures. A: The cochlea, basoapical direction, cat (based on data from cat, Liberman 1982). B: The inferior colliculus, from dorsal to ventral, cat (Snyder et al. 1990). C: The primary auditory cortex, caudorostral direction, cat (Merzenich et al. 1975; Reale and Imig, 1980).

amplitude modulated (AM) and frequency modulated (FM) sinusoidal stimuli (Langner and Schreiner 1988; Schreiner and Langner 1988), which are closer to natural sounds than pure tones. So-called modulation transfer functions are a measure of the synchronized response (vector strength or modulation depth) as a function of modulation frequency. The peak response of the given neuron to a modulation occurs at the best modulation frequency. In the central nucleus of the IC there is an orderly representation of best modulation frequencies perpendicular to the cochleotopic map of CFs (Langner and Schreiner 1988; Schreiner and Langner 1988). This is a possible way to transform the temporal code to a place code in the IC. Cortical responses to AM sinusoidal stimuli are weak for modulation frequencies about 20 Hz (Fig. 6.3; Eggermont 1993). The reason for a decrease in temporal response properties in the centripetal pathway is not well understood. However, an increase in the duration of inhibition following the onset response (postexcitatory inhibition) is the most probable cause of this phenomenon.

2.2.4 Binaural Hearing

Speech understanding in a noisy environment strongly relies on the ability to localize sound sources in space and to focus the attention to these sources selectively. This situation applies to a cocktail party, where it is possible to selectively switch attention and listen to different speakers despite a background noise. The source of broadband signals can be localized in space using three cues:

1. By determining the time difference in the arrival of the sound front at the ears (interaural time difference). For periodic stationary (low frequency) stimuli interaural phase differences can be used for localization of the sound source in space.

2. By determining the interaural sound level difference. It is a consequence of the head shadow at one ear when the sound source is lateralized. This holds, however, only for stimulus wavelengths that are less than the dimension of the head (this corresponds in humans to frequencies of more than ~500 Hz).

3. In contrast to the binaural cues mentioned above, sound localization can also be (in part) performed using monaural information. The pinna works as a direction-dependent frequency filter. For known broadband acoustic stimuli the perceived spectrum can be compared to the known reference spectrum of that or a similar stimulus stored in memory. The direction of the sound source can be learned from such comparisons using the direction-dependent filter characteristics of the pinna.

The neuronal representation of the binaural time- and level-difference cues can already be seen in the first nuclei with bilateral input, in the olivary

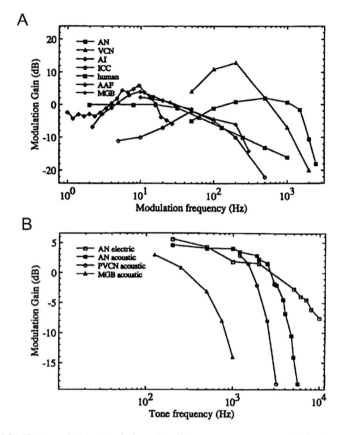

FIGURE 6.3. Temporal characteristics of auditory responses recorded in the auditory pathway of cats. A: Amplitude-modulation-transfer functions with low- or band-pass characteristics. B: Synchronization to continuous unmodulated tones. A modulation gain of zero dB corresponds with a synchronization index 0.5 (from Eggermont 1993, with permission). The temporal characteristics of the responses deteriorate in the centripetal direction. Note that the frequency range is expanded and the highest synchronization is measured with electrical nerve stimulation (Dynes and Delgutte 1992). AAF: anterior auditory field; AI: primary auditory cortex; AN: auditory nerve; ICC: central part of the inferior colliculus; MGB: medial geniculate body; PVCN: posteroventral cochlear nucleus; VCN: ventral cochlear nucleus.

complex (Caird and Klinke 1983; Joris and Yin 1995, 1998; Brand et al. 2002). Neurons in the IC can also be activated from both ears and, together with the lateral lemniscus, participate in the process of sound localization (Phillips and Brugge 1985). Eventually, the primary auditory cortex also has an important role in binaural processing (Middlebrooks et al. 1980; Reale and Brugge 2000).

3. Central Auditory Responses to Electrical Stimulation

The aim of electrical stimulation of the auditory nerve is to mimic the natural conditions as closely as possible. Current coding strategies implement both place coding and temporal coding; however, nonlinear cochlear interactions (masking and combination tones) are neglected in the coding strategies of cochlear implant speech processors (for details see Wilson, Chapter 2). Cochlear implant stimulation can excite two groups of elements in the cochlea:

1. The hair cells directly or indirectly; perceptions evoked by functional hair cells are called *electrophonic hearing*.
2. The spiral ganglion neurons, either at the primary afferents or at the central axons; perceptions evoked by stimulation of the spiral ganglion are called *electroneural hearing*.

The differentiation of these two effects is already possible at the level of the auditory nerve (Abbas and Miller, Chapter 5). Electrophony shows up as decreased thresholds for electrical stimulation in the high-frequency portion of the cochlea and preserved tuning similar to that in acoustic stimulation (Kiang and Moxon 1972; McAnally and Clark 1994). It leads to a bimodal poststimulus time histogram in auditory nerve fiber recordings with pulsatile electrical stimulation (Hartmann et al. 1984b), as two action potential are evoked by the electrical pulse: the first one is generated by electroneural stimulation, the second one, with a delay of 1.0 to 2.0 ms, by electrophonic stimulation. Below we discuss how electrophony influences the central auditory responses to electrical stimulation. Afterward we analyze the characteristics of electroneural hearing in the central auditory system.

3.1 Electrophony

Electrophony substantially influences the activity in the central auditory system in response to sinusoidal electrical stimulation of the cochlea with frequencies >10 kHz and also in response to pulsatile electrical stimulation. Amplitude-intensity functions of local field potentials in the IC and auditory cortex have a large dynamic range in electrically stimulated hearing animals (~30 dB, pulsatile stimulation, IC: Lusted and Simmons 1988; sinusoidal stimulation, auditory cortex: Popelar et al. 1995). Amplitude-intensity functions had a shallow slope at low stimulus intensities with a transition point to a steep slope at high intensities. After deafening, the dynamic range decreased to 3 to 10 dB, the shallow portion of the amplitude-intensity function disappeared, and response thresholds increased. Thus, the shallow portion of the amplitude-intensity functions was generated by electrophonic effects. They also lead to lower thresholds and a larger dynamic range in electrically stimulated hearing cochleas.

As a rule, in animal experiments the electrophonic effect is eliminated by destruction of hair cells to approach the responses elicited by electrical stimulation in cochlear implanted (deaf) humans. This is in most cases achieved by application of ototoxic agents in experimental animal studies.

3.2 Thresholds

Electrical stimulation of a node of Ranvier with a monopolar cathodic extracellular electrode leads to a depolarization with opening of fast voltage-dependent sodium channels generating action potentials. The threshold for spike initialization depends on the local electrical field produced by current injection and its relation to the position of excitable elements (Ranck 1975; Basser and Roth 2000).

To selectively stimulate a subpopulation of auditory never fibers, the stimulation electrodes have to be positioned as close as possible to the fibers. Experiments with electrical stimulation at the round window demonstrated no electrical spatial tuning in acutely deafened cats (Hartmann et al. 1984a; Hartmann and Klinke, 1990). The whole population of auditory nerve fibers responded to the stimulus without significant differences in threshold. This can be explained by current spread over the whole cochlea with high current densities in round window stimulation.

Bipolar stimulation with a pair of electrodes positioned inside the scala tympani near the modiolar wall revealed a spatial tuning in responses of auditory nerve fibers (Stypulkowski and van den Honert 1984; van den Honert and Stypulkowski 1984). Longitudinal arrangements of the stimulation electrodes (along the longitudinal axis of the scala tympani) revealed less sharp spatial tuning than radial arrangement (perpendicular to the longitudinal axis of the scala tympani).

Electrical spatial tuning curves[1] of primary afferents with mono-, bi-, and tripolar configuration of the intracochlear stimulation electrodes showed spatial tuning (Fig. 6.4A; Kral et al. 1998). Lowest thresholds were found with monopolar stimulation. Sharpest tuning curves were observed for the tripolar configuration (20.8 ± 5.2 dB/mm cochlear space in tripolar vs. 8.5 ± 4.5 dB/mm in bipolar stimulation and 3.1 ± 2.5 dB/mm in monopolar stimulation for the Nucleus 22 implant; Kral et al. 1998). With tripolar stimulation the electrical response area of auditory nerve fibers was nearly as narrow as one critical band (Abbas and Miller, Chapter 5) over a larger range of stimulus intensities.

---

[1] Electrical spatial tuning curves represent electrical stimulation thresholds of an auditory nerve fiber as a function of the electrode position in the cochlea (distance from the round window). They are determined for one given fiber and different stimulation sites in the cochlea. Spatial tuning curves referring to sound source position are not discussed in this chapter.

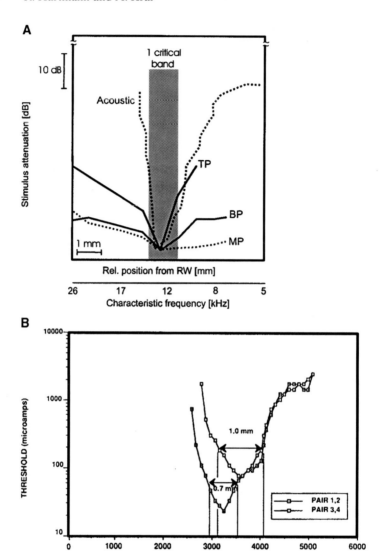

FIGURE 6.4. Place coding as revealed by electrical stimulation of the auditory nerve. The auditory pathway demonstrates cochleotopy also with electrical stimulation. A: Spatial tuning curves measured on a single auditory nerve fiber with cochlear implant stimulation and different stimulation configurations (modified, based on data from Kral et al. 1998). Curves are plotted relative to lowest threshold. Tripolar stimulation shows the most restricted tuning curve. Black dotted line is the tuning curve measured with acoustical stimulation. The gray bar marks one critical band. Only the tuning curve with tripolar stimulation restricts the stimulation to one critical band over a larger intensity range. B: Spatial threshold curves of several multiunit clusters encountered during penetration through the inferior colliculus. Preserved cochleotopy with electrical stimulation is shown by the case of two different cochlear stimulation sites (from Snyder et al. 1990, with permission).

C

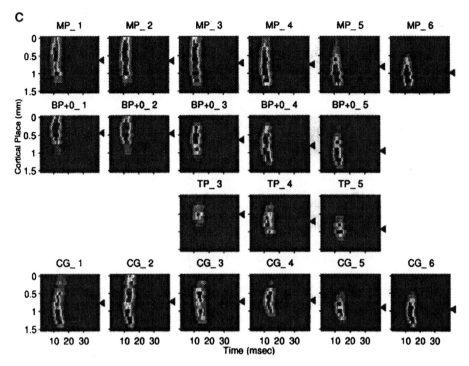

FIGURE 6.4. C: Cortical spatiotemporal maps of multiunit responses to cochlear implant stimulation (guinea pig, from Bierer and Middlebrooks 2002, with permission). MP: monopolar stimulation, number designates the active electrode; BP: bipolar stimulation; TP: tripolar stimulation; CG: monopolar stimulation with common ground. The most restricted spatiotemporal maps are found with tripolar stimulation; however, all stimulation modes demonstrate a cochleotopy.

Is cochleotopic organization with cochlear implant stimulation also conserved in the central auditory system? Despite complex neural interactions and a high convergence and divergence in the auditory pathway, cochleotopy can be found in the acoustically stimulated central auditory system. It is not obvious that such a network shows cochleotopic organization also with electric stimulation. The complex neural processing might be significantly affected by the lack of stochasticity and high synchronicity of the electrically evoked action potentials (Abbas and Miller, Chapter 5; also see section 3.5).

According to Glass and collaborators, neural activity in the CN of deafened cats with bipolar intracochlear stimulation shows no clear spatial tuning (Glass 1983, 1984; Clopton and Glass 1984). These results were possibly due to the small sample of recorded cells and the fact that there was no control of the position of the recorded cells within the cochleotopic map of the CN. Single units in the cat central nucleus of the inferior colliculus, in contrast, showed V-shaped spatial tuning curves with bipolar electrical stimulation (Merzenich and Reid 1974). The authors electrically stimulated the deafened ear and performed, at the same time, an acoustic control of the characteristic frequency of the investigated binaural neurons through acoustic stimulation at the other (hearing) ear. With a large population of multiunit recordings in the ICc of acutely deafened cats, spatial threshold curves[2] were also measured with a custom-made cochlear implant (Snyder et al. 1990). The width of these threshold curves was ~1 mm at 6 dB above the minimum threshold (Fig. 6.4B).

The deoxyglucose marking technique also revealed cochleotopic organization of the cochlear nucleus and inferior colliculus with electrical stimulation (Ryan et al. 1990; Saito et al. 1999). Electrical intracochlear stimulation was monopolar and bipolar at four different cochlear positions. With low current levels (50 μA at 1 kHz) monopolar stimulation evoked activity concentrated within a layer inside the nuclei. With bipolar stimulation at 50 to 100 μA, clear position-dependent activity in the IC could be distinguished. With higher current intensity, almost the whole volume of IC became activated. These data were confirmed for the IC of electrically stimulated cats (Brown et al. 1992). Electrical stimulation was monopolar and bipolar at different cochlear positions, and evaluation was performed with the 2-deoxyglucose method. In these experiments the recorded cochleotopic organization of the IC was compared to "acoustical" tonotopy through stimulation at the nonimplanted ear. This study provided further evidence on cochleotopic organization of electrically evoked activity in the IC.

Cortical local field potentials in response to short electric pulsatile stimulation of the auditory nerve (Woolsey and Walzl 1942) provided the first evidence of a cochleotopic organization in the auditory cortex with an electrically stimulated auditory nerve. Cochleotopic organization of the auditory cortex was further demonstrated by the dependence of the thresholds of local field potentials on the position of the stimulating electrode in the cochlea. The threshold functions differed for different cortical recording sites in a manner consistent with a cochleotopic organization of field A1 in cats (Popelar et al. 1995; Hartmann et al. 1997a,b). Optical recordings in the primary auditory cortex of electrically stimulated animals also revealed a cochleotopy (Dinse et al. 1997a; Taniguchi et al. 1997).

Excitation patterns of CI-evoked cortical multiunit activity was measured simultaneously at 16 positions with a 16-channel recording electrode

---

[2] Spatial threshold curves were measured from several multiunits during a penetration through the central nucleus of the inferior colliculus. Stimulation was performed at the same cochlear position for the given threshold curve. In that respect they differ from the electrical spatial tuning curves as defined in Kral et al. (1998), which were computed from one single auditory nerve fiber with stimulation at different cochlear positions.

inserted into layers III/IV approximately parallel to the cortical surface (guinea pig, Bierer and Middlebrooks 2002; Fig. 6.4C). Monopolar stimulation activated more than a 1-mm-thick rostrocaudal stripe[3] along the auditory cortex. Changing the intracochlear stimulation electrode changed the "center of gravity" of activated areas consistent with a cochleotopic organization of the cortex. Bipolar stimulation evoked smaller activated areas in the cortex, whereas tripolar stimulation evoked the smallest areas.

In summary, a cochleotopic organization has been demonstrated in the central auditory system using different methods and different animal models. Monopolar configuration allows low stimulation currents, but activates the auditory system broadly. However, changing the position of the stimulation electrode in the cochlea shifts the position of the excited neurons in the central auditory system even in monopolar stimulation, providing some place information. Bipolar stimulation with narrow electrode spacing activates a smaller population of neurons in the auditory system. Tripolar stimulation produces the most focused neural activity.

3.3 Psychophysical and Electrophysiological Thresholds

A strong correlation between different electrophysiological measures of threshold intensity and behavioral thresholds has been provided for different species and stimulation modes. However, already at the level of the auditory nerve the electrophysiological thresholds were higher than behavioral thresholds (Parkins and Colombo 1987; Pfingst 1988).

Electrically evoked auditory brainstem responses (EABR) can be recorded from the vertex of the head. The shape of responses to pulsatile stimulation through a cochlear implant corresponds well to the shape of auditory-evoked brainstem responses recorded with acoustic click stimulation (Fig. 6.5). The correlation of EABR thresholds and psychophysical hearing thresholds is statistically significant (guinea pig: Miller et al. 1995a,b; cat: Smith et al. 1994; Beitel et al. 2000a,b; human: Abbas and Brown 1991a,b; Shallop et al. 1991; Brown et al. 1994, 1999; Truy et al. 1998; Firszt et al. 1999). Nevertheless, thresholds determined objectively by auditory brainstem responses are ~5–6 dB higher than psychophysical thresholds. Generally, the EABR thresholds are a better measure of comfortable levels than psychophysical threshold levels. Additionally, large differences between the EABR thresholds and psychophysical thresholds were demonstrated in individual subjects.

Different factors contribute to the variability in EABR thresholds in relation to hearing thresholds. The number of surviving spiral ganglion cells influence the EABR threshold, but is not a sufficient explanation for the

---

[3] One octave corresponds to ~0.5 mm cortical distance in guinea pigs, but to ~1 mm in cats. One critical band corresponds to ~1/3 of an octave.

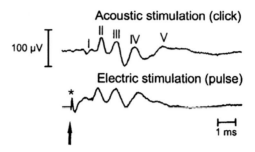

FIGURE 6.5. Auditory brainstem response of a cat. A: Acoustic stimulation (100-μs click duration, 80 dB SPL peak equivalent). B: Bipolar electric cochlear stimulation of a congenitally deaf cat (bipolar stimulation with a charge-balanced pulse, 300 μA peak equivalent, 200 μs/phase, repetition 13/s). Arrow: Stimulus presentation. Asterisk: artifact of electrical stimulus (based on data from Hartmann et al. 1997a).

variability (Miller et al. 2000). The position of the cochlear implant in the scala tympani also affects EABR thresholds. The difference between the highest and lowest thresholds in different distances of the CI from the modiolar wall was 9 dB (Shepherd et al. 1993). A localization close to the modiolus showed the lowest thresholds.[4] These effects, however, do not explain the difference between EABR thresholds and hearing thresholds.

The lowest multiunit threshold in the IC of cochlear-implanted cats was only 1 dB higher than behavioral thresholds (Beitel et al. 2000a,b; Vollmer et al. 2001) but EABR thresholds varied more and were ~6 dB higher than hearing thresholds in the same studies. The EABR thresholds become detectable only if the evoked neuronal activity is well above the lowest multiunit threshold. Obviously, EABRs cannot be used to assess the lowest multiunit thresholds and strongly depend on the synchronization of the neural activity at suprathreshold levels. One factor contributing to the variability of the relation of EABR and hearing thresholds therefore could be the synchronization of the neural activity (see section 4).

Thresholds of cortical middle-latency evoked potentials correlated with behavioral thresholds in promontory-stimulated patients (Kileny and Kemink 1987). However, the difference of psychophysical and electrophysiological thresholds was large (−4 dB to +3 dB). In single-channel cochlear implants the postoperative electrophysiological and psychophysical thresholds were within ±1 dB (Miyamoto 1986). Follow-up studies showed that electrophysiological and psychophysical thresholds in individual subjects can vary substantially with multichannel implants (Shallop et al. 1990). Consistent with this result, epidurally recorded cortical middle-

---

[4] Experiments and model calculations indicate that for optimum spatial selectivity the scala tympani should be filled with Silastic carrier as widely as possible to minimize current shunts between the electrodes (Abbas and Miller, Chapter 5).

latency responses in guinea pigs correlated with behavioral thresholds; however, differences up to ±5 dB were observed in awake and anesthetized animals (Miller et al. 2001). The awake animals showed only slightly larger variability in thresholds than anesthetized ones. Despite the effect of anesthesia on amplitude and latency of middle-latency responses (McGee et al. 1983; Crowther et al. 1990; Kral et al. 1999), only minor effects of anesthesia on thresholds have been reported (McGee et al. 1983). The lowest thresholds of single- and multiunits in the auditory cortex were <1 dB higher than behavioral thresholds (cats: Beitel et al. 2000a,b; Vollmer et al. 2001) and were up to 10 dB lower than the lowest middle-latency response thresholds recorded epidurally (guinea pigs: Miller et al. 2001).

Eventually, the criteria used for defining the electrophysiological thresholds can significantly affect the absolute threshold values. Additionally, psychophysical thresholds are known to be dependent on the spatiotemporal integration, and consequently different stimulation modes yield different results (Pfingst et al. 1991; Pfingst and Morris 1993; McKay, Chapter 7). Therefore, the selected threshold criterion and the stimulation mode can affect the absolute difference in psychophysical and neurophysiological thresholds in a given investigation.

In summary, neurophysiological thresholds to CI stimulation are up to 5 to 10 dB higher than hearing thresholds, depending on the method of stimulation, recording, averaging, and criteria for threshold definition. The lowest multiunit thresholds represent the best approximation of the hearing thresholds.

3.4 Dynamic Range

The dynamic range in CI stimulation is essentially similar in the auditory nerve and in the central auditory system. The dynamic range is very small for pulsatile stimulation (3–6 dB) and only slightly larger for low-frequency (~100 Hz) sinusoidal stimulation (10–12 dB), both for multiunit responses and for local field potentials. For stimulation with higher sinusoidal frequencies, the dynamic range decreases to 3 dB (IC: Snyder et al. 1995; Vischer et al. 1997; Snyder et al. 2000; auditory cortex: Raggio and Schreiner 1994; Hartmann et al. 1997a; human data: Zeng and Galvin 1999; Zeng and Shannon 1999). Both the amplitude-intensity and rate-intensity functions have a saturating shape. The dynamic range of electrically evoked responses (in dB) is on average ~2.5 times smaller than with acoustically evoked multiunit responses (Fig. 6.6; Raggio and Schreiner 1994; Schreiner and Raggio 1996). Raggio and Schreiner found indications that inhibition is more effectively activated with electrical stimulation of a normally developed auditory system than with acoustical stimulation; the firing rate of cortical single units was more strongly negatively correlated with their latency in electrical stimulation. A possible reason could be the high synchronization of responses in electrically stimulated subjects (see below).

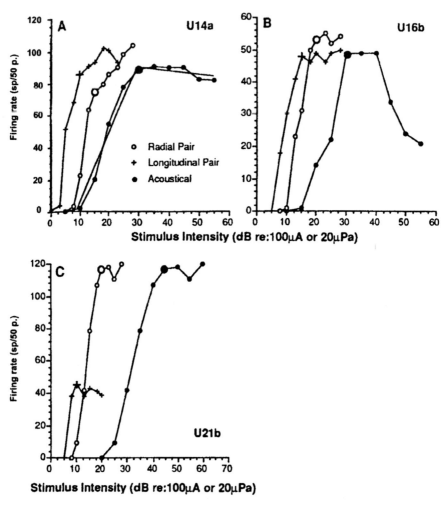

FIGURE 6.6. Rate-intensity functions of cortical multiunits (field A1, cat) for electrical stimulation of the deafened contralateral ear and acoustical stimulation of the ipsilateral ear (from Raggio an Schreiner 1994, with permission). The responses to electrical stimulation have a compressed dynamic range and more monotonic rate-intensity functions than responses to acoustical stimulation.

3.5 Temporal Code

The auditory nerve reacts to suprathreshold electrical stimulation by generating action potentials with very short latencies (in the order of 0.3 ms). With monopolar sinusoidal stimulation at low frequencies (e.g., 100 Hz) action potentials synchronize to the cathodic phase of the stimulus. With increasing current levels, one to three spikes per period can be evoked (Hartmann et al. 1984b). Spontaneous activity is absent in the auditory nerve of deaf cats. Consequently, the period histograms show strong synchronization to the electrical stimulus with synchronization indices >0.9 (Hartmann et al. 1984b). In comparison to maximum synchronization to pure tones in hearing cats, the electrical phase-lock is much stronger and covers a frequency range, up to 12 kHz (Fig. 6.7).

A short acoustic click generates poststimulus time histograms with many peaks (a "ringing" pattern) in auditory nerve fibers. The poststimulus time histograms for pulsatile electrical stimulation of a deaf cochlea show only one sharp peak. This strong synchronization ("hypersynchronization") to the stimulus with a very small stochastic jitter of the responses is a major problem for cochlear implants (Abbas and Miller, Chapter 5).

Complex signals like steady-state vowels, when used as electrical stimulus in a one-channel analog coded speech processor ("Vienna" type), lead to intensity-dependent action potential patterns in the auditory nerve (Knauth et al. 1994). Just above threshold the spikes locked to the fundamental frequency. With increasing level, two to three spikes per period were locked depending on strong F1 (first formant) or F2 (second formant) components. The interspike intervals coded the higher formant frequencies F1 and F2 only partially. The temporal code thus cannot unambiguously code the higher formants in electrical stimulation. Good speech comprehension in one-channel cochlear implant subjects has to rely on context-related pattern-completition mechanisms in the higher auditory centers (Warren 1970; Kral 2000). Multichannel systems complement the temporal information with place information to eliminate this problem. Nonetheless, the time structure still provides the cardinal information in these devices (Dorman et al. 1997).

Intracellular recordings from stellate and bushy cells in the anteroventral cochlear nucleus of rats with electrical stimulation of the auditory nerve (Paolini and Clark 1998) showed a short excitatory postsynaptic potential $(3.6 \pm 2.3 \, ms)$ with short latency (~1 ms) and fast rise time (<1 ms). The excitatory potential was followed by a hyperpolarization in most neurons. The hyperpolarization was either of short duration (<10 ms) or long duration (>19 ms). However, in 25% of the neurons, the hyperpolarization was absent. Consistent with this result, strong phase-locking to sinusoidal electrical stimulation was measured in the cochlear nucleus of deafened cats (Glass 1983, 1984). The synchronization indices declined with increasing frequency to a minimum at 6 kHz and increased again at 12 kHz. This unusual rise of the synchronization indices between 6 and 12 kHz may be caused by a problem of spike detection in recordings with large electrical artifacts.

The time resolution of neuronal responses to electrical cochlear stimulation has been extensively investigated in the IC (Snyder et al. 1995, 2000). The neurons responded to unmodulated sinusoidal signals and pulse trains almost one to one with every stimulus cycle (Fig. 6.8) up to frequencies of ~80 Hz. At higher frequencies an adaptation of the responses was observed. At the end of a 320-ms pulse train, fewer spikes were evoked than at the

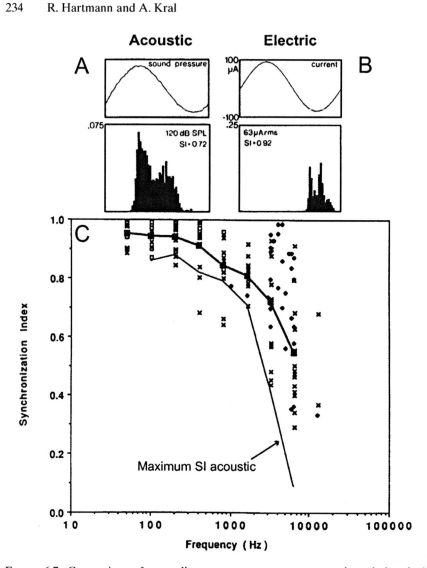

FIGURE 6.7. Comparison of cat auditory nerve responses to acoustic and electrical sinusoidal stimulation. A: Acoustic monitor of a sinusoidal stimulus (frequency 100 Hz) and the corresponding period histogram. High-intensity stimulation (120 dB SPL) leads to phase-locked responses in the period histogram. B: Electrical monitor of a sinusoidal stimulus (100 Hz) and the corresponding period histogram of the response. Stimulation current $63 \mu A_{rms}$. Electrical stimulation yields a higher synchronization of the responses to the stimulus. C: Group data of synchronization indices (with electrical stimulation) for all fibers recorded. Thick line represents means of the synchronization indices at the given stimulus frequency, thin line the maximum synchronization index obtained with acoustical stimulation at the characteristic frequency (CF) of the fibers. Electrical stimulation results in higher synchronization indices (based on data from Hartmann and Klinke 1990).

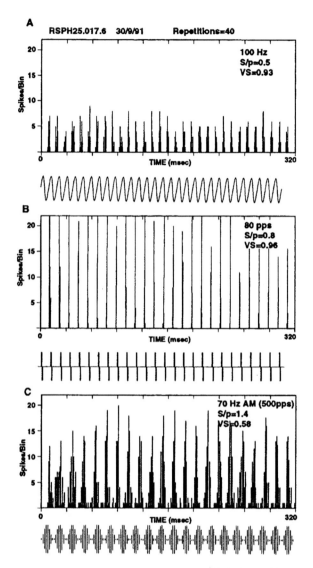

FIGURE 6.8. Responses of "sustained" neurons in cat inferior colliculus to electrical stimulation. A: Sinusoidal stimuli. B: Biphasic pulses 200 μA/phase. C: Amplitude-modulated sinusoidal stimuli (from Snyder et al. 1995, with permission). Neurons with sustained responses can lock to the phase of electrical stimuli with similar maximum following frequencies as with acoustic stimulation.

beginning. Above pulse rates of 130 pulses per second (pps) only the first pulse of the train evoked action potentials. The neurons' ability to follow unmodulated and AM-modulated pulse trains was divided into a slow-, medium-, and fast-responding groups. The mean maximum frequency to which neurons could entrain[5] (maximum following frequency) for unmodulated pulses was 104 pps. Responses to higher pulse rates were possible in case of AM-modulated signals (carrier up to 600 pps). Most of the neurons (70%) showed a low-pass response for modulation frequencies up to 40 Hz. The remaining 30% had band-pass functions with a maximum between 34 and 42 Hz. Additionally, amplitude modulation around carrier threshold intensity leads to complex response patterns depending on the modulation depth.

In the primary auditory cortex only little temporal information on the structure of the original stimulus is left. Because neurons in the auditory cortex can be activated by short signals like single pulses and short duration sinusoidal stimuli but less well by sustained unmodulated signals, most experiments concentrated on stimulation with single pulses with a long repetition time of 0.5 to 2 s (Hartmann et al. 1997a; Klinke et al. 1999, 2001; Kral et al. 2000, 2001, 2002). In acutely deafened cats, a short biphasic pulse evoked local field potentials with latencies between 8 and 20 ms. Also single- and multiunit responses in A1 showed an onset response (on average a single action potential) in this time window (Raggio and Schreiner 1994; Schreiner and Raggio 1996). The onset response was followed by a silent period and a long latency response in the 150-ms range. The responses with latencies over 50 ms have been shown to be clearly dependent on the anesthesia level (Kral et al. 1999) and seem to be influenced by a long-lasting inhibition (accommodation) following the onset response in the majority of cortical units. This effect prohibits entrainment of cortical activity to high pulse repetition rates.

In higher pulse rates the first pulse always elicited a synchronized spike, and the response to the following pulses of the train showed increasing jitter (Fig. 6.9). At higher repetition rates (>10 pps) the response pattern to the later pulses of the train became more and more stochastic (Schreiner and Raggio 1996; Kadner and Scheich 2000). In comparison to the response of neurons from cat auditory cortex to acoustic clicks (Eggermont 2002), a higher synchronization of the electrically evoked spikes was found (Schreiner and Raggio 1996). The coding of repetition rate by electrical cochlear stimulation in the primary auditory cortex of cats looked similar to the responses in the IC. The maximum following frequency of neurons was lower in the cortex than in the IC. Schreiner and Raggio (1996), compared maximum discharge rates of cortical single units between 5 and 20 pps using bipolar stimulation with radial and longitudinal electrode pairs in acutely deafened ears. The evoked discharge rate was dependent on the electrode orientation and spacing (Fig. 6.6). The maximum following frequency was slightly (~3 Hz) higher for electrical than acoustical stimulation.

---

[5] Entrainment is defined as the mean number of action potentials generated in response to a pulse train divided by the number of pulses in the train. At high repetition rates each pulse is no longer responded to by an action potential, and the entrainment decreases.

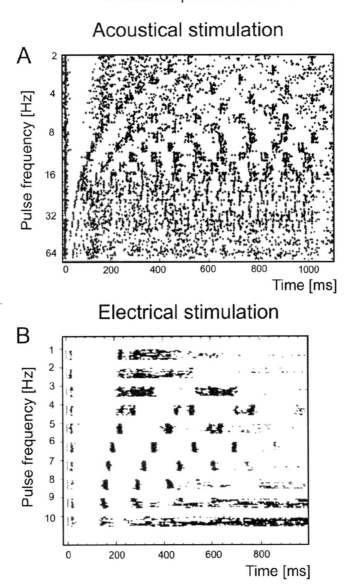

FIGURE 6.9. Locking of cortical responses to acoustical and electrical stimuli. A: Pulse frequency following response in the primary auditory cortex of the cat with acoustic click stimulation (from Eggermont 2002, with permission). B: Recording of a unit in the primary auditory cortex of a gerbil with electrical pulse stimulation of the auditory nerve (from Fiseifis and Scheich 1995, with permission).

Also a ~20% higher spike entrainment (spikes/pules) was determined for electrical stimulation. These data have been interpreted as indicative of an earlier onset of cortical postexcitatory inhibition in electrically stimulated ears. This results, with unchanged duration of the inhibition, in earlier termination of these inhibitory potentials. In consequence, higher maximum following frequencies with electrical stimulation are possible.

3.6 Brainstem Implants

From a theoretical point of view the central auditory system can also be used for stimulation with a neuroprosthetic device. It is known that in the sensory system specific nerve/cell stimulation leads to a corresponding sensation. In the visual system electrical stimulation of the visual cortex in blind subjects leads to punctuate light sensations called phosphenes (Brindley et al. 1972). However, stimulation of the temporal gyrus (not the primary auditory cortex) by Dobelle and colleagues (1973) showed no analog auditory sensations (like sounds with periodicity pitch or noisy sensations) that could be used for coding strategies in specialized signal processors. Early studies tested the superior olivary complex as a potential site for electrical stimulation; rats could be conditioned to electrical stimulation of this structure (Berard et al. 1983). However, from the surgical and neurophysiological points of view, the cochlear nucleus is more suitable for electrical stimulation: an auditory brainstem implant (ABI) (Eisenberg et al. 1987; Shannon and Otto 1990; Brackmann et al. 1993; Shannon et al. 1993) was developed as a prosthesis for deaf patients with atrophy of the auditory nerve and for patients after dissection of the auditory nerve due to acoustic neurinomas. The EABRs were tested as a useful tool for assisting in placement of the implant (Waring 1998).

The ABI has to be able to selectively stimulate spatially restricted populations of neurons and, at the same time, remain in place for long periods of time. The latter precondition is important for preventing damage or stimulation of nonauditory pathways in the brainstem. An additional complication is the high diversity of cell types with very different functional properties in the brainstem, highly contrary to the situation in auditory nerve. Many of these functionally different cells are simultaneously stimulated by the same electrical stimulus.

In theory, the optimal place for a brainstem implant is the AVCN, where the properties of the neurons are similar to those of the auditory nerve ("primary-like"). In practice, the dorsal part of the cochlear nucleus has been stimulated safely via surface ABI electrode arrays and a modified cochlear implant speech processor (Shannon et al., Chapter 8; Shannon and Otto 1990; Brackmann et al. 1993; Shannon et al. 1993; Otto et al. 2002). A disadvantage of DCN stimulation is that only a subpopulation of cells can be reached with surface electrodes. They mostly stimulate the dorsal stria of the DCN as well as activate intrinsic inhibitory circuits. This was tested in several animal experiments with surface electrodes (Shore et al. 1990). Consequently, patients initially describe their perceptions as muffled, like "beeps and boops" or like trying to talk with the mouth closed, with improvement in quality of perceptions with use of the device (Otto et al. 2002). To introduce a better controlled cochleotopic stimulation, penetrating electrodes for microstimulation were tested in animal experiments. The most important questions concern safety aspects. Insertion of needle electrode arrays was tested in the cochlear nucleus of cats (McCreery et al. 1992, 1994) and guinea pigs (El-Kashlan et al. 1991; Liu et al. 1997; Rosahl et al. 2001). The main differences between surface and insertion electrodes were the higher thresholds and lower dynamic ranges for the surface position. With insertion electrodes in the PVCN (McCreery et al. 1998), a cochleotopic organization was demonstrated by stimulating four positions inside the CN and recording multiunit activity and local field potentials in penetration tracks in the IC. The uptake of 2-deoxyglucose in the contralateral SOC, IC, and ipsi- and contralateral LL was higher with the penetrating electrodes (El-Kashlan et al. 1991). From evaluation of the neurophysiological data, penetrating electrode arrays stimulate more specifically the auditory pathway in the CN than surface electrodes. Consequently, this type of electrode appears suitable for future ABI, if its safety can be guaranteed for human subjects. However, even the surface ABI helps facilitate spoken communication (Shannon et al., Chapter 8).

3.7 Summary on Cochlear Implant Stimulation

Electrical stimulation of the auditory nerve can provide both a place code as well as the temporal code to the central auditory system. Currently, the place code, to a certain extent, can be exploited with cochlear implants. Tripolar configurations can even restrict the excitation pattern to one critical band over a considerable intensity range (Fig. 6.4A). Such restricted excitation patterns are not substantially degraded in the afferent auditory pathway, so that the cortical activation pattern evoked by tripolar configuration is substantially more restricted than the one evoked by monopolar stimulation. The temporal code is exactly transmitted by electrical stimulation, even better than in hearing ears. The lack of spontaneous activity in deaf ears combined with the hypersynchronized activity substantially reduces the dynamic range with electrical stimulation.

4. Profound Hearing Loss and Central Auditory System

Cochlear implant stimulation in clinical practice is often performed on an auditory system that has been profoundly deaf for a long period of time. *Profound sensorineural hearing loss* deprives the auditory system of its inputs, and, as will be shown below, affects its morphological and functional properties (see Leake and Rebscher, Chapter 4). Profound hearing loss can affect the whole hearing range of sound frequencies (total hearing loss) or only a part of it (circumscribed hearing loss). Additionally, hearing loss can be binaural or monaural. These different conditions have different impacts on the central auditory system.

A *circumscribed cochlear lesion* deprives the central cochleotopic representations (maps) of a part of their inputs. Consequently, circumscribed cochlear lesions initiate a reorganization in the central representations of the cochlea. The functional parts of the cochlea, mainly those adjoining the lesion, become overrepresented at the cost of the representation of the "deaf" parts of the cochlea (Robertson and Irvine 1989; Rajan et al. 1993). A competition of neighboring cochlear representations for neuronal targets are supposed to initiate the expansion: the representation of more activated input streams expands replacing less active or inactive streams. Reorganization of sensory maps is supposed to be based on long-term potentiation and long-term depression of synaptic efficiencies (Buonomano and Merzenich 1998). Reorganization of sensory maps after circumscribed sensory lesions (*injury-induced plasticity*) differs from plastic reorganization induced by learning (Recanzone et al. 1993). Learning involves controlled activation of neuromodulatory systems like the cholinergic system (Bakin and Weinberger 1996; Weinberger 1997; Kilgard and Merzenich 1998). Involvement of these structures in injury-related plasticity has not been reported.

Circumscribed cochlear lesions induce changes also in the cochlear nucleus (Kaltenbach et al. 1992; Rajan and Irvine 1998). However, in contrast to the changes in more central parts of the auditory system (IC, auditory cortex), the former are not based on neuronal plasticity. Rather, they are caused by originally inhibited prelesion inputs from neighboring cochlear regions.

Several interesting facts can be deduced from the studies on circumscribed cochlear lesions. First, the lack of plastic reorganizations in the peripheral auditory system and its presence in the more central parts lead to the conclusion that central mechanisms are responsible for the effect of injury-induced plasticity. Second, the capacity for neuronal plasticity is larger in the more central parts of the auditory system. Third, it has been shown that this type of plasticity is larger in young (immature) animals (Harrison et al. 1991), indicating a sensitive period for injury-induced plasticity in the auditory system.

Total bilateral deafness does not induce a disequilibrium between auditory inputs to the central cochlear maps, as all inputs are inactive. None of the inputs has an advantage in the competition for neuronal targets. Reorganizations of cochlear maps, as found in circumscribed hearing loss, therefore cannot take place in the same way in total deafness. Total deafness, however, has other profound effects on the central auditory system. The "unused" auditory system undergoes degenerative changes. More importantly, neuronal postnatal development is supposed to depend on "experience" (sensory input). Maturation of the central auditory system extends well into postnatal age in mammals. It is not completed before the age of 5 months in cats and before 12 years in humans, possibly extending into adolescence. Total auditory deprivation from birth leads to a "naive" auditory system, as normal developmental processes are affected by the total absence of auditory input.

4.1 Postnatal Auditory Maturation

4.1.1 Normal Morphological Development

The human auditory system matures during postnatal life. The myelination of the afferent auditory pathway matures during the first 2 to 4 years postnatally (p.n.). The development of axon diameter and myelin sheath can even be traced into adolescence (Yakovlev and Lecour 1967; Paus et al. 1999). Dendritic development of neurons in the brainstem is initiated around the 24th fetal week (Moore et al. 1997a); however, the development also extends into the perinatal and postnatal periods. The structure of auditory brainstem neurons, as revealed by changes in microtubule-associated protein 2 (a molecule that stabilizes dendritic process by promoting assembly of microtubules), is not terminated before 6 months after birth (Moore et al. 1998). The human auditory cortex, on the other hand, shows development up to the 12th year p.n. (Moore and Guan 2001; Moore 2002). Postnatal development of human neocortex includes formation of new dendrites and deletion of already existing ones (Conel 1939–1967). Deletion of existing dendrites is massive and takes place between the 2nd and the 6th year, extending to adolescence. Correspondingly also the number of synapses in the human neocortex demonstrates postnatal development. A massive formation of new synapses can be observed during the first 2 to 4 years of life. After the 4th year of life many synapses gradually disappear in all investigated cortical regions (including auditory cortex), so that only 50% of the maximum counts are found in adolescent brains ("synaptic pruning," Huttenlocher 1984; Huttenlocher and Dabholkar 1997; recent review of synaptogenesis theory in Benson et al. 2001). Synaptic pruning has been causally connected to sensory experience in birds (Wallhäuser and Scheich 1987). In mammals the evidence of a connection of sensorial experience and pruning is less direct and only known for the visual system. In cats, pruning takes place between the 5th and the 12th week postnatally in the visual cortex (Cragg 1975a; O'Kusky 1985; Schoop et al. 1997; for monkeys compare O'Kusky and Colonnier 1982a–c;). This process is affected by sensory deprivation in the early postnatal period. Cats that were dark-reared (or whose visual nerve was mechanically destroyed) had significantly fewer cortical synapses (Cragg 1975b; O'Kusky 1985; Schoop et al. 1997). This effect was more pronounced in infragranular layers of the cortex. Consequently, the theory of synaptogenesis assumes that synapses that are not activated by sensory inputs are pruned, and the activated ones are stabilized in the early postnatal development.

4.1.2 Normal Functional Development

Functional development of the auditory system continues postnatally. The human cochlea and brainstem start functioning at the 28th fetal week, as shown by auditory brainstem responses and acousticomotor reflexes (Moore et al. 1995). Nevertheless, the human auditory brainstem evoked response becomes mature only by 2 years after birth (Ponton et al. 1992). Cortical evoked potentials can be recorded already in neonates. Their shape is very variable in childhood and shows significant developmental changes until 12 to 14 years of age, possibly extending into adolescence (Barnet and Ohlrich 1975; Eggermont 1985, 1988, 1992a; Sharma et al. 1997; Ponton et al. 2000a). Mismatch-negativity of cortical evoked responses has been demonstrated in neonates, and also shows maturation in the first months after birth, with most rapid maturation rates between 6 and 12 months p.n. (Cheour et al. 2000).

   In cats, the cochlea achieves functional maturity during the second week after birth (Brugge 1992). Functional development in the cochlear nucleus, however, proceeds until the fourth week after birth (Brugge et al. 1978). Responses in the cat IC mature (in term of frequency tuning and periodicity coding) on days 26 to 30 p.n. (Aitkin and Moore 1975; Moore and Irvine 1979). The cortical local field potentials and multiunit responses in cortical field A1 reach threshold levels of less than 100 dB SPL only on day 10 p.n. (König et al. 1972; Brugge 1992). Functional development in cat primary auditory cortex proceeds, however, until the fifth month p.n., shortly before the animal becomes sexually mature (Eggermont 1991, 1996).

   What is the effect of complete hearing loss on the auditory system? The effects may include synaptic changes together with more extensive structural reorganization of the affected system. Hearing loss can additionally interfere with normal auditory development if onset of deafness takes place during the normal auditory development. Therefore, two different conditions for effects of hearing loss are discussed here: adult-onset deafness and congenital/neonatal deafness.

4.2 Central Effects of Hearing Loss in Adults

Investigation of long-term effects of hearing loss is complicated by methodological difficulties as a complete hearing loss cannot be reversibly made in mammals. Reversibility is important, as it would provide the opportunity to test central auditory responses to sound after periods of complete deprivation. Ear plugging (conductive hearing loss) leads to reversible threshold shifts, but does not cause complete hearing loss due to preserved bone conduction. The increase in hearing threshold after ear plugging is between 20 and 40 dB and exceeds 40 dB only in exceptional cases (Nixon and von Gierke 1959). Consequently, complete hearing loss can be induced only by destruction of the sensory structures (sensory, cochlear deafness). Sensory deafness is irreversible in mammals. To investigate the functional properties of the auditory system following complete sensory hearing loss, electrical stimulation of the auditory nerve becomes necessary.

Can we find central auditory changes as a consequence of complete hearing loss that occurs late in life? The auditory system is no doubt plastic even in adults (Buonomano and Merzenich 1998); however, there are few data on functional consequences of complete hearing loss.

One central effect of total bilateral deafness induced in adult animals is the central down-regulation of inhibition, as demonstrated on guinea pigs and rats (Bledsoe et al. 1995). Complete bilateral deafness was induced by intracochlear application of neomycin. The auditory system was stimulated by pulsatile electrical stimulation through an intracochlear electrode (monopolar configuration). The responses in the central nucleus of the IC were investigated both electrophysiologically and by a c-*fos* technique. The number of units showing inhibitory responses to electrical cochlear stimulation decreased from 43% in hearing animals to less than 5% in animals that were deaf for 21 days. After 21 days of deafness, the number of *fos*-immunoreactive neurons increased by a factor of about 2. On the other hand, the amount of potassium-stimulated γ-aminobutyric acid (GABA) release was markedly suppressed in animals deafened for 30 days. Consequently, after several weeks of deafness the adult auditory midbrain shows signs of disinhibition. Indeed, an increased amplitude of local field potentials was found in animals after induction of temporary threshold shift by acoustic overexposure (Willott and Henry 1974; Willott and Lu 1982; Syka and Rybalko 2000).

Indications of disinhibition as a consequence of deafness have been found also in the primary auditory cortex (Raggio and Schreiner 1999). Adult cats were deafened by subcutaneous application of kanamycin and aminooxyacetic acid 2 weeks before the neurophysiological experiment (in deaf animals, 2 weeks of deafness before the experiment) and compared to adult animals deafened by intracochlear application of neomycin or by mechanical damage right at the beginning of the experiment (control animals, normally hearing before the experiment). Stimulation was performed by cochlear implants in both groups. The authors compared multiunit responses measured in middle cortical layers of field A1 at the hemisphere contralateral to the implanted ear. The study revealed a rudimentary cochleotopy in both groups (see section 4.3). The control animals had higher minimum thresholds than the deaf animals. A cortical "activation area" was defined as the cortical area comprising units with a threshold within 6 dB of the minimum recorded threshold of the given animal and stimulation configuration. This area was about twice as large in the deaf animals as in controls. Decreased thresholds and increased cortical activated areas in adult-deafened animals thus demonstrated a "hypersensitiv-ity" of the auditory cortex as a consequence of deafness (compare also Woolf et al. 1983; Sasaki et al. 1980).

Some human subjects who have become deaf in adulthood (postlingual deafness) perform surprisingly well in speech recognition with cochlear implants even after decades of complete deafness. Late cortical evoked potentials (P300) and mismatch negativity recorded a few months after cochlear implantation in postlingually deaf subjects have characteristics similar to those obtained from hearing controls with corresponding acoustic stimuli (Kraus et al. 1993; Micco et al. 1995). Hence, if deafness is complete and bilateral, the adult auditory system is affected by functional degenera-tive changes only to a minor extent and quickly learns to adapt to the arti-ficial electrical stimulation. The disinhibition demonstrated in numerous studies after adult onset of deafness seems counterbalanced after auditory input is resumed.

4.3 Central Effects of Neonatal Hearing Loss

Manipulation of hearing during development has been performed on a number of animal models. Altricial animals (e.g., ferrets, cats) can neither see nor hear at birth (Kral et al. 2001; Syka 2002). They provide the possi-bility to manipulate the auditory input before hearing onset. Deprivation studies have been performed with different deafening techniques, bearing different respective advantages and disadvantages.

4.3.1 Deprivation Models

Mechanical cochlear ablation was performed to investigate effects of audi-tory deprivation well before the onset of hearing function (Kitzes et al. 1995; Russell and Moore 1999; Gabriele et al. 2000). The main drawback of the ablation model is a concomitant complete withdrawal of all neu-rotrophic factors released from the cochlea. Additionally, cochlear implant stimulation is not possible in this model, so that functional studies are limited to unilateral deafness. Spontaneous activity is completely normal before and completely absent after ablation, which most probably does not match the pathophysiology of "natural" hearing loss.

Another deafness model involves congenitally deaf species (e.g., con-genitally deaf cats: Larsen and Kirchhoff 1992; Heid et al. 1998; congeni-tally deaf mice: Steel and Bock 1980; Kiernan and Steel 2000; or congenitally deaf dalmatians: Niparko and Finger 1997). Many congenitally deaf animals show slow degeneration of the spiral ganglion (e.g., congeni-tally deaf cats: Heid et al. 1998), comparable to human congenital (perina-tal) deafness (Felix and Hoffmann 1985; Vasama and Linthicum 2000). However, they often have small litters and low reproduction in the colony.

A third model is animals deafened neonatally by ototoxic drugs (Leake-Jones et al. 1982; Xu et al. 1993). Cochlear sensitivity to ototoxic agents is known to develop in line with hearing function (Carlier and Pujol 1980; Raphael et al. 1983; Shepherd and Martin 1995), providing the possibility to deafen animals before they gain auditory experience. These animals, however, suffer from a pronounced loss of spiral ganglion cells (Leake et al. 1999) due to direct toxic effects of ototoxic drugs. Degeneration of the auditory nerve results in different degrees of activation of the auditory system by cochlear electrostimulation, depending on the number of surviving neurons (Schwartz et al. 1993; Araki et al. 2000). Auditory nerve degeneration can set up denervation effects, caused by the physical absence of the auditory nerve fibers in addition to the absence of neural activity. However, this model does not impose limits on the number of experimental animals.

4.3.2 Morphological Changes in the Auditory Brainstem and Midbrain

Dystrophic changes in the brainstem following binaural neonatal (congenital) auditory deprivation have been reported by several laboratories. A decrease of somatic area of neurons in the brainstem of auditory-deprived animals has been demonstrated repeatedly (Trune 1982; Nordeen et al. 1983; Hultcrantz et al. 1991; Matsushima et al. 1991; Saada et al. 1996; Niparko and Finger 1997; Heid 1998; Hardie and Shepherd 1999; reviewed in Shepherd and Hardie 2001). However, no change in neuronal cell counts has been reported in the brainstem of neonatally deafened and congenitally deaf animals. This is surprising in light of studies demonstrating that if auditory deprivation (by cochlear ablation or block of action potential conduction) is performed before hearing onset, the neuronal cell counts in the cochlear nucleus are reduced (Born and Rubel 1988; Pasic and Rubel 1989; Tierney et al. 1997; Mostafapour et al. 2000, 2002). This effect can be explained by the fact that the susceptibility of the cochlea to insults (ototoxic substances and possibly also some genetically caused cochlear pathologies) emerges only with the hearing onset (Carlier and Pujol 1980; Raphael et al. 1983; Shepherd and Martin 1995). Both the neonatally deafened and congenitally deaf animals possibly develop degenerative changes in the cochlea first at the time of the hearing onset and not before that time (Shepherd and Hardie 2001). Consequently, at the time of cochlear degeneration the number of neurons in the cochlear nucleus is already established and does not decrease afterward. The absence of auditory experience in the central auditory system, however, has been provided in all these deafness models. There was no postnatal period with recordable auditory brainstem responses for stimuli up to 115 dB SPL in congenitally deaf cats (Heid et al. 1998). Similarly, in neonatally deafened cats there were no auditory brainstem responses for stimuli up to 108 dB SPL (Vollmer et al. 1999) or up to 92 dB SPL (Hardie and Shepherd 1999).

No data are available about neuronal activity in the auditory nerve of young congenitally deaf or neonatally deafened cats. However, the presence of spontaneous activity in young animals (before cochlear degeneration or deafening) is likely. Spontaneous activity in the auditory nerve could suffice to prevent neuronal death in cochlear nucleus in congenitally deaf animals or neonatally deafened animals. Dying hair cells could even induce an increase in activity of the auditory nerve.

Changes in the end bulbs of Held have been demonstrated in the cochlear nucleus of congenitally deaf cats (Larsen and Kirchhoff 1992; Ryugo et al. 1997, 1998): reduction in the number of terminal ramifications, in the density of synaptic vesicles, and an increase in the size of the synaptic area and in pre- and postsynaptic densities. Consequently, not only dystrophic but also hypertrophic, possibly compensatory, processes, take place after early deprivation. The dendritic trees are reduced after auditory deprivation in the cochlear nucleus (Trune 1982; Conlee and Parks 1983). Similar dystrophic changes were also found in the brainstem of deaf humans with adult onset of deafness (Moore et al. 1997b).

Synaptic numbers in the midbrain of neonatally deafened animals were reduced (Hardie et al. 1998), demonstrating that synaptogenesis is affected by auditory deprivation. However, basic interconnections between brainstem and midbrain nuclei are preserved and demonstrate a nucleotopy in congenitally deaf cats (Heid et al. 1997). The projections from the thalamus to the primary auditory cortex are also relatively normal in neonatally deafened cats (Stanton and Harrison 2000). Consequently, despite certain morphological deficits the gross interconnection pattern in the subcortical auditory pathway is preserved in naive deaf animals.

Extensive morphological studies have been undertaken on the brainstem and midbrain of monaurally deprived animals (gerbils, ferrets, and rats), mainly by cochlear ablation (Kitzes and Semple 1985; Moore and Kitzes 1985; Reale et al. 1987; McMullen and Glaser 1988; McMullen et al. 1988; Moore et al. 1993; Moore 1994; Kitzes et al. 1995; Russell and Moore 1995). These studies demonstrated that unilateral deprivation leads to a reorganization of binaural projections to the superior olive and inferior colliculus. The ablated side loses projections to the investigated nuclei. Instead, the number of projections from the intact side increases, indicating a replacement of the silenced projections by the active ones. Bilateral ablations do not lead to such extensive rewiring in the auditory system (Moore 1990; Silverman and Clopton 1977). The structure of dendritic trees in superior olivary nuclei also depends on acoustic input: monaural deprivation leads to reorganization of the dendritic trees with a reduction of dendritic length at the side of reduced cochlear activity (Feng and Rogowski 1980; Rogowski and Feng 1981; Deitch and Rubel 1984).

In summary, more extensive morphological changes can be demonstrated with monaural deprivation than with binaural deprivation. The reason can be that monaural deprivation causes an unbalance between active and inactive inputs to neurons receiving binaural projections. Binaural inputs are in competition for neuronal targets (neurotrophic factors synthetized by the targets). If one input shows significantly smaller activity, it will be replaced by more active inputs.

4.3.3 Functional Changes in the Auditory Brainstem and Midbrain

In very young gerbils Kotak and Sanes (1997) demonstrated fast and profound effects of unilateral cochlear ablation on the functionality of synapses in the auditory brainstem. The authors ablated the cochlea on day 7 p.n. and investigated the brainstem using the brain-slice technique 1 to 6 days after the manipulation. Synaptic transmission was markedly decreased after cochlear ablation on the deprived side of the cochlear nucleus. A follow-up study demonstrated a different effect in the midbrain: following a 5-day-long deafness (induced on day 9 p.n.) in young gerbils, there was an increase of excitatory postsynaptic currents and a decrease in postsynaptic inhibitory currents in the IC, indicating a disinhibition in the midbrain (Vale and Sanes 2002).

In waking neonatally deafened and congenitally deaf cats, electrical stimulation of the auditory nerve evoked pinna and head orientation reflexes (Snyder et al. 1991; Klinke et al. 1999; Kral et al. 2002). These brainstem reflexes thus develop independent of auditory experience. Auditory midbrain activity evoked by electrical stimulation demonstrated only small functional deficits after neonatal bilateral deprivation. The most prominent difference between deaf and control (hearing, acutely deafened) animals was the shape of wave IV of the brainstem evoked response (Hardie and Shepherd 1999). Thresholds of auditory-evoked brainstem responses were increased after auditory deprivation in neonatally deafened cats (Snyder et al. 1990; Hardie and Shepherd 1999). In contrast to this finding, single-unit thresholds in the IC were lower in neonatally deafened cats than in hearing acutely deafened controls (Snyder et al. 1991). This contradiction indicates that synchronization of the neuronal activity in the central auditory system decreased with deprivation. Indeed, an increased temporal jitter of the responses has been demonstrated in the IC of neonatally deafened cats (Shepherd et al. 1999). This effect could also be responsible for the high variability of EABR thresholds in comparison to hearing thresholds (see section 3.3).

Electrical spatial threshold curves in the central nucleus of the IC were not affected by auditory deprivation (Snyder et al. 1990). The topographic representation and the cochleotopic gradient were not affected by auditory deprivation either (Snyder et al. 1991). Other functional parameters of the units in the IC (e.g., rate-intensity functions, types of peristimulus time histograms) were also not significantly affected by auditory deprivation.

In summary, in contrast to morphological data on auditory-deprived animals, functional properties of IC neurons show only minor deficits. The basic functional properties in the auditory brainstem and midbrain are not extensively affected by auditory deprivation. This, however, does not hold for the properties of binaural interaction. The development of binaural inhibitory-excitatory interactions critically depends on auditory experience in the olivary complex (Brand et al. 2002; Kapfer et al. 2002). Consequently, the ability to localize the sound source could be significantly altered in auditory-deprived animals, as also evidenced from the owl studies (Knudsen 1999).

4.3.4 Functional Changes in the Primary Auditory Cortex

The primary auditory cortex of adult, congenitally deaf cats, can be activated by electrical stimulation of the auditory nerve (Hartmann et al. 1997b; Shepherd et al. 1997). Both local field potentials and multiunit activity in the contralateral A1 field can be recorded in response to electrical stimulation of the auditory nerve. Single units in these cats displayed a short latency response to pulsatile stimulation (shortest latency ~8 ms) with a successive suppression of spontaneous activity. The latencies were comparable to normal-hearing cats when electrically stimulated (Raggio and Schreiner 1994). A rudimentary cochleotopy was demonstrated in field A1 of congenitally deaf cats (Hartmann et al. 1997b). The pronounced degeneration of the auditory nerve fibers by pharmacological deafening could have contributed to a more smeared cochleotopy in neonatally deafened cats (Raggio and Schreiner 1999). Two regions of lower multiunit thresholds separated by a high-threshold ridge were found in the auditory cortex of neonatally deafened cats (Raggio and Schreiner 1999), which possibly correspond to two cortical high-amplitude spots in local field potentials recorded in congenitally deaf cats (see Fig. 6.15). Cortical areas activated by the stimulation were larger and the lowest cortical thresholds lower in neonatally deafened animals compared to hearing acutely deafened controls (Raggio and Schreiner 1999), indicating a similar cortical "hypersensitivity" at threshold stimulation intensities as in adult-onset of deafness.

Most studies on cerebral cortex, including plasticity studies, have been carried out in "middle cortical layers" (layers III and IV). Cortical layers differ in their functional properties. Plasticity is high in the cortex of young animals; however, only layer III remains plastic into adulthood in primary sensory cortex (Singer 1995; Kaczmarek et al. 1997). The effects of deprivation have to be investigated in all cortical layers.

To address both this question, synaptic activity in the primary auditory cortex of congenitally deaf animals was investigated in a layer-specific manner (Kral et al. 2000). Current source densities were computed from field potentials recorded in different cortical layers. These current source densities provide a measure of extracellular components of synaptic currents (Mitzdorf 1985). Far-field influences are eliminated with this technique.

Stimulation was performed using charge-balanced biphasic pulses applied in a monopolar configuration through a cochlear implant. Comparisons were performed with stimulation at a high intensity (10 dB over the lowest individual cortical thresholds, in saturation of the amplitude-intensity functions of field potentials).

In accordance with the theory of synaptogenesis, adult congenitally deaf cats showed significantly smaller mean synaptic currents when compared to hearing animals (Fig. 6.10), as a consequence of either more extensive synaptic pruning in the inactive auditory cortex or degenerative processes (Kral et al. 2000). Earliest synaptic currents had longer latencies in naive animals than in hearing cats. Synaptic currents were small with latencies above 30 ms in deaf cats. Infragranular synaptic currents were substantially reduced by deprivation.

Reduced synaptic activity in infragranular cortical layers implies reduced feedback activation of subcortical structures, most importantly the auditory thalamus, as these layers are the source of corticothalamic projections (layer VI; Wong and Kelly 1981; Mitani and Shimokouchi 1985; de Venecia et al. 1998). Corticothalamocortical loops may be involved in short-term memory (Edeline 1999; Steriade 1999). Projections to more peripheral nuclei are also affected in naive animals: layer V is the source of afferent projections to the IC (Andersen et al. 1980; Mitani et al. 1983; Kral et al. 2000). Corticocortical interactions are also in part mediated by infragranular layers. Low synaptic activity after 30 ms poststimulus further indicates functional impairments in corticocortical interactions (Winguth and Winer 1986). The highly coordinated interaction of the primary auditory cortex with other cortical and subcortical centers seems to be disrupted after auditory deprivation. The postnatal development of the cerebral cortex of cats starts in infragranular layers (Friauf and Shatz 1991; Friauf et al. 1990), which are also the first to become mature (Huttenlocher and Dabholkar 1997). They are downstream in the information processing within the cortical column; they receive inputs from supragranular layers and layer IV. This might explain why they are most affected by deprivation.

The auditory system remains in an immature state also in prelingually deaf humans (e.g., Ponton et al. 1996a,b; Firszt et al. 2002a,b). Right after cochlear implantation deaf children showed P_1 waves in cortical evoked potentials. However, group data showed that P_1 latency had not matured normally during the deafness period. Deaf children had longer P_1 latency than age-matched hearing children. Additionally, later responses like waves N_{1b} or P_2 were either absent or had longer latencies in deaf children. These data agree with the animal studies mentioned above. (For cortical maturation with CI stimulation see section 4.4.)

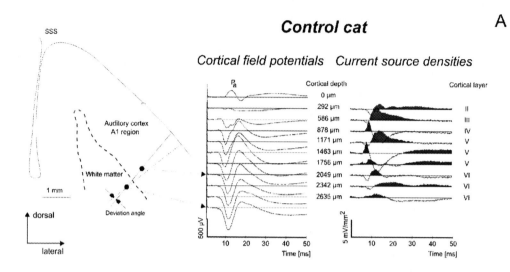

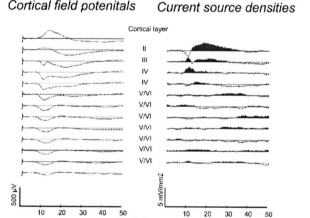

FIGURE 6.10. Deficits in synaptic activity in the primary auditory cortex of congenitally deaf cats. A: A profile of local field potentials recorded during a penetration through the primary auditory cortex with a microelectrode in a control (hearing, acutely deafened) cat. Drawing on the left is the histological reconstruction of the electrode penetration. The depth profiles of the current source densities computed from the field potentials are shown on the right. B: Similar data taken from a naive congenitally deaf cat. Mean current source densities are significantly smaller and show a different activation pattern in congenitally deaf cats (modified from Kral et al. 2000).

In summary, cochleotopic organization of the auditory system is at least partly developed in naive deaf animals, although it appears rudimentary in the auditory cortex. Synchronization of responses is slightly but significantly decreased in the brainstem and midbrain after auditory deprivation. Decreased cortical thresholds to cochlear implant stimulation in neonatally deafened animals, together with large cortical areas activated by the stimulation at threshold intensity, indicate a central "hypersensitivity" to stimulation. A decrease in synaptic currents in primary auditory cortex in congenitally deaf cats in response to intense stimulation (10 dB above threshold), on the other hand, indicates that in deaf animals functional intracortical synapses were either not formed or were lost in comparison to hearing controls. The decrease in synaptic activation was most pronounced in infragranular cortical layers and with longer latencies (>30 ms). The changes caused by congenital auditory deprivation imply substantial deficits in transmission of neuronal activity from primary auditory cortex to higher cortical areas and back to subcortical nuclei. Similarly, electrophysiological data suggest that the auditory system remains in an immature state in deaf children.

4.4 Auditory Experience After Periods of Deafness

Funtional deficits in the auditory system after long-term deafness are supposed to be the consequence of lack of activity in the auditory system. Such theory can be further tested by providing long-term deaf subjects with auditory input and demonstrating reversibility of the deficits due to hearing experience. This is possible with electrical stimulation through a cochlear implant. Furthermore, it is of clinical interest to investigate the ability of the auditory system to learn to appropriately process the artificial electrical stimuli. Eventually, it is important to find out whether or not the adaptation to artificial stimulation is restricted to certain developmental periods of the auditory system.

4.4.1 Subcortical Developmental Plasticity—Morphology

The studies of subcortical reorganization of the auditory system with chronic electrostimulation rely mainly on data from neonatally deafened cats (Leake-Jones and Rebscher 1983; Xu et al. 1993). The chronic electrostimulation of the auditory system was performed using pulsatile stimulation in a single-channel strategy (bipolar stimulation, most apical electrodes of the implant). The animals were implanted at the age of 2 to 5 months and attached to a stimulator for 1 to 4 hours/day. They were stimulated either with pulse trains of a given repetition rate (30 Hz in the initial studies, over 300 Hz in the more recent studies) or by amplitude- or frequency-modulated pulse trains. Duration of stimulation was up to 3 to 10 months, 2 or 6 dB over behaviorally determined hearing thresholds (Beitel et al. 2000a,b).

The total duration of auditory experience is an important factor in reversing the effects of auditory deprivation in the cochlear nucleus. Hultcrantz et al. (1991) investigated the cochlear nucleus of chronically stimulated neonatally deafened cats. Stimulation through the cochlear implant was initiated at the age of 3.0 to 4.5 months. The animals were stimulated 1 hour a day, 5 days a week, for a period of 3 months (total stimulation time: 60 hours). The study did not demonstrate any effects of chronic electrostimulation on the anteroventral cochlear nucleus (cross-sectional somatic neuronal area, spherical cell densities, or cochlear nucleus volume). Consequently, the 60 hours' stimulation (over 3 months) was not sufficient to reverse the deprivation-induced dystrophic changes in the cochlear nucleus. Lustig et al. (1994) repeated the study with a longer period of chronic stimulation (4 hours/day, implantation age: 2 to 4 months, stimulation at 2 dB above the EABR threshold, bipolar stimulation, stimulation duration 3 months, total stimulation time: up to 240 hours). Again, the effect of electrostimulation on the cross-sectional somatic area was not significant. Longer stimulation (implantation age ~3 months, monopolar stimulation with biphasic pulses at a repetition rate of 100 pps, stimulation 16 hours/day, 3 to 4 months, total stimulation time: 1100–1600 hours), however, led to reductions of shrinkage of neuronal somata in the anteroventral cochlear nucleus of neonatally deafened cats (Matsushima et al. 1991; for rodents compare Chouard et al. 1983). Data from congenitally deaf animals implanted at a comparable age confirmed the results (Heid et al. 2000). When stimulated over 1 to 2 months, 24 hours a day, 7 days a week (implantation age ~3 months, monopolar chronic stimulation, compressed analog coding strategy, total stimulation time: 700–1400 hours) the animals showed significant increases of cross-sectional neuronal somatic areas in the anteroventral cochlear nucleus, but only after 1400 hours of stimulation. In contrast, the superior olive showed significantly larger somatic areas than in naive animals already after 700 hours of auditory experience. Consequently, auditory experience in young age reverses the effects of auditory deprivation on cells in the brainstem; however, the stimulation has to be longer than 700 hours in cats.

Corresponding human data are sparse. No effect of chronic electrostimulation on dystrophic changes of neuronal somata has been found in the cochlear nucleus of cochlear implant patients (Moore et al. 1997b; Chao et al. 2002). Unfortunately, the data are not completely comparable to those obtained in animals studies, as the patients had adult onset of deafness. Additionally, the sample of brains investigated so far was strongly biased by the age of patients (e.g., a mean of 78 years in the study of Chao et al. 2002) and a relatively short duration of cochlear implant stimulation (a few years).

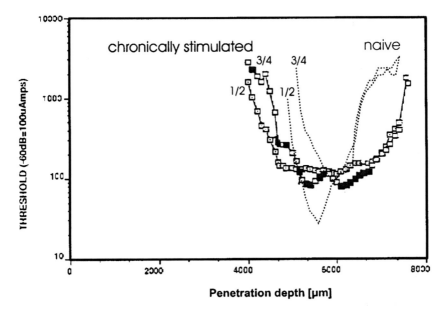

FIGURE 6.11. Chronic electrostimulation with a single-channel device smears the cochleotopic organization in the inferior colliculus (modified from Snyder et al. 1990, with permission). Dotted lines: spatial threshold curves reconstructed from a neonatally deafened (naive) cat. Black lines: spatial threshold curves reconstructed from an animal chronically stimulated over several months with a single-channel device. Bipolar stimulation with electrodes 1/2 or 3/4. After electrostimulation the threshold curves became broader and only weak signs of cochleotopy could be demonstrated.

4.4.2 Subcortical Developmental Plasticity—Physiology

Snyder et al. (1990) demonstrated extensive effects on the midbrain in chronically electrostimulated neonatally deafened cats (implantation age 2–4 months, pulsatile stimulation 6 dB above EABR threshold 1 hour/day, 5 days/week, constant repetition rate 30 pps, or 2 dB above brainstem evoked response, 4 hours/day, 5 days/week, for a period of up to 3 months). The IC representation of chronically stimulated parts of the cochlea (electrical spatial threshold curves) was greatly expanded (Fig. 6.11). The expansion included changes in neuronal thresholds and thus represented a plastic adaptation to the artificial one-channel stimulation. A follow-up study (Snyder et al. 1991) found that the occurrence of purely inhibitory responses in the IC increased after chronic electrostimulation. The minimum latency of responses decreased slightly but significantly in chronically stimulated animals (by ~1 ms). The response latency became smaller than that in hearing or neonatally deafened animals. A long-latency response (mean latency ~75 ms in normal controls) was found in 10% to 16% units of both naive cats and hearing controls. Occurrence of long-latency responses increased to 40% in chronically stimulated animals. The latency of the long-latency response increased after chronic electrostimulation to a mean latency of ~97 ms. The authors suggested that this long-latency response could be the consequence of a top-down projection from higher auditory structures. In conclusion, not only were some deficits of naive animals reversed or prevented by chronic electrostimulation, but also some specific adaptations to the electrical stimulus, going beyond the deficit-reversal, were demonstrated.

Increasing the frequency of chronic electrostimulation up to 300 Hz and the use of frequency- and amplitude-modulated sweeps in the chronic electrostimulation (Snyder et al. 1995; Vollmer et al. 1999) substantially affected the temporal response properties of neurons in the IC; the maximum following frequency of neurons in the inferior colliculus increased from ~200 to ~600 pps in comparison to unstimulated animals. These data consequently demonstrate a high plasticity of the auditory system also in the temporal domain of the responses.

In contrast to single-channel stimulation, no expansion of cochlear representations could be found in the IC with two-channel stimulation if stimulation was presented to the active channels in an interleaved mode (Leake et al. 2000). However, if the stimulation was presented to both channels simultaneously, the representations in the IC were very similar to those in one-channel stimulation.

The subcortical pathway also develops with chronic electrostimulation in prelingually (congenitally) deaf humans (Gordon et al. 2002). Electrically evoked ABRs were present in all of the investigated children before chronic auditory stimulation. With auditory experience through cochlear implants, the latencies of electrically evoked ABRs decreased from the time of initial activation throughout the first year of cochlear implant use. That corresponds to some of the results of animal studies discussed above.

4.4.3 Cortical Developmental Plasticity—Physiology

Auditory experience through cochlear implants leads to extensive reorganization at the auditory cortex in congenitally deaf cats (Klinke et al. 1999, 2001; Kral et al. 2001, 2002). The animals were provided with a custom-made cochlear implant at the age of 2 to 5 months p.n. Testing of electrical hearing thresholds in cochlear-implanted animals was performed using thresholds of pinna orientation reflexes (Ehret 1985).

For chronic electrostimulation with biologically meaningful stimuli (monopolar stimulation, 24 hours/day, 7 days/week, 1 to 6 months), a portable signal processor with a single-channel compressed analog stimulation strategy was used. Animals carried the processor in a jacket on their back and could move freely. The processor allowed stimulation by all ambient sounds above 65 dB SPL in the frequency range of 100 to 8000 Hz. Klinke et al. (1999) showed that such stimulation provides auditory experience to congenitally deaf cats and that these animals can be successfully conditioned to acoustic stimuli. Implanted animals were raised in an acoustically enriched animal house environment. The animals were also classically conditioned to acoustic stimuli; they had to pick up a reward at a specified location in the room after presentation of tone stimuli (Klinke et al. 1999). The response approached 80% to 100% after 2 weeks of training in most of the animals.

Following months of auditory experience cortical local field potentials in implanted animals had larger amplitudes than in both naive congenitally deaf cats and hearing controls. Long-latency responses (wave P_1, latency ~150 ms), which are rarely seen in naive deaf cats, were recorded reproducibly with large amplitudes in chronically stimulated cats (Fig. 6.12). The long-latency response has been previously described in the auditory cortex of hearing cats (Eggermont 1992b, Dinse et al. 1997b). It appears later in postnatal development than middle-latency responses (cat: König et al.

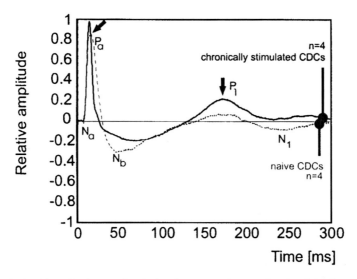

FIGURE 6.12. Chronic electrostimulation increases the amplitude of the P_1 wave and decreases the latency of local field potentials. Shown are grand mean averages taken from age-matched naive animals and chronically electrostimulated animals. Individual field potentials were normalized to maximum (P_a) amplitude of chronically stimulated congenitally deaf cats and consequently averaged. Differences in P_a latencies between animals lead to an amplitude <1 in naive congenitally deaf cats. Compare Klinke et al. (1999) and Kral et al. (2002).

1972; Eggermont 1996; monkey: Auther and Hackett 2001) and reaches mature latency in the 5th postnatal month in the cat, shortly before sexual maturity (Eggermont 1996). These facts indicate that auditory experience is important for the development of long-latency responses.

The long-latency responses are based on a rebound from cortical inhibition (Eggermont 1992b) influenced by corticothalamic loops (Contreras et al. 1996; auditory cortex: Cotillon and Edeline 2000). Corticothalamic loops are thought essential for short-term memory (Steriade 1999) and for subsequent processing in the higher order cortical auditory areas, as the primary auditory cortex has reciprocal connections with higher order auditory areas. Higher order auditory areas in turn project to cortical multimodal areas (Pandya and Yeterian 1985). Deficits in the functionality of thamalo-cortico-thalamic loops therefore can affect activity in higher order cortical areas and auditory short-term memory (see Pisoni and Cleary, Chapter 9). Maturation of the long-latency responses is thus of crucial importance for auditory cognitive functions. In addition to reduction of long-latency responses in the naive auditory cortex, most pronounced deficits in synaptic activity have been described in infragranular layers, which are a source of corticothalamic projections (see above). Taken together, these data provide an additional indication for deficits in corticothalamic loops in congenitally deaf cats and their reversibility after auditory experience with cochlear implants. When considering the results of the above studies on long-latency responses in the IC (Snyder et al. 1991), it appears very probable that loops down to the IC (its external nucleus) could also participate in this process. An alternative explanation of the effect of auditory experience on long-latency responses could be the restoration of a normal inhibitory function in the auditory cortex.

The complexity of single- and multiunit responses increased after chronic electrostimulation in congenitally deaf cats (Klinke et al. 1999; Kral et al. 2001). Based on poststimulus time histograms, different classes of responses could be differentiated in chronically stimulated animals, while mainly one was found in naive animals (Fig. 6.13). Long-latency responses of different types were found in 43% of the investigated units in chronically stimulated animals compared to 1% in naive deaf cats. The rate-intensity functions of single units in the cortex of naive animals in response to biphasic electric pulses corresponded to rate-intensity functions recorded in the auditory nerve (Hartmann et al. 1984b). In chronically stimulated animals, single-unit rate-intensity functions in the cortex were less uniform, including strongly nonmonotonic functions (Fig. 6.14). The more complex response characteristics in chronically stimulated cats indicate more complex cortical processing of input activity in chronically stimulated animals; an identical stimulus is responded to by different units in different ways, indicating that different units acquire distinct functions in the analysis of the input in chronically stimulated animals.

Current source density analyses in chronically stimulated animals (Klinke et al. 1999) indicate synaptic activation of infragranular layers comparable to normal hearing cats. This suggests that a maturation of the circuitry in the primary auditory cortex takes place under chronic electrical stimulation. The largest increases in synaptic currents in comparison to naive animals were found in supragranular layers II/III (Klinke et al. 1999), where cortical plasticity is greatest (Kaczmarek et al. 1997). To a certain extent, the auditory cortex can obviously achieve experience-dependent maturation under electrical stimulation through a cochlear implant if implantation takes place early.

Concomitantly with an increase in amplitudes of field potentials following long-term electrostimulation, there was an expansion of the activated cortical area (Fig. 6.15; Klinke et al. 1999; Kral et al. 2002). The activated cortical area was defined as the cortical region showing middle-latency (10–80 ms) responses above 300 µV with stimulation at 10 dB above the lowest cortical threshold. This region expanded substantially with increasing duration of auditory experience. The region with large long-latency responses in the field potentials also expanded. Largest long-latency responses were found in roughly the same cortical position as large middle-latency responses. These expansions of the activated area represent an adequate functional adaptation to the artificial one-channel stimulation strategy. They most likely correspond to earlier data on expansions of spatial threshold curves in the IC of chronically stimulated naive cats (Snyder et al. 1990).

The underlying mechanisms of the above expansions have not been elucidated yet. Considering the long time-course and the substantial extent of the expansions of cortical maps in the early-implanted animals, two mechanisms may be involved in plasticity:

1. Several smaller reorganizations of the cochleotopic maps in the subcortical structures of the auditory system (Snyder et al. 1990), which in sum yield a large expansion in cortical representation (Jones 2000). However, suppression of the somatosensory cortex leads to immediate expansions of neuronal receptive fields in the ventroposterior thalamus (Ergenzinger et al. 1998; Florence et al. 2000). The reorganization of subcortical maps can thus also be a top-down effect of cortical reorganization (for evidence of such effect in the auditory system, see Gao and Suga 2000).

2. A substantial expansion mainly of the thalamocortical projections and/or of corticocortical connections caused by enlargement of dendritic trees or axonal sprouting in the auditory cortex.

The expansions of the activated areas seem to reflect the basic funtional plasticity of the cerebral cortex, as described after lesions in somatosensory, visual, and motor systems (Dodson and Mohuiddin 2000; Buonomano and

naive

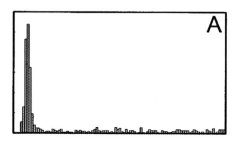

chronically stimulated

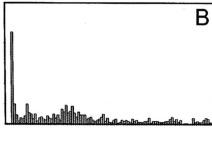

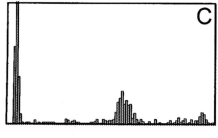

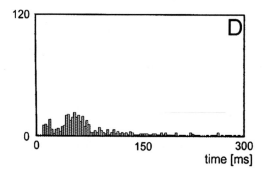

spikes

120

0

0   150   300

time [ms]

Merzenich 1998). Expansions of biologically more significant and/or more utilized cortical representations have also been reported in the auditory cortex (Robertson and Irvine 1989; Irvine and Robertson 1990; Irvine and Rajan 1993; Recanzone et al. 1993; Ohl and Scheich 1997; Weinberger 1998; Rauschecker 1999; Pantev and Lutkenhoner 2000).

Cortical responses to the "untrained" ear were additionally investigated in unilaterally chronically stimulated congenitally deaf cats. The ear that was not chronically electrostimulated ("untrained ear") was provided with a second cochlear implant during the final neurophysiological experiment. Despite months of unilateral auditory experience, cortical responses to stimulation of the "untrained" ear were not suppressed in animals implanted before their 5th months of life (Kral et al. 2001, 2002). This finding is of potential clinical importance: early-implanted prelingually deaf children could benefit from implantation of the unstimulated ear even if this takes place much later in life.

In summary, chronic electrostimulation in congenitally deaf cats leads to restoration of functional properties in the primary auditory cortex. The complexity of multiunit response types increases, the synaptic currents increase in all cortical layers and develop a pattern comparable to hearing controls, and the cortical field potentials become comparable to hearing controls. Additionally, a plastic adaptation to the artificial one-channel stimulation was found, and the cortical activated areas expanded. All these data demonstrate reversibility of the functional deficits found in naive animals.

4.4.4 Developmental Plasticity Is Age Dependent

Expansions of cortical activated areas following chronic electrostimulation mentioned above can be used as a measure of cortical plasticity and the reorganization capability of the auditory cortex. Using such an approach, it is possible to investigate the effect of age of implantation on cortical plasticity. Activated cortical areas expanded with duration of electrostimulation

---

FIGURE 6.13. Chronic electrostimulation increases the complexity of response types in the auditory cortex of congenitally deaf cats. A: The majority of units (99%) in A1 field of naive animals responded to electrical sinusoidal stimulation (437 Hz, 5-ms duration, 10 dB above threshold, repetition rate ~2 Hz) with a simple onset response. After chronic electrostimulation still 56% of the units responded with an onset response (A), often followed by an increased firing rate over up to 100 ms. However, 43% of the units responded with an onset response and a long-latency response (B, C). Rarely, units with no onset (D) but delayed response were found (1% of all units). (From Klinke et al. 1999, with permission).

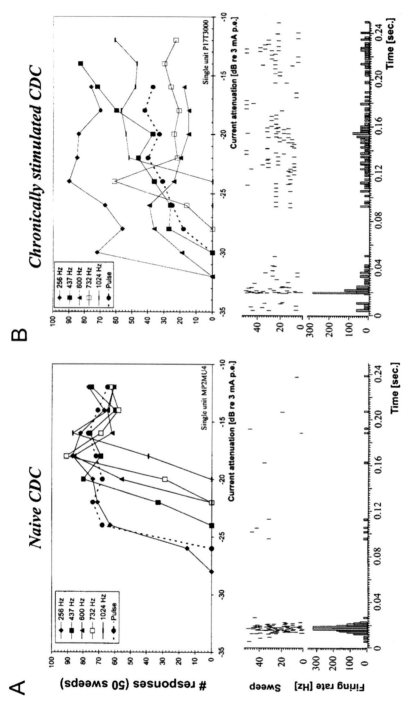

FIGURE 6.14. Chronic electrostimulation increases the complexity of cortical responses to cochlear implant stimulion. A: Rate-intensity functions of cortical units in response to stimulation with sinusoidal stimulation (6-ms duration) or single biphasic pulses (200μs/phase). The spike events with poststimulus time histogram for pulsatile stimulation at 10dB above threshold are shown for the same unit below. B: Same data for a unit from a chronically stimulated animal. Rate-intensity functions show a large diversity, larger dynamic range, and the poststimulus time histogram demonstrates a long-latency response. (From Kral et al. 2001, with permission).

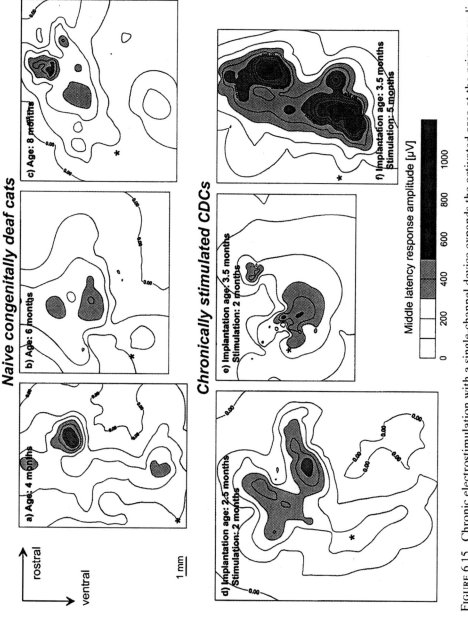

FIGURE 6.15. Chronic electrostimulation with a single-channel device expands the activated area at the primary auditory cortex (congenitally deaf cats, Kral et al. 2002). Shown are contour plots of the P_a amplitudes at the auditory cortex. Chronically stimulated animals show larger activated areas than age-matched naive animals.

up to the implantation age of 5 months. With higher implantation age, the activated areas expanded to lesser extent (Kral et al. 2001, 2002), indicating decreasing plasticity with increasing age. Additionally, after months of auditory experience in early-implanted animals, the latency of wave P_a (Fig. 6.12) was significantly shorter than in age-matched naive animals. However, P_a latency was no longer significantly different from naive animals if chronic electrostimulation was initiated after the 5th month of age (Kral et al. 2002). Similarly, the amplitude of the P_1 wave was smaller in animals implanted after their 5th month of age. These data demonstrate the existence of a sensitive period in plasticity of the auditory cortex.

The responses in the cortex ipsilateral to the chronically implanted ear showed a stronger dependence on the implantation age than the contralateral cortex. In theory one can assume that the amplitude of the postsynaptic depolarization is critical for initiation of plastic changes, i.e., there is an absolute plasticity threshold. Sensitive periods would then be the consequence of an age-dependent increase in the plasticity threshold. The given threshold will be more likely reached if the given neuron receives more excitatory projections (inputs) from the stimulated ear. Consequently, as the ipsilateral cortex receives less projections from the implanted ear, it will be more susceptible to an age-dependent decrease of plasticity. The presented data, therefore, are consistent with such a hypothesis. Whether the observed sensitive period is critical (i.e., whether the effects comparable to those due to chronic electrostimulation in early-implanted animals can be achieved in late-implanted animals after auditory experience exceeding 4000 hours) remains to be answered. However, the study demonstrates that the earlier the implantation takes place, the better the results that can be achieved.

Other data also indicate the existence of a sensitive period in the auditory system: a high-frequency hearing loss due to administration of ototoxic agents leads to larger reorganization of the tonotopic map in the feline auditory cortex if the cochlear hearing loss takes place neonatally (Harrison et al. 1991). The tonotopic organization of the rodent auditory cortex can be disrupted by unpatterned acoustical stimulation only within a critical period early in life (Zhang et al. 2002). The duration of the sensitive period of about 5 months in cats discussed here correlates with the time of functional maturation of the cat auditory cortex, which is achieved in the 5th month of life (Eggermont 1996).

Results of investigations on sensitive periods in *subcortical* auditory structures do not allow a clear conclusion. Seldon et al. (1996) used the 2-deoxyglucose method to investigate the volume of activated tissue in the IC of chronically stimulated neonatally deafened cats. The stimulation was initiated at different ages (from 3 months on). It was a single-channel stimulation, applied for 16 hours/day, over a period of 6 months. A large variability of the extent of the activated region was found. No significant effects of age at implantation could be identified. These data have been comple-mented by electrophysiological recordings in the IC (Moore et al. 2002). The authors deafened adult normal-hearing cats and chronically stimulated the animals over 4 to 6 months using a stimulation protocol comparable to similar studies on neonatally deafened cats (Snyder et al. 1990, 1991). When comparing the results of the chronically stimulated animals that were deaf-ened as adults to chronically stimulated cats deafened neonatally, no dif-ferences were found. Nonetheless, the animals differed not only in the age at implantation: the adult group was deafened shortly before implantation, and thus electrical stimulation was performed on an auditorily mature animal and not on a naive but adult animal (see section 4.2). A recent report on c-*fos* activity in the cochlear nucleus and IC of neonatally deafened, chronically electrostimulated rats demonstrated a sensitive period in both the cochlear nucleus and IC (Hsu et al. 2001); the number of immuno-labeled neurons after chronic electrostimulation of late-implanted animals was smaller than that of early-implanted animals.

In humans, the morphological postnatal development of the auditory cortex is characterized by two phases (Conel 1939–1967; Huttenlocher and Dabholkar 1997), including rapid synaptogenesis that is mainly genetically determined (up to 2–4 years postnatally), and slow synaptic maturation (stabilization or elimination, from the 2–4 years postnatally to adolescence).

Evoked potentials in response to electrical stimulation through cochlear implants revealed a sensitive period in the human auditory system. Two main developmental steps can be differentiated. First, in prelingually deaf children implanted after the fourth year, the latencies of evoked potentials demonstrated maturation of comparable rate as in hearing controls (Ponton et al. 1996b; 2000b; Ponton and Eggermont 2001). The maturational delays in latencies already present at the time of implantation were, nevertheless, not compensated for by chronic electrostimulation. In contrast, in prelin-gually deaf children implanted before 4 years of age, the maturational delay disappeared after 6 months of auditory experience (Sharma et al. 2002a–c). This indicates the existence of a sensitive period within the first 4 years of life. This developmental boundary corresponds well to the analysis of speech competence in prelingually deaf cochlear implant users; the speech competence is worse if patients are implanted after their fourth year of age (Fryauf-Bertschy et al. 1997; Tyler et al. 1997; Pisoni and Cleary, Chapter 9). Thus, cortical maturation is possible even in children implanted after the fourth year of age, but their maturational delay is not compensated and their speech understanding reaches lower scores than in children implanted before the fourth year. Second, there is another temporal landmark around 12 years of age (Ponton et al. 1996a; Ponton and Eggermont 2001) when cortical evoked potentials mature. This is correlated with the maturation of cortical neuronal morphology (Moore and Guan 2001). Possibly, cortical

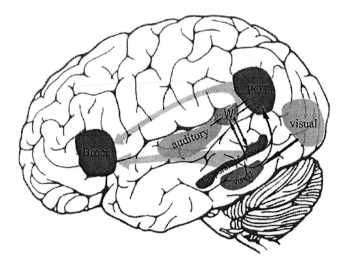

FIGURE 6.16. Schematic illustration of main areas activated during speech recognition in the dominant hemisphere of the human brain. Arrows indicate possible interconnections. Syntactic-semantic areas are Broca's area, parieto-occipito-temporal junction (POT) and areas in the inferior posterior temporal lobe. Phonetic-phonologic analyses take place in Wernicke's area (W). Exact location of activated areas in a given subject with the given task is variable.

functional maturation is limited if first hearing experience sets in after this age.

4.5 Higher Order (Association) Auditory Cortex and Cochlear Implant Stimulation

The effect of deprivation and delayed auditory experience on the association auditory cortex has not yet been addressed experimentally. Nevertheless, there are data indicative of a reorganization of higher order sensory cortex as a consequence of auditory deprivation in humans. The reports are limited to positron emission tomography, as evoked potentials can only indirectly indicate the locus of changes they demonstrate. Functional magnetic resonance imaging has been possible in cochlear-implanted subjects (e.g., Parving et al. 1995; Lazeyras et al. 2002); however, due to the technical complications of using strong magnetic fields with cochlear implants, few results have been obtained with this method so far.

The present model of cortical activation by language (presented acoustically or visually) is shown in Figure 6.16 (e.g., Price 2000). Giraud et al. (2001a,b) investigated the specificity of cortical regions activated by

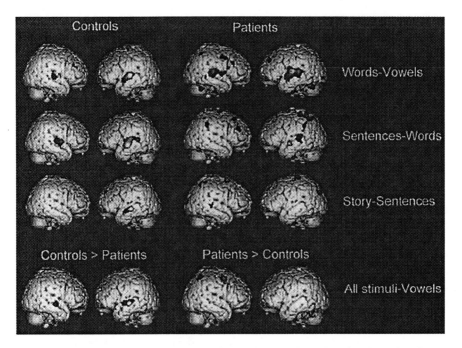

FIGURE 6.17. Cortical activation of hearing subjects and cochlear-implanted subjects by speech signals (position emission tomography study). (From Giraud et al. 2000, with permission).

cochlear implant stimulation in subjects with adult-onset hearing loss after periods of deafness (Fig. 6.17). The data indicated that some of the areas that, using the given battery of stimuli, showed speech specificity in normal hearing controls did not show this specificity in electrically stimulated patients after periods of deafness (e.g., Wernicke's area). Additionally, the patients recruited other areas, not activated by hearing subjects (left dorsal occipital cortex). Cochlear-implanted subjects activated more the pre-cuneus and the parahippocampal gyrus, whereby hearing controls activated more the left inferior frontal, left posterior inferior temporal, and left and right temporoparietal junctions. These data provided evidence that cochlear-implanted subjects process speech differently than hearing controls; most probably, some of the processing is automatic in hearing controls but involves conscious effort in deaf. Changes in the cortical circuitry by deprivation can also be responsible for the differences. Additionally, the data raise the question whether responses to speech stimuli applied through a cochlear implant are comparable to processing of the mother tongue presented acoustically. Maybe cortical responses to speech presented through cochlear implants should be compared to processing of a second language in hearing controls. One interesting finding was the activation of the visual cortex by speech stimuli in cochlear-implanted subjects (Giraud et al. 2001c), which was interpreted as imagination of lip movements during speaking to help understand the speech by visual analysis of the image of the speaker's lips.

Naito et al. (1995, 1997), addressed the question of cortical activation by electrical stimulation of the auditory nerve in pre- or postlingually implanted subjects. The primary auditory cortex was activated by electrical stimulation of the auditory nerve in both pre- and postlingually deaf subjects right after implantation to a similar extent as the primary auditory cortex of hearing subjects by sounds. However, when higher-order (association) cortex was compared between these groups, prelingually deaf patients showed substantially smaller activation compared to hearing or postlingually deaf patients. These data are particularly interesting, as they correspond well with animal studies showing that the primary auditory cortex can be activated in congenitally deaf or neonatally deafened cats (Hartmann et al. 1997a; Shepherd et al. 1997; Raggio and Schreiner 1999). However, the processing of input is deficient and the transmission of neuronal activity to higher cortical areas and back to subcortical nuclei is most probably compromised (Kral et al. 2000). This would explain why the auditory association cortex is much less activated in prelingually deaf patients than in hearing controls.

Prelingually deaf individuals show increased capacity in visual motion processing, indicative of cross-modal reorganization following congenital deafness (Bavelier and Neville 2002). The site of corresponding cross-modal reorganization is most probably not the primary, but rather the higher order cortex. No cross-modal reorganization of primary auditory cortex was found in adult congenitally deaf cats (Kral et al. 2003). Also in humans there is no consistent evidence of cross-modal reorganization of primary auditory cortex (compare Finney et al. 2001 with Nishimura et al. 1999; Petitto et al. 2000; Bernstein et al. 2002). However, higher order auditory cortex does process visual activity in the prelingually deaf (Nishimura et al. 1999; Finney et al. 2001; Petitto et al. 2001; Bernstein et al. 2002). Consequently, cross-modal reorganization is extensive, but seems limited to the higher order auditory cortex. This could explain the findings of Naito and colleagues that cochlear implants activate the primary auditory cortex; the higher order auditory areas, however, are activated only deficiently. Lee et al. (2001) demonstrated that the spontaneous glucose metabolism (i.e., in absence of auditory stimulation) in the auditory (higher order) cortex increases with age in prelingually deaf individuals. The amount of glucose metabolism in the higher order auditory cortex negatively correlates with postoperative speech perception scores during postoperative hearing training. We hypothetize that the increase in spontaneous glucose metabolism is an expression of the cross-modal reorganization of the higher order auditory cortex.

Then these data, together with the animal data on primary auditory cortex, would be consistent with the hypothesis of a decoupling between primary and secondary auditory cortex in congenitally deaf. Cochlear implantation has to overcome this decoupling.

5. Summary

Hearing loss leads to "hypersensitivity" in the adult auditory system, in part by disinhibition. The central disinhibition seems to be compensated for very soon after initiation of cochlear implant stimulation. This may explain the benefit that some postlingually deaf patients obtain from cochlear implants even after many years of deafness.

The auditory systems need auditory experience for maturation. Therefore, congenital (prelingual) hearing loss affects the development of the auditory system. The primary auditory cortex can be activated by auditory input after congenital deafness, but the activity is not appropriately processed and relayed to the auditory association cortex and language cortical areas. The top-down corticothalamic streams are also affected. This deprivation-induced deficit is reversible, but full reversibility is possible only if auditory stimulation is initiated early. In humans this corresponds to the age of 3 to 4 years after birth. This age correlates with the onset of elimination of synapses (synaptic pruning, 2–4 years) in the auditory cortex. To achieve optimal results, deaf children therefore should be implanted before the age of 2 to 4 years.

Acknowledgments. We thank Drs. R. Klinke and J. Smolders for critical comments and valuable suggestions on an earlier version of this manuscript. The work was supported by the grant SFB 269 C1 from Deutsche Forschungs-gemeinschaft.

Abbreviations

A1	primary auditory cortex
AAF	anterior auditory field
AII	secondary auditory cortex
ABI	auditory brainstem implant
AM	amplitude modulation
AN	auditory nerve
CDC	congenitally deaf cat
CF	characteristic frequency
CI	cochlear implant

CN   cochlear nucleus
  AVCN   anteroventral cochlear nucleus
  DCN   dorsal cochlear nucleus
  PVCN   postero-ventral cochlear nucleus
E-ABR   electrically evoked brainstem response
FM   frequency modulation
FTC   frequency threshold curve
Hz   Hertz
LL   lateral lemniscus
IC   inferior colliculus
  ICc   central part IC
  ICx   external part IC
  ICd   dorsal part IC
MGB   medial geniculate body
MNTB   medial nuclei of trapezoid body
MU   multi units (neurons)
p.n.   post natal (after birth)
pps   pulses per second
SI   synchronization index
SOC   superior olivary complex
  LSO   lateral part
  MSO   medial part
SPL   sound pressure level
SU   single unit (neuron)

References

Abbas PJ, Brown CJ (1991a) Electrically evoked auditory brainstem response: refractory properties and strength-duration functions. Hear Res 51:139–147.

Abbas PJ, Brown CJ (1991b) Electrically evoked auditory brainstem response: growth of response with current level. Hear Res 51:123–137.

Aitkin LM, Moore DR (1975) Inferior colliculus. II. Development of tuning characteristics and tonotopic organization in central nucleus of the neonatal cat. J Neurophysiol 38:1208–1216.

Andersen RA, Knight PL, Merzenich MM (1980) The thalamocortical and corticothalamic connections of AI, AII, and the anterior auditory field (AAF) in the cat: evidence for two largely segregated systems of connections. J Comp Neurol 194:663–701.

Araki S, Kawano A, Seldon HL, Shepherd RK, Funasaka S, Clark GM (2000) Effects of intracochlear factors on spiral ganglion cells and auditory brainstem response after long-term electrical stimulation in deafened kittens. Otolaryngol Head Neck Surg 122:425–433.

Auther LL, Hackett TA (2001) Postnatal development of auditory pathways in primates. I. Auditory evoked response maturation. Assoc Res Otolaryngol 24:24.

Bakin JS, Weinberger NM (1996) Induction of a physiological memory in the cerebral cortex by stimulation of the nucleus basalis. Proc Natl Acad Sci USA 93:11219–11224.

Barnet AB, Ohlrich ES (1975) Auditory evoked potentials during sleep in normal children from ten days to three years of age. Electroencephalogr Clin Neurophysiol 39:29–41.

Basser PJ, Roth BJ (2000) New currents in electrical stimulation of excitable tissues. Annu Rev Biomed Eng 2:377–397.

Bavelier D, Neville HJ (2002) Cross-modal plasticity: where and how? Nat Rev Neurosci 3:443–452.

Beitel RE, Snyder RL, Schreiner CE, Raggio MW, Leake PA (2000a) Electrical cochlear stimulation in the deaf cat: comparisons between psychophysical and central auditory neuronal thresholds. J Neurophysiol 83:2145–2162.

Beitel RE, Vollmer M, Snyder RL, Schreiner CE, Leake PA (2000b) Behavioral and neurophysiological thresholds for electrical cochlear stimulation in the deaf cat. Audiol Neurootol 5:31–38.

Benson DL, Colman DR, Huntley GW (2001) Molecules, maps and synapse specificity. Nat Rev Neurosci 2:899–909.

Berard DR, Coleman WR, Berger LH (1983) Electrical stimulation of the superior olivary complex can produce cortical evoked potential and behavioral discrimination correlates of pitch perception in the rat. Int J Neurosci 18:87–95.

Bernstein LE, Auer ET Jr, Moore JK, Ponton CW, Don M, Singh M (2002) Visual speech perception without primary auditory cortex activation. Neuroreport 13:311–315.

Bierer JA, Middlebrooks JC (2002) Auditory cortical images of cochlear-implant stimuli: dependence on electrode configuration. J Neurophysiol 87:478–492.

Bledsoe SC, Nagase S, Miller JM, Altschuler RA (1995) Deafness-induced plasticity in the mature central auditory system. Neuroreport 7:225–229.

Born DE, Rubel EW (1988) Afferent influences on brainstem auditory nuclei of the chicken: presynaptic action potentials regulate protein synthesis in nucleus magnocellularis neurons. J Neurosci 8:901–919.

Brackmann DE, Hitselberger WE, Nelson RA, Moore J, et al. (1993) Auditory brainstem implant: I. Issues in surgical implantation. Otolaryngol Head Neck Surg 108:624–633.

Brand A, Behrend O, Marquardt T, McAlpine D, Grothe B (2002) Precise inhibition is essential for microsecond interaural time difference coding. Nature 417:543–547.

Brindley GS, Donaldson PE, Falconer MA, Rushton DN (1972) The extent of the region of occipital cortex that when stimulated gives phosphenes fixed in the visual field. J Physiol 225:57P–58P.

Brown M, Shepherd RK, Webster WR, Martin RL, Clark GM (1992) Cochleotopic selectivity of a multichannel scala tympani electrode array using the 2-deoxyglucose technique. Hear Res 59:224–240.

Brown CJ, Abbas PJ, Fryauf-Bertschy H, Kelsay D, Gantz BJ (1994) Intraoperative and postoperative electrically evoked auditory brainstem responses in nucleus cochlear implant users: implications for the fitting process. Ear Hear 15:168–176.

Brown CJ, Hughes ML, Lopez SM, Abbas PJ (1999) Relationship between EABR thresholds and levels used to program the CLARION speech processor. Ann Otol Rhinol Laryngol Suppl 177:50–57.

Brugge JF (1992) Development of the lower auditory brainstem of the cat. In: Romand R, ed. Development of Auditory and Vestibular Systems 2. Amsterdam: Elsevier Science Publishers B.V., pp. 173–296.

Brugge JF, Javel E, Kitzes LM (1978) Signs of functional maturation of peripheral auditory system in discharge patterns of neurons in anteroventral cochlear nucleus of kitten. J Neurophysiol 41:1557–1559.

Buonomano DV, Merzenich MM (1998) Cortical plasticity: from synapses to maps. Annu Rev Neurosci 21:149–186.

Caird DM, Klinke R (1983) Processing of binaural stimuli by cat superior olivary complex neurons. Exp Brain Res 52:385–399.

Carlier E, Pujol R (1980) Supra-normal sensitivity to ototoxic antibiotic of the developing rat cochlea. Arch Otorhinolaryngol 226:129–133.

Chao TK, Burgess BJ, Eddington DK, Nadol JB (2002) Morphometric changes in the cochlear nucleus in patients who had undergone cochlear implantation for bilateral profound deafness (1). Hear Res 174:196–205.

Cheour M, Leppanen PHT, Kraus N (2000) Mismatch negativity (MMN) as a tool for investigating auditory discrimination and sensory memory in infants and children. Clin Neurophysiol 111:4–16.

Chouard CH, Meyer B, Josset P, Buche JF (1983) The effect of the acoustic nerve chronic electric stimulation upon the Guinea pig cochlear nucleus development. Acta Otolaryngol (Stockh) 95:639–645.

Clopton BM, Glass I (1984) Unit responses at cochlear nucleus to electrical stimulation through a cochlear prosthesis. Hear Res 14:1–11.

Conel JL (1939–1967) The Postnatal Development of Human Cerebral Cortex, vol. I–VIII. Cambridge, MA: Harvard University Press.

Conlee JW, Parks TN (1983) Late appearance and deprivation-sensitive growth of permanent dentrites in the avian cochlear nucleus (Nuc. Magnocellularis). J Comp Neurol 217:216–226.

Contreras D, Destexhe A, Sejnowski TJ, Steriade M (1996) Control of spatiotemporal coherence of a thalamic oscillation by corticothalamic feedback. Science 274:771–774.

Cotillon N, Edeline JM (2000) Tone-evoked oscillations in the rat auditory cortex result from interactions between the thalamus and reticular nucleus. Eur J Neurosci 12:3637–3650.

Cragg BG (1975a) The development of synapses in the visual system of the cat. J Comp Neurol 160:147–166.

Cragg BG (1975b) The development of synapses in kitten visual cortex during visual deprivation. Exp Neurol 46:445–451.

Crowther JA, Miller JM, Kileny PR (1990) Effect of anesthesia on acoustically evoked middle latency response in guinea pigs. Hear Res 43:115–120.

de Venecia RK, Smelser CB, McMullen NT (1998) Parvalbumin is expressed in a reciprocal circuit linking the medial geniculate body and auditory neocortex in the rabbit. J Comp Neurol 400:349–362.

Deitch GS, Rubel EW (1984) Afferent influences on brainstem auditory nuclei of the chicken: tune course and specific of dentritic atrophy following deafferentation. J Comp Neurol 229:66.

Delgutte B, Kiang NY (1984) Speech coding in the auditory nerve: I. Vowel-like sounds. J Acoust Soc Am 75:866–878.

Dinse HR, Godde B, Hilger T, Reuter G, et al. (1997a) Optical imaging of cat auditory cortex cochleotopic selectivity evoked by acute electrical stimulation of a multi-channel cochlear implant. Eur J Neurosci 9:113–119.

Dinse HR, Kruger K, Akhavan AC, Spengler F, Schoner G, Schreiner CE (1997b) Low-frequency oscillations of visual, auditory and somatosensory cortical neurons evoked by sensory stimulation. Int J Psychophysiol 26:205–227.

Dobelle WH, Stensaas SS, Mladejovsky MG, Smith JB (1973) A prosthesis for the deaf based on cortical stimulation. Ann Otol Rhinol Laryngol 82:445–463.

Dodson HC, Mohuiddin A (2000) Response of spiral ganglion neurones to cochlear hair cell destruction in the guinea pig. J Neurocytol 29:525–537.

Dorman MF, Loizou PC, Rainey D (1997) Speech intelligibility as a function of the number of channels of stimulation for signal processors using sine-wave and noise-band outputs. J Acoust Soc Am 102:2403–2411.

Dynes SBC, Delgutte B (1992) Phase-locking of auditory-nerve discharges to sinusoidal electric stimulation of the cochlea. Hear Res 58:79–90.

Edeline JM (1999) Learning-induced physiological plasticity in the thalamo-cortical sensory systems: a critical evaluation of receptive field plasticity, map changes and their potential mechanisms. Prog Neurobiol 57:165–224.

Eggermont JJ (1985) Evoked potentials as indicators of auditory maturation. Acta Otolaryngol (Stockh) 421:41–47.

Eggermont JJ (1988) On the rate of maturation of sensory evoked potentials. Electroencephalogr Clin Neurophysiol 70:293–305.

Eggermont JJ (1991) Maturational aspects of periodicity coding in cat primary auditory cortex. Hear Res 57:45–56.

Eggermont JJ (1992a) Development of auditory evoked potentials. Acta Otolaryngol 112:197–200.

Eggermont JJ (1992b) Stimulus induced and spontaneous rhythmic firing of single units in cat primary auditory cortex. Hear Res 61:1–11.

Eggermont JJ (1993) Functional aspects of synchrony and correlation in the auditory nervous system. Concepts Neurosci 4:105–129.

Eggermont JJ (1996) Differential maturation rates for response parameters in cat primary auditory cortex. Auditory Neurosci 2:309–327.

Eggermont JJ (2002) Temporal modulation transfer functions in cat primary auditory cortex: separating stimulus effects from neural mechanisms. J Neurophysiol 87:305–321.

Ehret G (1985) Behavioural studies on auditory development in mammals in relation to higher nervous system functioning. Acta Otolaryngol Suppl 421:31–40.

Eisenberg LS, Maltan AA, Portillo F, Mobley JP, House WF (1987) Electrical stimulation of the auditory brainstem structure in deafened adults. J Rehabil Res Dev 24:9–22.

El-Kashlan HK, Niparko JK, Altschuler RA, Miller JM (1991) Direct electrical stimulation of the cochlear nucleus: surface vs. penetrating stimulation. Otolaryngol Head Neck Surg 105:533–543.

Ergenzinger ER, Glasier MM, Hahm JO, Pons TP (1998) Cortically induced thalamic plasticity in the primate somatosensory system. Nat Neurosci 1:226–229.

Felix H, Hoffmann V (1985) Light and electron microscopic investigation of cochlear nerve specimens from profoundly deaf patients. Acta Otolaryngol Suppl 423:67–72.

Feng AS, Rogowski BA (1980) Effects of monaural and binaural occlusion on the morphology of neurons in the medial superior olivary nucleus of the rat. Brain Res 189:530–534.

Finney EM, Fine I, Dobkins KR (2001) Visual stimuli activate auditory cortex in the deaf. Nat Neurosci 4:1171–1173.

Firszt JB, Rotz LA, Chambers RD, Novak MA (1999) Electrically evoked potentials recorded in adult and pediatric CLARION implant users. Ann Otol Rhinol Laryngol Suppl 177:58–63.

Firszt JB, Chambers RD, Kraus AN, Reeder RM (2002a) Neurophysiology of cochlear implant users I: effects of stimulus current level and electrode site on the electrical ABR, MLR, and N1-P2 response. Ear Hear 23:502–515.

Firstz JB, Chambers RD, Kraus N (2002b) Neurophysiology of cochlear implant users II: comparison among speech perception, dynamic range, and physiological measures. Ear Hear 23:516–531.

Fiseifis S, Scheich H (1995) Cochlear-prosthesis-induced activity in Gerbils auditory cortex. In: Elsner N, Menzel R, eds. Göttingen Neurobiology Report 23. Stuttgart-New York: Thieme, 317.

Florence SL, Hackett TA, Strata F (2000) Thalamic and cortical contributions to neural plasticity after limb amputation. J Neurophysiol 83:3154–3159.

Friauf E, Shatz CJ (1991) Changing patterns of synaptic input to subplate and cortical plate during development of visual cortex. J Neurophysiol 66:2059–2071.

Friauf E, McConnell SK, Shatz CJ (1990) Functional synaptic circuits in the subplate during fetal and early postnatal development of cat visual cortex. J Neurosci 10:2601–2613.

Fryauf-Bertschy H, Tyler RS, Kelsay DM, Gantz BJ, Woodworth GG (1997) Cochlear implant use by prelingually deafened children: the influences of age at implant and length of device use. J Speech Lang Hear Res 40:183–199.

Gabriele ML, Brunsobechtold JK, Henkel CK (2000) Plasticity in the development of afferent patterns in the inferior colliculus of the rat after unilateral cochlear ablation. J Neurosci 20:6939–6949.

Gao E, Suga N (2000) Experience-dependent plasticity in the auditory cortex and the inferior colliculus of bats: role of the corticofugal system. Proc Natl Acad Sci USA 97:8081–8086.

Giraud AL, Truy E, Frackowiak RS, Gregoire MC, Pujol JF, Collet L (2000) Differential recruitment of the speech processing system in healthy subjects and rehabilitated cochlear implant patients. Brain 123:1391–1402.

Giraud AL, Price CJ, Graham JM, Frackowiak RS (2001a) Functional plasticity of language-related brain areas after cochlear implantation. Brain 124:1307–1316.

Giraud AL, Truy E, Frackowiak R (2001b) Imaging plasticity in cochlear implant patients. Audiol Neurootol 6:381–393.

Giraud AL, Price CJ, Graham JM, Truy E, Frackowiak RS (2001c) Cross-modal plasticity underpins language recovery after cochlear implantation. Neuron 30:657–663.

Glass I (1983) Tuning characteristics of cochlear nucleus units in response to electrical stimulation of the cochlea. Hear Res 12:223–238.

Glass I (1984) Phase-locked responses of cochlear nucleus units to electrical stimulation through a cochlear implant. Exp Brain Res 55:386–390.

Gordon KA, Papsin BC, Harrison RV (2002) Auditory brainstem and midbrain development after cochlear implantation in children. Ann Otol Rhinol Laryngol Suppl 189:32–37.

Hackett TA, Stepniewska I, Kaas JH (1998) Thalamocortical connections of the parabelt auditory cortex in macaque monkeys. J Comp Neurol 400:271–286.

Hardie NA, Shepherd RK (1999) Sensorineural hearing loss during development: morphological and physiological response of the cochlea and auditory brainstem. Hear Res 128:147–165.

Hardie NA, Martsimcclintock A, Aitkin LM, Shepherd RK (1998) Neonatal sensorineural hearing loss affects synaptic density in the auditory midbrain. Neuroreport 9:2019–2022.

Harrison RV, Nagasawa A, Smith DW, Stanton S, Mount RJ (1991) Reorganization of auditory cortex after neonatal high frequency cochlear hearing loss. Hear Res 54:11–19.

Hartmann R, Klinke R (1990) Response characteristics of nerve fibers to patterned electrical stimulation. In: Miller JM, Spelman FA, eds. Cochlear Implants. Models of the Electrically Stimulated Ear. New York: Springer-Verlag, pp. 135–160.

Hartmann R, Topp G, Klinke R (1984a) Electrical stimulation of the cat cochlea—discharge pattern of single auditory fibres. Adv Audiol 1:18–29.

Hartmann R, Topp G, Klinke R (1984b) Discharge patterns of cat primary auditory fibers with electrical stimulation of the cochlea. Hear Res 13:47–62.

Hartmann R, Shepherd RK, Heid S, Klinke R (1997a) Response of the primary auditory cortex to electrical stimulation of the auditory nerve in the congenitally deaf white cat. Hear Res 112:115–133.

Hartmann R, Klinke R, Heid S (1997b) Functional organization of the auditory cortex in the congenitally deaf white cat. In: Syka J, ed. Acoustical Signal Processing in the Central Auditory System. New York: Plenum Press, pp. 561–575.

Heid S (1998) Morphologische Befunde am peripheren und zentralen auditorischen System der kongenital gehörlosen weißen Katze. [PhD thesis]. J. W. Goethe University, Frankfurt am Main.

Heid S, Jahnsiebert TK, Klinke R, Hartmann R, Langner G (1997) Afferent projection patterns in the auditory brainstem in normal and congenitally deaf white cats. Hear Res 110:191–199.

Heid S, Hartmann R, Klinke R (1998) A model for prelingual deafness, the congenitally deaf white cat—population statistics and degenerative changes. Hear Res 115:101–112.

Heid S, Kral A, Ladewig V, Hartmann R, Klinke R (2000) Morphology of the cochlea and the spiral ganglion in cochlear-implanted congenitally deaf cats. Assoc Res Otolaryngol 23:229.

Hsu WC, Campos-Torres A, Portier F, Lecain E, et al. (2001) Cochlear electrical stimulation: influence of age of implantation on Fos immunocytochemical reactions in inferior colliculi and dorsal cochlear nuclei of the rat. J Comp Neurol 438:226–238.

Hultcrantz M, Snyder R, Rebscher S, Leake P (1991) Effects of neonatal deafening and chronic intracochlear electrical stimulation on the cochlear nucleus in cats. Hear Res 54:272–280.

Huttenlocher PR (1984) Synapse elimination and plasticity in developing human cerebral cortex. Am J Ment Defic 88:488–496.

Huttenlocher PR, Dabholkar AS (1997) Regional differences in synaptogenesis in human cerebral cortex. J Comp Neurol 387:167–178.

Irvine DRF, Rajan R (1993) Plasticity in the frequency organization of auditory cortex of adult mammals with restricted cochlear lesions. Biomed Res 14:55–59.

Irvine DRF, Robertson D (1990) Reorganization of frequency representation in auditory cortex of guinea pigs with partial unilateral deafness. Inf Proc Mammal Audit Tact Sys 56:253–266.

Jones EG (2000) Cortical and subcortical contributions to activity-dependent plasticity in primate somatosensory cortex. Annu Rev Neurosci 23:1–37.

Joris PX, Yin TCT (1995) Envelope coding in the lateral superior olive. 1. Sensitivity to interaural time differences. J Neurophysiol 73:1043–1062.

Joris PX, Yin TCT (1998) Envelope coding in the lateral superior olive. III. Comparison with afferent pathways. J Am Physiol Soc 79:253–269.

Kaas JH, Hackett TA, Tramo MJ (1999) Auditory processing in primate cerebral cortex. Curr Opin Neurobiol 9:164–170.

Kaczmarek L, Kossut M, Skangiel-Kramska J (1997) Glutamate receptors in cortical plasticity: molecular and cellular biology. Physiol Rev 77:217–255.

Kadner A, Scheich H (2000) Trained discrimination of temporal patterns: cochlear implants in gerbils. Audiol Neurootol 5:23–30.

Kaltenbach JA, Czaja JM, Kaplan CR (1992) Changes in the tonotopic map of the dorsal cochlear nucleus following induction of cochlear lesions by exposure to intense sound. Hear Res 59:213–223.

Kapfer C, Seidl AH, Schweizer H, Grothe B (2002) Experience-dependent refinement of inhibitory inputs to auditory coincidence-detector neurons. Nat Neurosci 5:247–253.

Kiang NYS, Moxon EC (1972) Physiological considerations in artificial stimulation of the inner ear. Ann Otol Rhinol Laryngol 81:714–730.

Kiernan AE, Steel KP (2000) Mouse homologues for human deafness. Adv Otorhinolaryngol 56:233–243.

Kileny PR, Kemink JL (1987) Electrically evoked middle-latency auditory potentials in cochlear implant candidates. Arch Otolaryngol Head Neck Surg 113:1072–1077.

Kilgard MP, Merzenich MM (1998) Cortical map reorganization enabled by nucleus basalis activity. Science 279:1714–1718.

Kitzes LM, Semple MN (1985) Single unit responses in the inferior colliculus: different consequences of contralateral and ipsilateral stimulation. J Neurophysiol 53:1483–1498.

Kitzes LM, Kageyama GH, Semple MN, Kil J (1995) Development of ectopic projections from the ventral cochlear nucleus to the superior olivary complex induced by neonatal ablation of the contralateral cochlea. J Comp Neurol 353:341–363.

Klinke R, Kral A, Heid S, Tillein J, Hartmann R (1999) Recruitment of the auditory cortex in congenitally deaf cats by long-term cochlear electrostimulation. Science 285:1729–1733.

Klinke R, Hartmann R, Heid S, Tillein J, Kral A (2001) Plastic changes in the auditory cortex of congenitally deaf cats following cochlear implantation. Audiol Neurootol 6:203–206.

Knauth M, Hartmann R, Klinke R (1994) Discharge pattern in the auditory nerve evoked by vowel stimuli—a comparison between acoustical and electrical stimulation. Hear Res 74:247–258.

Knudsen EI (1999) Mechanisms of experience-dependent plasticity in the auditory localization pathway of the barn owl. J Comp Physiol 185:305–321.

König N, Pujol R, Marty R (1972) A laminar study of evoked potentials and unit responses in the auditory cortex of the postnatal cat. Brain Res 36:469–473.

Kotak VC, Sanes DH (1997) Deafferentation weakens excitatory synapses in the developing central auditory system. Eur J Neurosci 9:2340–2347.

Kral A (2000) Temporal code and speech recognition. Acta Oto-Laryngol 120:529–530.

Kral A, Hartmann R, Mortazavi D, Klinke R (1998) Spatial resolution of cochlear implants: the electrical field and excitation of auditory afferents. Hear Res 121:11–28.

Kral A, Tillein J, Hartmann R, Klinke R (1999) Monitoring of anaesthesia in neurophysiological experiments. Neuroreport 10:781–787.

Kral A, Hartmann R, Tillein J, Heid S, Klinke R (2000) Congenital auditory deprivation reduces synaptic activity within the auditory cortex in a layer-specific manner. Cereb Cortex 10:714–726.

Kral A, Hartmann R, Tillein J, Heid S, Klinke R (2001) Delayed maturation and sensitive periods in the auditory cortex. Audiol Neurootol 6:346–362.

Kral A, Hartmann R, Tillein J, Heid S, Klinke R (2002) Hearing after congenital deafness: central auditory plasticity and sensory deprivation. Cereb Cortex 12:797–807.

Kral A, Schröder J-H, Klinke R, Engel AK (2003) Absence of cross-modal reorganization in the primary auditory cortex of congenitally deaf cats. Exp Brain Res 153:605–613.

Kraus N, Micco AG, Koch DB, McGee T, et al. (1993) The mismatch negativity cortical evoked potential elicited by speech in cochlear-implant users. Hear Res 65:118–124.

Langner G, Schreiner CE (1988) Periodicity coding in the inferior colliculus of the cat. I. Neuronal mechanisms. J Neurophysiol 60:1799–1822.

Larsen SA, Kirchhoff TM (1992) Anatomical evidence of synaptic plasticity in the cochlear nuclei of white-deaf cats. Exp Neurol 115:151–157.

Lazeyras F, Boex C, Sigrist A, Seghier ML, et al. (2002) Functional MRI of auditory cortex activated by multisite electrical stimulation of the cochlea. Neuroimage 17:1010–1017.

Leake PA, Hradek GT, Snyder RL (1999) Chronic electrical stimulation by a cochlear implant promotes survival of spiral ganglion neurons after neonatal deafness. J Comp Neurol 412:543–562.

Leake PA, Snyder RL, Rebscher SJ, Moore CM, Vollmer M (2000) Plasticity in central representations in the inferior colliculus induced by chronic single- vs. two-channel electrical stimulation by a cochlear implant after neonatal deafness. Hear Res 147:221–241.

Leake-Jones PA, Rebscher SJ (1983) Cochlear pathology with chronically implanted scala tympani electrodes. In: Parkins CW, Anderson SW, eds. Cochlear Prostheses. An International Symposium. Ann N Y Acad Sci 405:203–223.

Leake-Jones PA, Vivion MC, O'Reilly BF, Merzenich MM (1982) Deaf animal models for studies of a multichannel cochlear prosthesis. Hear Res 8:225–246.

Lee DS, Lee JS, Oh SH, Kim SK, et al. (2001) Cross-modal plasticity and cochlear implants. Nature 409:149–150.

Liberman MC (1982) The cochlear frequency map for the cat: labeling auditory nerve fibers of known characteristic frequency. J Acoust Soc Am 72:1441–1449.

Liu X, McPhee G, Seldon HL, Clark GM (1997) Histological and physiological effects of the central auditory prosthesis: surface versus penetrating electrodes. Hear Res 114:264–274.

Lusted HS, Simmons FB (1988) Comparison of electrophonic and auditory-nerve electroneural responses. J Acoust Soc Am 83(2):657–661.

Lustig LR, Leake PA, Snyder RL, Rebscher SJ (1994) Changes in the cat cochlear nucleus following neonatal deafening and chronic intracochlear electrical stimulation. Hear Res 74:29–37.

Matsushima JI, Shepherd RK, Seldon HL, Xu SA, Clark GM (1991) Electrical stimulation of the auditory nerve in deaf kittens—effects on cochlear nucleus morphology. Hear Res 56:133–142.

McAnally KI, Clark GM (1994) Stimulation of residual hearing in the cat by pulsatile electrical stimulation of the cochlea. Acta Otolaryngol (Stockh) 114:366–372.

McCreery DB, Yuen TGH, Agnew WF, Bullara LA (1992) Stimulation with chronically implanted microelectrodes in the cochlear nucleus of the cat—histologic and physiologic effects. Hear Res 62:42–56.

McCreery DB, Yuen TGH, Agnew WF, Bullara LA (1994) Stimulus parameters affecting tissue injury during microstimulation in the cochlear nucleus of the cat. Hear Res 77:105–115.

McCreery DB, Shannon RV, Moore JK, Chatterjee M (1998) Accessing the tonotopic organization of the ventral cochlear nucleus by intranuclear microstimulation. IEEE Trans Rehabil Eng 6:391–399.

McGee TJ, Ozdamar O, Kraus N (1983) Auditory middle latency responses in the guinea pig. Am J Otolaryngol 4:116–122.

McMullen NT, Glaser EM (1988) Auditory cortical responses to neonatal deafening: pyramidal neuron spine loss without changes in growth or orientation. Exp Brain Res 72:195–200.

McMullen NT, Goldberger B, Suter CM, Glaser EM (1988) Neonatal deafening alters nonpyramidal dendrite orientation in auditory cortex: a computer microscope study in the rabbit. J Comp Neurol 267:92–106.

Merzenich MM, Reid MD (1974) Representation of the cochlea within the inferior colliculus of the cat. Brain Res 77:397–415.

Merzenich MM, Knight PL, Roth GL (1975) Representation of cochlea within primary auditory cortex in the cat. J Neurophysiol 38:231–249.

Micco AG, Kraus N, Koch DB, McGee TJ, et al. (1995) Speech-evoked cognitive P300 potentials in cochlear implant recipients. Am J Otol 16:514–520.

Middlebrooks JC, Dykes RW, Merzenich MM (1980) Binaural response-specific bands in primary auditory cortex (AI) of the cat: topographical organization orthogonal to isofrequency contours. Brain Res 181:31–48.

Miller AL, Morris DJ, Pfingst BE (2000) Effects of time after deafening and implantation on guinea pig electrical detection thresholds. Hear Res 144:175–186.

Miller AL, Arenberg JG, Middlebrooks JC, Pfingst BE (2001) Cochlear implant thresholds: comparison of middle latency responses with psychophysical and cortical-spike-activity thresholds. Hear Res 152:55–66.

Miller CA, Woodruff KE, Pfingst BE (1995a) Functional responses from guinea pigs with cochlear implants. 1. Electrophysiological and psychophysical measures. Hear Res 92:85–99.

Miller CA, Faulkner MJ, Pfingst BE (1995b) Functional responses from guinea pigs with cochlear implants. 2. Changes in electrophysiological and psychophysical measures over time. Hear Res 92:100–111.

Mitani A, Shimokouchi M (1985) Neuronal connections in the primary auditory cortex: an electrophysiological study in the cat. J Comp Neurol 235:417–429.

Mitani A, Shimokouchi M, Nomura S (1983) Effects of stimulation of the primary auditory cortex upon colliculogeniculate neurons in the inferior colliculus of the cat. Neurosci Lett 42:185–189.

Mitani A, Shimokouchi M, Ftoh K, Nomura S, Mizuno N, Kudo M (1985) Morphology and laminar organization of electrophysiologically identified neurons in the primary auditory cortex in the cat. J Comp Neurol 235:430–430.

Mitzdorf U (1985) Current source-density method and application in cat cerebral cortex: investigation of evoked potentials and EEG phenomena. Physiol Rev 65:37–100.

Miyamoto RT (1986) Electrically evoked potentials in cochlear implant subjects. Laryngoscope 96:178–185.

Moore CM, Vollmer M, Leake PA, Snyder RL, Rebscher SJ (2002) The effects of chronic intracochlear electrical stimulation on inferior colliculus spatial representation in adult deafened cats. Hear Res 164:82–96.

Moore DR (1990) Auditory brainstem of the ferret: bilateral cochlear lesions in infancy do not affect the number of neurons projecting from the cochlear nucleus to the inferior colliculus. Brain Res Dev Brain Res 54:125–130.

Moore DR (1994) Auditory brainstem of the ferret—long survival cochlear removal progressively changes projections from the cochlear nucleus to the inferior colliculus. J Comp Neurol 339:301–310.

Moore DR, Irvine DR (1979) The development of some peripheral and central auditory responses in the neonatal cat. Brain Res 163:49–59.

Moore DR, Kitzes LM (1985) Projections from the cochlear nucleus to the inferior colliculus in normal and neonatally cochlea-ablated gerbils. J Comp Neurol 240:180–195.

Moore DR, King AJ, McAlpine D, Martin RL, Hutchings ME (1993) Functional consequences of neonatal unilateral cochlear removal. Prog Brain Res 97:127–133.

Moore JK (2002) Maturation of human auditory cortex: implications for speech perception. Ann Otol Rhinol Laryngol Suppl 189:7–10.

Moore JK, Guan YL (2001) Cytoarchitectural and axonal maturation in human auditory cortex. J Assoc Res Otolaryngol 2:297–311.

Moore JK, Perazzo LM, Braun A (1995) Time course of axonal myelination in the human brainstem auditory pathway. Hear Res 87:21–31.

Moore JK, Guan YL, Shi SR (1997a) Axogenesis in the human fetal auditory system, demonstrated by neurofilament immunohistochemistry. Anat Embryol (Berl) 195:15–30.

Moore JK, Niparko JK, Perazzo LM, Miller MR, Linthicum FH (1997b) Effect of adult-onset deafness on the human central auditory system. Ann Otol Rhinol Laryngol 106:385–390.

Moore JK, Guan YL, Shi SR (1998) MAP2 expression in developing dendrites of human brainstem auditory neurons. J Chem Neuroanat 16:1–15.

Morosan P, Rademacher J, Schleicher A, Amunts K, Schormann T, Zilles K (2001) Human primary auditory cortex: cytoarchitectonic subdivisions and mapping into a spatial reference system. Neuroimage 13:684–701.

Mostafapour SP, Cochran SL, DelPuerto NM, Rubel EW (2000) Patterns of cell death in mouse anteroventral cochlear nucleus neurons after unilateral cochlea removal. J Comp Neurol 426:561–571.

Mostafapour SP, Del Puerto NM, Rubel EW (2002) bcl-2 overexpression eliminates deprivation-induced cell death of brainstem auditory neurons. J Neurosci 22:4670–4674.

Naito Y, Okazawa H, Honjo I, Hirano S, et al. (1995) Cortical activation with sound stimulation in cochlear implant users demonstrated by positron emission tomography. Cognitive Brain Res 2:207–214.

Naito Y, Hirano S, Honjo I, Okazawa H, et al. (1997) Sound-induced activation of auditory cortices in cochlear implant users with post- and prelingual deafness demonstrated by positron emission tomography. Acta Otolaryngol 117:490–496.

Niparko JK, Finger PA (1997) Cochlear nucleus cell size changes in the dalmatian: model of congenital deafness. Otolaryngol Head Neck Surg 117:229–235.

Nishimura H, Hashikawa K, Doi K, Iwaki T, et al. (1999) Sign language "heard" in the auditory cortex. Nature 397:116.

Nixon CW, von Gierke HE (1959) Experiments on bone-conduction thresholds in a free sound field. J Acoust Soc Am 31:1121–1125.

Nordeen KW, Killackey HP, Kitzes LM (1983) Ascending projections to the inferior colliculus following unilateral cochlear ablation in the neonatal gerbil, Meriones unguiculatus. J Comp Neurol 214:144–153.

Ohl FW, Scheich H (1997) Learning-induced dynamic receptive field changes in primary auditory cortex of the unanaesthetized Mongolian gerbil. J Comp Physiol A 181:685–696.

O'Kusky JR (1985) Synapse elimination in the developing visual cortex: a morphometric analysis in normal and dark-reared cats. Brain Res 354:81–91.

O'Kusky J, Colonnier M (1982a) Postnatal changes in the number of astrocytes, oligodendrocytes, and microglia in the visual cortex (area 17) of the macaque monkey: a stereological analysis in normal and monocularly deprived animals. J Comp Neurol 210:307–315.

O'Kusky J, Colonnier M (1982b) Postnatal changes in the number of neurons and synapses in the visual cortex (area 17) of the macaque monkey: a stereological analysis in normal and monocularly deprived animals. J Comp Neurol 210:291–306.

O'Kusky J, Colonnier M (1982c) A laminar analysis of the number of neurons, glia, and synapses in the adult cortex (area 17) of adult macaque monkeys. J Comp Neurol 210:278–290.

Otte J, Schuknecht HF, Kerr AG (1978) Ganglion cell populations in normal and pathological human cochleae. Implications for cochlear implantation. Laryngoscope 88:1231–1246.

Otto SR, Brackmann DE, Hitselberger WE, Shannon RV, Kuchta J (2002) Multichannel auditory brainstem implant: update on performance in 61 patients. J Neurosurg 96:1063–1071.

Pandya DN, Sanides F (1973) Architectonic parcellation of the temporal operculum in rhesus monkey and its projection pattern. Z Anat Entwicklungsgesch 139:127–161.

Pandya DN, Yeterian EH (1985) Architecture and connections of cortical association areas. In: Peters A, Jones EG, eds. Cerebral Cortex, vol. 4: Association and Auditory Cortices. New York, London: Plenum Press, pp. 3–88.

Pantev C, Lutkenhoner B (2000) Magnetoencephalographic studies of functional organization and plasticity of the human auditory cortex. J Clin Neurophysiol 17:130–142.

Paolini AG, Clark GM (1998) Intracellular responses of the rat anteroventral cochlear nucleus to intracochlear electrical stimulation. Brain Res Bull 46:317–327.

Parkins CW, Colombo J (1987) Auditory-nerve single-neuron thresholds to electrical stimulation from scala tympani electrodes. Hear Res 31:267–285.

Parving A, Christensen B, Salomon G, Pedersen CB, Friberg L (1995) Regional cerebral activation during auditory stimulation in patients with cochlear implants. Arch Otolaryngol Head Neck Surg 121:438–444.

Pasic TR, Rubel EW (1989) Rapid changes in cochlear nucleus cell size following blockade of auditory nerve electrical activity in gerbils. J Comp Neurol 283:474–480.

Paus T, Zijdenbos A, Worsley K, Collins DL, et al. (1999) Structural maturation of neural pathways in children and adolescents: in vivo study. Science 283:1908–1911.

Petitto LA, Zatorre RJ, Gauna K, Nikelski EJ, Dostie D, Evans AC (2000) Speech-like cerebral activity in profoundly deaf people processing signed languages: implications for the neural basis of human language. Proc Natl Acad Sci USA 97:13961–13966.

Petitto LA, Katerelos M, Levy BG, Gauna K, Tetreault K, Ferraro V (2001) Bilingual signed and spoken language acquisition from birth: implications for the mechanisms underlying early bilingual language acquisition. J Child Lang 28:453–496.

Pfingst BE (1988) Comparisons of psychophysical and neurophysiological studies of cochlear implants. Hear Res 34:243–252.

Pfingst BE, Morris DJ (1993) Stimulus features affecting psychophysical detection thresholds for electrical stimulation of the cochlea. 2. Frequency and interpulse interval. J Acoust Soc Am 94:1287–1294.

Pfingst BE, Dehaan DR, Holloway LA (1991) Stimulus features affecting psychophysical detection thresholds for electrical stimulation of the cochlea. 1. Phase duration and stimulus duration. J Acoust Soc Am 90:1857–1866.

Phillips DP, Brugge JF (1985) Progress in neurophysiology of sound localization. Annu Rev Psychol 36:245–274.

Ponton CW, Eggermont JJ (2001) Of kittens and kids: altered cortical maturation following profound deafness and cochlear implant use. Audiol Neurootol 6:363–380.

Ponton CW, Eggermont JJ, Coupland SG, Winkelaar R (1992) Frequency-specific maturation of the 8th nerve and brainstem auditory pathway—evidence from derived auditory brainstem responses (ABRs). J Acoust Soc Am 91:1576–1586.

Ponton CW, Don M, Eggermont JJ, Waring MD, Masuda A (1996a) Maturation of human cortical auditory function: differences between normal-hearing children and children with cochlear implants. Ear Hear 17:430–437.

Ponton CW, Don M, Eggermont JJ, Waring MD, Kwong B, Masuda A (1996b) Auditory system plasticity in children after long periods of complete deafness. Neuroreport 8:61–65.

Ponton CW, Eggermont JJ, Kwong B, Don M (2000a) Maturation of human central auditory system activity: evidence from multi-channel evoked potentials. Clin Neurophysiol 111:220–236.

Ponton CW, Eggermont JJ, Don M, Waring MD, et al. (2000b) Maturation of the mismatch negativity: effects of profound deafness and cochlear implant use. Audiol Neurootol 5:167–185.

Popelar J, Hartmann R, Syka J, Klinke R (1995) Middle latency responses to acoustical and electrical stimulation of the cochlea in cats. Hear Res 92:63–77.

Popper AN, Fay RR (1992) The Mammalian Auditory Pathway: Neurophysiology. New York, Berlin, Heidelberg: Springer-Verlag.

Price CJ (2000) The anatomy of language: contributions from functional neuroimaging. J Anat 197 (pt 3):335–359.

Raggio MW, Schreiner CE (1994) Neuronal responses in cat primary auditory cortex to electrical cochlear stimulation. I. Intensity dependence of firing rate and response latency. J Neurophysiol 72:2334–2359.

Raggio MW, Schreiner CE (1999) Neuronal responses in cat primary auditory cortex to electrical cochlear stimulation. III. Activation patterns in short- and long-term deafness. J Neurophysiol 82:3506–3526.

Rajan R, Irvine DR (1998) Absence of plasticity of the frequency map in dorsal cochlear nucleus of adult cats after unilateral partial cochlear lesions. J Comp Neurol 399:35–46.

Rajan R, Irvine DR, Wise LZ, Heil P (1993) Effect of unilateral partial cochlear lesions in adult cats on the representation of lesioned and unlesioned cochleas in primary auditory cortex. J Comp Neurol 338:17–49.

Ranck JB (1975) Which elements are excited in electrical stimulation in mammalian central nervous system: a review. Brain Res 98:417–440.

Raphael Y, Fein A, Nebel L (1983) Transplacental kanamycin ototoxicity in the guinea pig. Arch Otorhinolaryngol 238:45–51.

Rauschecker JP (1999) Auditory cortical plasticity: a comparison with other sensory systems. Trends Neurosci 22:74–80.

Rauschecker JP, Tian B, Pons T, Mishkin M (1997) Serial and parallel processing in rhesus monkey auditory cortex. J Comp Neurol 382:89–103.

Reale RA, Brugge JF (2000) Directional sensitivity of neurons in the primary auditory (AI) cortex of the cat to successive sounds ordered in time and space. J Neurophysiol 84:435–450.

Reale RA, Imig TJ (1980) Tonotopic organization in auditory cortex of the cat. J Comp Neurol 192:265–291.

Reale RA, Brugge JF, Chan JCK (1987) Maps of auditory cortex in cats reared after unilateral cochlear ablation in the neonatal period. Dev Brain Res 34:281–290.

Recanzone GH, Schreiner CE, Merzenich MM (1993) Plasticity in the frequency representation of primary auditory cortex following discrimination training in adult owl monkeys. J Neurosci 13:87–103.

Robertson D, Irvine DRF (1989) Plasticity of frequency organization in auditory cortex of guinea pigs with partial unilateral deafness. J Comp Neurol 282:456.

Rogowski BA, Feng SA (1981) Normal postnatal development of medial superior olivary neurons in the Albino rat: a Golgi and Nissl study. J Comp Neurol 196:85–97.

Rosahl SK, Mark G, Herzog M, Pantazis C, et al. (2001) Far-field responses to stimulation of the cochlear nucleus by microsurgically placed penetrating and surface electrodes in the cat. J Neurosurg 95:845–852.

Russell FA, Moore DR (1995) Afferent reorganisation within the superior olivary complex of the gerbil: development and induction by neonatal, unilateral cochlear removal. J Comp Neurol 352:607–625.

Russell FA, Moore DR (1999) Effects of unilateral cochlear removal on dendrites in the gerbil medial superior olivary nucleus. Eur J Neurosci 11:1379–1390.

Ryan AF, Miller JM, Wang ZX, Woolf NK (1990) Spatial distribution of neural activity evoked by electrical stimulation of the cochlea. Hear Res 50:57–70.

Ryugo DK, Pongstaporn T, Huchton DM, Niparko JK (1997) Ultrastructural analysis of primary endings in deaf white cats: morphologic alterations in endbulbs of Held. J Comp Neurol 385:230–244.

Ryugo DK, Rosenbaum BT, Kim PJ, Niparko JK, Saada AA (1998) Single unit recordings in the auditory nerve of congenitally deaf white cats: morphological correlates in the cochlea and cochlear nucleus. J Comp Neurol 397:532–548.

Saada AA, Niparko JK, Ryugo DK (1996) Morphological changes in the cochlear nucleus of congenitally deaf white cats. Brain Res 736:315–328.

Sachs MB (1984) Neural coding of complex sounds. Annu Rev Physiol 46:261.

Saito H, Miller JM, Pfingst BE, Altschuler RA (1999) Fos-like immunoreactivity in the auditory brainstem evoked by bipolar intracochlear electrical stimulation: effects of current level and pulse duration. Neuroscience 91:139–161.

Sasaki CT, Kauer JS, Babitz L (1980) Differential [14C]2-deoxyglucose uptake after deafferentation of the mammalian auditory pathway—a model for examining tinnitus. Brain Res 194:511–516.

Schoop VM, Gardziella S, Muller CM (1997) Critical period-dependent reduction of the permissiveness of cat visual cortex tissue for neuronal adhesion and neurite growth. Eur J Neurosci 9:1911–1922.

Schreiner CE, Langner G (1988) Periodicity coding in the inferior colliculus of the cat. II. Topographical organization. J Neurophysiol 60:1823–1840.

Schreiner CE, Raggio MW (1996) Neuronal responses in cat primary auditory cortex to electrical cochlear stimulation. II. Repetition rate coding. J Neurophysiol 75:1283–1300.

Schwartz DR, Schacht J, Miller JM, Frey K, Altschuler RA (1993) Chronic electrical stimulation reverses deafness-related depression of electrically evoked 2-deoxyglucose activity in the guinea pig inferior colliculus. Hear Res 70:243–249.

Seldon HL, Kawano A, Clark GM (1996) Does age at cochlear implantation affect the distribution of 2-deoxyglucose label in cat inferior colliculus? Hear Res 95:108–119.

Shallop JK, Beiter AL, Goin DW, Mischke RE (1990) Electrically evoked auditory brainstem responses (EABR) and middle latency responses (EMLR) obtained from patients with the nucleus multichannel cochlear implant. Ear Hear 11:5–15.

Shallop JK, VanDyke L, Goin DW, Mischke RE (1991) Prediction of behavioral threshold and comfort values for Nucleus 22-channel implant patients from electrical auditory brainstem response test results. Ann Otol Rhinol Laryngol 100:896–898.

Shannon RV, Otto SR (1990) Psychophysical measures from electrical stimulation of the human cochlear nucleus. Hear Res 47:159–168.

Shannon RV, Fayad J, Moore J, Lo WW, et al. (1993) Auditory brainstem implant: II. Postsurgical issues and performance. Otolaryngol Head Neck Surg 108:634–642.

Sharma A, Kraus N, McGee TJ, Nicol TG (1997) Developmental changes in P1 and N1 central auditory responses elicited by consonant-vowel syllables. Electroencephalogr Clin Neurophysiol 104:540–545.

Sharma A, Dorman M, Spahr A, Todd NW (2002a) Early cochlear implantation in children allows normal development of central auditory pathways. Ann Otol Rhinol Laryngol Suppl 189:38–41.

Sharma A, Dorman And MF, Spahr AJ (2002b) A sensitive period for the development of the central auditory system in children with cochlear implants: implications for age of implantation. Ear Hear 23:532–539.

Sharma A, Dorman MF, Spahr AJ (2002c) Rapid development of cortical auditory evoked potentials after early cochlear implantation. Neuroreport 13:1365–1368.

Shepherd RK, Hardie NA (2001) Deafness-induced changes in the auditory pathway: implications for cochlear implants. Audiol Neurootol 6:305–318.

Shepherd RK, Martin RL (1995) Onset of ototoxicity in the cat is related to onset of auditory function. Hear Res 92:131–142.

Shepherd RK, Hatsushika S, Clark GM (1993) Electrical stimulation of the auditory nerve—the effect of electrode position on neural excitation. Hear Res 66:108–120.

Shepherd RK, Hartmann R, Heid S, Hardie N, Klinke R (1997) The central auditory system and auditory deprivation: experience with cochlear implants in the congenitally deaf. Acta Otolaryngol Suppl 532:28–33.

Shepherd RK, Baxi JH, Hardie NA (1999) Response of inferior colliculus neurons to electrical stimulation of the auditory nerve in neonatally deafened cats. J Neurophysiol 82:1363–1380.

Shore SE, Wiler JA, Anderson DJ (1990) Evoked vertex and inferior colliculus responses to electrical stimulation of the cochlear nucleus. Ann Otol Rhinol Laryngol 99:571–576.

Silverman MS, Clopton BM (1977) Plasticity of binaural interaction. 1. Effect of early auditory deprivation. J Neurophysiol 40:1266–1274.

Singer W (1995) Development and plasticity of cortical processing architectures. Science 270:758–764.

Smith DW, Finley CC, VandenHonert C, Olszyk VB, Konrad KEM (1994) Behavioral and electrophysiological responses to electrical stimulation in the cat. 1. Absolute thresholds. Hear Res 81:1–10.

Snyder RL, Rebscher SJ, Cao K, Leake PA, Kelly K (1990) Chronic intracochlear electrical stimulation in the neonatally deafened cat. 1. Expansion of central representation. Hear Res 50:7–34.

Snyder RL, Rebscher SJ, Leake PA, Kelly K, Cao K (1991) Chronic intracochlear electrical stimulation in the neonatally deafened cat. 2. Temporal properties of neurons in the inferior colliculus. Hear Res 56:246–264.

Snyder R, Leake P, Rebscher S, Beitel R (1995) Temporal resolution of neurons in cat inferior colliculus to intracochlear electrical stimulation: effects of neonatal deafening and chronic stimulation. J Neurophysiol 73:449–467.

Snyder RL, Vollmer M, Moore CM, Rebscher SJ, Leake PA, Beitel RE (2000) Responses of inferior colliculus neurons to amplitude-modulated intracochlear electrical pulses in deaf cats. J Neurophysiol 84:166–183.

Stanton SG, Harrison RV (2000) Projections from the medial geniculate body to primary auditory cortex in neonatally deafened cats. J Comp Neurol 426:117–129.

Steel KP, Bock GR (1980) The nature of inherited deafness in deafness mice. Nature 288:159–161.

Steriade M (1999) Coherent oscillations and short-term plasticity in corticothalamic networks. Trends Neurosci 22:337–345.

Stypulkowski PH, van den Honert C (1984) Physiological properties of the electrically stimulated auditory nerve. I. Compound action potential recordings. Hear Res 14:205–223.

Syka J (2002) Plastic changes in the central auditory system after hearing loss, restoration of function, and during learning. Physiol Rev 82:601–636.

Syka J, Rybalko N (2000) Threshold shifts and enhancement of cortical evoked responses after noise exposure in rats. Hear Res 139:59–68.

Taniguchi I, Horikawa J, Hosokawa Y, Nasu M (1997) Optical imaging of neural activity in auditory cortex induced by intracochlear electrical stimulation. Acta Oto-Laryngol 532:83–88.

Tierney TS, Russell FA, Moore DR (1997) Susceptibility of developing cochlear nucleus neurons to deafferentation-induced death abruptly ends just before the onset of hearing. J Comp Neurol 378:295–306.

Trune DR (1982) Influence of neonatal cochlear removal on the development of mouse cochlear nucleus: number, size, and density of its neurons. J Comp Neurol 209:409–424.

Truy E, Gallego S, Chanal JM, Collet L, Morgon A (1998) Correlation between electrical auditory brainstem response and perceptual thresholds in Digisonic cochlear implant users. Laryngoscope 108:554–559.

Tyler RS, Fryauf-Bertschy H, Kelsay DM, Gantz BJ, Woodworth GP, Parkinson A (1997) Speech perception by prelingually deaf children using cochlear implants. Otolaryngol Head Neck Surg 117:180–187.

Vale C, Sanes DH (2002) The effect of bilateral deafness on excitatory and inhibitory synaptic strength in the inferior colliculus. Eur J Neurosci 16:2394–2404.

van den Honert C, Stypulkowski PH (1984) Physiological properties of the electrically stimulated auditory nerve. II. Single fiber recordings. Hear Res 14:225–244.

Vasama JP, Linthicum FH Jr (2000) Idiopathic sudden sensorineural hearing loss: temporal bone histopathologic study. Ann Otol Rhinol Laryngol 109:527–532.

Vischer MW, Bajo VM, Zhang JS, Calciati E, Haenggeli CA, Rouiller EM (1997) Single unit activity in the inferior colliculus of the rat elicited by electrical stimulation of the cochlea. Audiology 36:202–227.

Vollmer M, Snyder RL, Leake PA, Beitel RE, Moore CM, Rebscher SJ (1999) Temporal properties of chronic cochlear electrical stimulation determine temporal resolution of neurons in cat inferior colliculus. J Neurophysiol 82:2883–2902.

Vollmer M, Beitel RE, Snyder RL (2001) Auditory detection and discrimination in deaf cats: psychophysical and neural thresholds for intracochlear electrical signals. J Neurophysiol 86:2330–2343.

Wallhäuser E, Scheich H (1987) Auditory imprinting leads to differential 2-deoxyglucose uptake and dendritic spine loss in the chick rostral forebrain. Dev Brain Res 31:29–44.

Waring MD (1998) Refractory properties of auditory brainstem responses evoked by electrical stimulation of human cochlear nucleus: evidence of neural generators. Evoked Potential 108:331–344.

Warren RM (1970) Perceptual restoration of missing speech sounds. Science 167:392–393.

Webster DB (1992) An overview of mammalian auditory pathways with an emphasis on humans. In: Webster DB, Popper AN, Fay RR, eds. Springer Handbook of Auditory System: Neuroanatomy. New York: Springer-Verlag, pp. 1–21.

Webster DB, Popper AN, Fay RR (1992) The Mammalian Auditory Pathway: Neuroanatomy. New York: Springer.

Weinberger NM (1997) Learning-induced receptive field plasticity in the primary auditory cortex. Semin Neurosci 9: 59–67.

Weinberger NM (1998) Physiological memory in primary auditory cortex: characteristics and mechanisms. Neurobiol Learn Memory 70:226–251.

Willott JF, Henry KR (1974) Auditory evoked potentials: developmental changes of threshold and amplitude following early acoustic trauma. J Comp Physiol Psychol 86:1–7.

Willott JF, Lu SM (1982) Noise induced hearing loss can alter neural coding and increase excitability in the central nervous system. Science 216:1331–1332.

Winguth SD, Winer JA (1986) Corticocortical connections of cat primary auditory cortex (AI): laminar organization and identification of supragranular neurons projecting to area AII. J Comp Neurol 248:36–56.

Wong D, Kelly JP (1981) Differentially projecting cells in individual layers of the auditory cortex: a double-labeling study. Brain Res 230:362–366.

Woolf NK, Sharp FR, Davidson TM, Ryan AF (1983) Cochlear and middle ear effects on metabolism in the central auditory pathway during silence: A2-deoxyglucose study. Brain Res 274:119–127.

Woolsey CN, Walzl EM (1942) Topical projections of nerve fibers from local regions of the cochlea to the cerebral cortex. Bull Johns Hopkins Hosp 71: 315–343.

Xu S-A, Shepherd RK, Chen Y, Clark GM (1993) Profound hearing loss in the cat following the single co-administration of kanamycin and ethacrynic acid. Hear Res 70:205–215.

Yakovlev PL, Lecour A-R (1967) The myelogenic cycles of regional maturation of the brain. In: Minkowski A, ed. Regional development of the brain in early life. Oxford: Blackwell Scientific, pp. 3–70.

Zeng FG, Galvin JJ 3rd (1999) Amplitude mapping and phoneme recognition in cochlear implant listeners. Ear Hear 20:60–74.

Zeng FG, Shannon RV (1999) Psychophysical laws revealed by electric hearing. NeuroReport 10:1931–1935.

Zhang LI, Bao S, Merzenich MM (2002) Disruption of primary auditory cortex by synchronous auditory inputs during a critical period. Proc Natl Acad Sci USA 99:2309–2314.

7
Psychophysics and Electrical Stimulation

Colette M. McKay

1. Introduction

Psychophysics is the study of sensory perception in response to physical stimuli. In the context of electrical stimulation with a cochlear implant, it is the study of the effect of the electrical stimulation patterns on perceived sound qualities. Perceptual qualities that can be measured with psychophysical techniques range from simple ones such as pitch, timbre, and loudness, to more complex percepts such as the categorical perception of the phonemes that constitute a speech signal. The basic psychophysical abilities that any sensory system must facilitate can be categorized into four types: detection of a stimulus; discrimination of two different stimuli; identification of a stimulus; and the scaling of a stimulus (which involves being able to rank similar stimuli along a particular dimension, for example loudness). In the sense of hearing, all of these psychophysical abilities are important for the accurate perception of the sounds that are significant in our everyday lives. In addition, we must be able to perceptually segregate one sound from another when there is more than one sound source.

Psychophysical research with cochlear implants has two main goals. The first is to understand the peripheral and central mechanisms underlying auditory processing. In cochlear implants, electrical stimulation bypasses the normal peripheral processing of the external and middle ears, and the subsequent processing of cochlear mechanics and hair-cell transduction, and thus allows peripheral and central processing effects to be separated. By measuring how the sound percept changes when the electrical stimulation pattern is changed, it is possible to test hypotheses for how the central auditory system processes peripheral neural inputs. In addition, cochlear implants provide an opportunity to extend similar research in acoustics, by allowing stimulus patterns to be created at the peripheral neural input stage that would be difficult with acoustic stimulation (e.g., spatial and temporal cues for pitch perception can be separately manipulated more easily with electrical stimulation). Finally, we can study the effects of pathology on central auditory processing, whether due to degraded neural input or abnormal central processing.

The second and more immediately practical goal of cochlear implant psychophysics is to optimally encode acoustic information into the electrical signal. In other words, we need to find out how to manipulate the pattern of electrical stimulation so that an implantee can perceive the same information about the sound as a normal-hearing person would perceive. As well as understanding the simple effects of stimulus parameters on loudness and pitch, we need to understand how to achieve more complex goals such as understanding speech, being able to separate a speech signal from another sound, perceiving the harmonic relationships present in music, localizing sound sources, and many other auditory achievements that normal-hearing people take for granted. Much of the current psychophysical research is directed at finding more effective ways of transforming complex acoustic sounds into electrical stimulation patterns so that complex listening tasks such as speech perception are facilitated. Psychophysical research is thus an important component of speech perception research and speech processing research.

In the following section, the simple percepts evoked by electrical stimulus patterns on a single electrode are described, and in later sections the discussion is extended to more complex multielectrode electrical stimuli and binaural electrical stimuli. Some common psychophysical test procedures are also introduced and briefly described at appropriate places throughout the chapter (mostly in section 2).

2. Perception of Stimuli Using a Single Electrode Position

2.1 Spatial and Temporal Frequency Analysis in Normal Hearing

One of the main functions of the normal cochlea is to separate the different acoustic frequencies in a sound along a place dimension. For example, two pure tones of different frequencies lead to a maximum movement of the basilar membrane in the cochlea at two different distances from the stapes. The hair cells located near the points of maximum movement will be maximally stimulated, and the acoustic nerve fibers that innervate these hair cells will be activated. Thus, the two tones activate two distinct (but sometimes overlapping) sets of nerve fibers, which transmit information to the central auditory pathways from two distinct places in the cochlea. The fact that different frequencies activate different nerve fibers, and that different places within a neural structure (corresponding to different groups of nerve cells) will be activated by different frequencies is called *tonotopic organization*. This tonotopic organization is preserved in all stages of the auditory pathway from the cochlea to the auditory cortex (see Leake and Rebscher, Chapter 4; Hartmann and Kral, Chapter 6).

The frequency of a tone is also represented by the temporal pattern of activity in auditory nerve fibers. When a person with normal hearing listens to a pure tone, the basilar membrane in the cochlea vibrates with the same frequency as the sound pressure waves. The acoustic nerve fibers that are activated by the hair cells thus tend to fire at a particular phase of the basilar membrane movement, provided that the frequency is below about 4 kHz. This phenomenon of normal neural behavior is called *phase locking* (Abbas and Miller, Chapter 5). Although any particular neuron does not usually fire on every cycle of basilar membrane movement, it always fires near a particular phase of the cycle. When the outputs of a group of activated nerve fibers are recorded, these outputs cluster in time, resulting in a temporal neural firing pattern that reflects the frequency of the incoming sound wave. Thus, there are two different ways that the frequency of an incoming sound wave is coded in the peripheral part of the normal auditory system: as spatial information (What part of the cochlea is excited?); and as temporal information (What timing patterns do the firing nerve fibers exhibit?).

It is beyond the scope of this chapter to discuss the various theories that relate pitch perception to spatial or temporal frequency coding in normal hearing. It suffices here to say that over most of the audible frequency range both aspects of auditory processing appear to be used, with the relative importance of the two types of processing depending on the signal and listening tasks. At frequencies below 50 Hz, the whole basilar membrane is oscillating up and down in phase, with no clear point of maximum amplitude, and so place coding is unlikely to play a role for these low frequencies. In contrast, for frequencies above about 4000 to 5000 Hz, the phase locking of nerve fibers to the input frequency is no longer possible, and so temporal coding is unlikely at frequencies higher than 5000 Hz.

The pitch of a pure tone increases as the frequency increases. For frequencies between 50 and 4000 Hz, this pitch change can be related to a shift to more basal positions and/or a decrease in period of temporal patterns in the neural activity. Of course, most ordinary environmental sounds are not pure tones, but consist of many pure-tone and noise components. For these more complex sounds, there is not the same simple relationship between the frequency of the sound and the pitch perceived by the listener. However, components of a complex sound that differ in frequency activate different places along the cochlea, and temporal analysis of the neural activity at these different cochlear places is carried out. The function of the inner ear, to separate different frequency components for analysis, is critical to our ability to distinguish and identify different sounds in the environment. *Frequency resolution* is our ability to perceptually separate the frequency components within a complex sound, and is discussed in section 3.

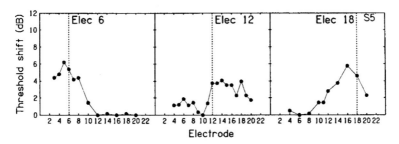

FIGURE 7.1. Forward-masking patterns from masking stimuli on three different electrodes (6, 12, and 18). The vertical axes represent the difference between the masked and unmasked thresholds (in dB of pulse duration change) for each probe stimulus. The pattern of probe-stimulus threshold shift across the electrode array is quite different for the three cases. (Data adapted from Cohen et al. 1996a.)

2.2 Electrode Position and Place Pitch

When a person has a profound cochlear hearing loss, there are too few remaining hair cells to translate the movement of the basilar membrane into place-specific nerve activity. The important function of the cochlea, to distinguish and separate different frequencies in a sound, is lost. With a multiple-electrode cochlear implant, we can activate neurons at different distances along the cochlea by stimulating at different electrodes positions (see Abbas and Miller, Chapter 5; Hartmann and Kral, Chapter 6).

Psychophysically, place-specific patterns of neural excitation can be inferred from *forward-masking patterns* (Cohen et al. 1996a; Chatterjee and Shannon 1998). In a forward-masking measurement, a masking stimulus (for example, a pulse train on a particular electrode) is immediately followed (within about 100 microseconds) by a short-duration probe stimulus. The detection threshold of the probe stimulus is measured with and without the masking stimulus, and the threshold shift caused by the masker is compared for a series of probes on all available electrode positions. The pattern of threshold shift across electrode position is thought to be related to the excitation pattern produced by the masker, although the exact relationship between excitation patterns and forward-masking patterns is still a matter for research. Figure 7.1 shows three forward-masking patterns produced by masking electrodes at different cochlear locations. It can be seen that the masking patterns are quite distinct, supporting the notion that the three electrodes activate different sets of neurons.

Fortunately, due to the fact that the surviving nerve fibers in implantees are tonotopically organized, the activation of distinct neural populations by different electrodes usually leads to a place-pitch phenomenon analogous to that in normal hearing. If the electrodes near the base of the cochlea are

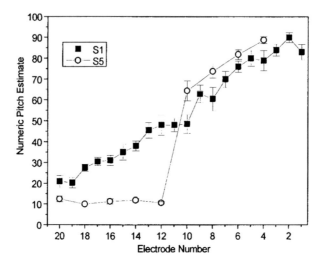

FIGURE 7.2. Data for two subjects in a pitch-scaling experiment, in which the subjects were asked to provide a numerical value that represented the pitch. The data of S1 show the usual pitch progression from high to low as the electrode position moves from basal to apical, but the data of S5 show a broad area in the apical region where pitch does not change with electrode position. (Adapted from Cohen et al. 1996b.)

activated, the person hears a high-pitched sound. As the electrode position is moved further down the cochlea toward the apex, the person hears progressively lower-pitched sounds.

Figure 7.2 shows the result of a *pitch-scaling* experiment with two listeners (S1 and S5) who are users of a multiple-electrode implant. In a pitch-scaling task, the person is asked to listen to a set of sounds, and then assign a number to each one that represents how high the pitch sounded. The pattern of pitch change with electrode position seen for S1 is typical of that seen among implantees, and is also the pattern expected from place-pitch theories. That is, as electrode position moves from apical to basal, the estimated pitch of the resultant sound increases systematically from low pitched to high pitched. However, this pattern of pitch change with electrode position is not always as ideal as for S1. A less typical, but not rare, case is shown in the data of S5, who gave pitch estimates that did not consistently increase when electrode position was moved in the basal direction. Occasionally, place-pitch reversals may occur in which more apical electrodes lead to higher, instead of lower, pitch (Nelson et al. 1995; Collins et al. 1997). There are many reasons why the pitch may not change as consistently as desired with the position of the electrodes. These include effects attributable to the electrode array itself (such as possible kinks in the array or unusual positioning relative to the cochlear structures) and effects attributable to the pathology of deafness (such as a limited quantity of surviving nerve cells, a nonuniform distribution of these cells, or other pathological processes such as new bone growth within the cochlea).

Another way of measuring the pitch change with electrode position is with a *pitch-ranking task*. In this case, the listener hears a pair of sounds, and is asked to say whether the first or second one has the higher pitch. The pairs of sounds are presented a number of times in random order. This is an example of a *two-alternative forced-choice task*. The percentage of times the listener said one particular sound was higher in pitch than the other can be converted to measure called the *perceptual distance*, usually denoted by d'. For example, if one sound was considered higher in pitch than the other 50% of the time, then the perceptual distance would be zero because 50% is the score predicted from chance guessing with two alternatives. This would be expected to occur if the two sounds were indistinguishable. A perceptual distance of 1 represents the difference between two sounds that are just distinguishable, and is often used as the criterion for the threshold of discrimination. The theory relating psychophysical discrimination tasks and perceptual distances is referred to as *signal detection theory*.[1] If one starts with the pair of electrodes at one end of the array, and measures the perceptual distance between stimuli on adjacent electrodes, one can plot the cumulated d' as the electrode position changes. Such a plot would look very similar to those in Figure 7.2, except that the range of data on the vertical scale would represent the perceptual difference between stimuli at each end of the electrode array (Nelson et al. 1995).

If two electrodes activate essentially the same set of auditory neurons, then stimulation on the two electrodes is difficult to distinguish perceptually (if the stimuli only differ in electrode position). This can be tested directly in an *electrode-discrimination task*. Electrode discrimination is usually tested with a *multiple-interval forced-choice task*. For example, in a four-interval forced-choice task, the listener hears four sounds and is asked to choose which was the different one. The different, or test, stimulus occurs in one of the four intervals, selected randomly, and the reference stimulus occurs in the remaining three intervals. In this case, the test and reference stimuli differ only in electrode position. In contrast to the pitch-ranking task, the discrimination task allows the use of any perceptual difference, including loudness, to detect the different stimulus. Thus it is important to carefully loudness balance the two stimuli first, and/or randomly vary the loudness across the four intervals, if one is interested only in the discrimination of percepts that depend on electrode position.

---

[1] The reader is referred to chapters 7 and 8 in Gelfand (1998) for a summary of psychoacoustic methods and signal detection theory. See also SHAR volume 3, *Human Psychophysics* (Yost et al. 1993).

Pitch-scaling and pitch-ranking tasks (but not electrode-discrimination tasks) are based on one important assumption: that as electrode position is changed (and all other aspects of the stimulus remain constant) the sound that the listener experiences changes along a single perceptual dimension called pitch. This appears to be a valid assumption in general, but may not always be valid for individual subjects and electrodes. A symptom of this problem sometimes shows itself in a pitch-ranking experiment, when one obtains an inconsistency called a transitivity violation: electrode A is considered higher in pitch than electrode B, and electrode B is higher than electrode C, but electrode A is considered lower in pitch than electrode C. This can happen when there is a perceptual change along more than one dimension and the listener is attempting to rank the change along a single dimension. For example, when Collins et al. (1997) tested 11 subjects on a pitch ranking task, three of these subjects showed a high level of transitivity violation.

One way to investigate whether electrode position affects the sound along a single dimension or multiple dimensions is to do a *multidimensional scaling* experiment.[2] In this experiment, the listener hears all possible pairs of sounds from a set of stimuli (all with different electrode positions in this case), and is asked to rate the dissimilarity of each pair of sounds along a scale from "exactly the same" to "the most different." The matrix of dissimilarities is analyzed to obtain a plot called a "stimulus space." In the stimulus space, the stimuli are represented by symbols whose separation in the space represents how different they are perceptually. Figure 7.3 shows three examples of stimulus spaces produced from dissimilarity judgments in cochlear-implant subjects (two subjects are from the set of seven subjects presented by Henshall and McKay 2001). The numbers represent stimuli on different electrodes, with numbers increasing as the electrode position goes from basal to apical. On the left is a typical example. This subject heard all the stimuli changing systematically along a single dimension (represented by the distance around the curve) as the electrode position was moved from basal to apical. Other subjects show the same one-dimensional perceptual change with electrode position, but may have particular electrodes that are "out of order." This can sometimes happen at the ends of the array: electrodes at the basal end may be outside the cochlea or a long distance from the neurons they activate; at the apical end the array may be kinked due to a poor insertion process. The middle panel is an example of such a case. On the right is an example in which the assumption of a one-dimensional perceptual change with electrode position is not valid. Although the electrode position increases relatively consistently from right to left in the stimulus space (perhaps illustrating a place-pitch dimension),

---

[2] The theory and application of multidimensional scaling are discussed in Schiffman et al. (1981).

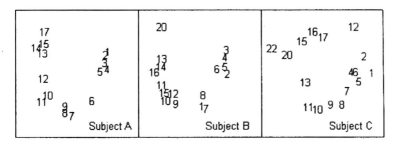

FIGURE 7.3. Stimulus spaces for three subjects, derived from a multidimensional scaling experiment using stimuli differing in electrode position. The numbers represent the electrode positions. The distances separating the numbers in the stimulus spaces represent the relative dissimilarity of the percepts. Subject A shows all stimuli arranged systematically around a single curve, demonstrating a one-dimensional perceptual change with electrode position. Subject B shows a similar one-dimensional pattern related to electrode position, but electrodes 1, 2, and 15 are in the wrong order, and many groups of electrodes produce extremely similar percepts. Subject C is a case where a single perceptual dimension does not describe the data well. Although the electrode position increases relatively consistently from right to left in the stimulus space (perhaps illustrating a place-pitch dimension), there is a second perceptual dimension that causes a dissimilarity of electrodes 12, 15, 16, and 17 from their close neighbors in the electrode array.

there is a second perceptual dimension that causes a dissimilarity of electrodes 12, 15, 16, and 17 from their close neighbors in the electrode array. Perhaps the current paths from these electrodes also produce nonauditory sensations (by activating nonauditory nerve fibers), or are different from those of neighboring electrodes in extent or direction due to pathological changes in the cochlea.

As described in Shannon et al., Chapter 8, there is a lot of individual variability in how well people can understand speech using a cochlear implant, and some of this variability is due to how well the concept of a systematic place-pitch relationship with electrode position applies in the individual case. As mentioned above, there are many factors that can limit the way place pitch changes with electrode position in individuals, and these can be either due to the positioning of the electrode array, or due to individual patterns of nerve cell survival or pathological changes in the cochlea.

Another factor that can degrade the perception of electrode place in complex electrical stimuli is unwanted interactions of place pitch with stimulation parameters other than electrode place. For example, as the loudness is changed (by altering the current or pulse duration), the place-pitch percept may also change. This phenomenon has not been extensively studied, but has been reported by several researchers (e.g., Shannon 1983a; Townshend et al. 1987). The size and direction of this covariance of pitch and loudness is highly dependent on the subject and the electrode used, and is most likely due to changes in the spatial shape of the neural excitation pattern as level is changed. Even when loudness is kept constant, the pulse duration used may affect the place-pitch percept evoked by a single electrode. McKay and McDermott (1999) found that subjects could discriminate stimulation with different pulse durations, even when the loudness was equalized by adjusting the current.

In summary, the concept of place pitch in cochlear implants is a generally valid one, and makes an important contribution to the ability of the implantee to recognize different sounds. As electrode position changes along the cochlea from the round window or basal end to the apical end, the auditory percept that the person experiences also changes along a perceptual dimension from high pitched to low pitched. Thus, using multielectrode implants, we can partially restore one of the most important functions of the cochlea: to separate different frequencies spatially so that they activate different nerve cells.

2.3 Perception of Temporal Information on a Single Electrode

2.3.1 Perception of Rate of Stimulation

In electrical stimulation, as in acoustic stimulation, the auditory nerve fibers exhibit phase locking to the stimulus temporal pattern. In fact, electrophysiology studies have generally shown that electrically stimulated nerve fibers are more tightly phase locked to the stimulus than for acoustic stimulation (Abbas and Miller, Chapter 5). It is therefore not surprising that cochlear implantees hear a pitch associated with periodic temporal features of the electrical signal. The simplest and most studied temporal parameter is the pulse rate in a constant-current pulse train. If the pulse rate is varied from about 50 Hz to about 300 Hz, the implantee hears the pitch of the sound increasing as the rate increases. At about 300 Hz, the rate pitch becomes less salient. As the rate is increased above about 300 Hz, the pitch percept becomes dominated solely by the place-pitch percept of the electrode used, and the implantee can no longer distinguish changes in rate. Implantees vary considerably in their ability to perceptually distinguish different rates of stimulation, and also vary in the upper limit to the range of rates that they can distinguish. These abilities can even vary considerably over different electrode positions in the same subject. Figure 7.4 shows the rate-discrimination ability of two implant subjects. It can be seen that subject PS found it difficult to discriminate rates above 225 Hz on electrode 5 and above 160 Hz on electrode 15, whereas subject BV could discriminate rates up to 330 Hz.

Studies with musically capable implantees have shown that, provided rates are kept within the limits of salient rate pitch, changes of rate produce

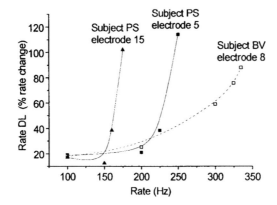

FIGURE 7.4. Data for two subjects from our laboratory, showing how their rate discrimination ability changes with the rate of stimulation. Note that the rate at which discrimination becomes difficult varies between the two subjects and also between the two electrodes for subject PS.

similar musical pitch intervals as would be expected from changing the frequency of pure tones or the fundamental frequency of complex sounds in normal hearing (Pijl and Schwarz 1995; McDermott and McKay 1997).

It is interesting to ask the question, What is the relationship between place pitch and rate pitch: are these two independent percepts of the same sound, or are they somehow amalgamated via central processing to make a single one-dimensional pitch percept? The evidence so far collected from psychophysical studies suggests that percepts related to spatial and temporal cues are independent. That is, a pulse train on a particular electrode evokes two simultaneous percepts, one related to the rate of stimulation and the other related to the electrode position, both of which may be described by subjects as a pitch percept. Some researchers suppose that the percept associated with cochlear place may be more properly described as timbre. However, implant subjects readily apply the pitch label to place-of-stimulation cues (as in pitch ranking or scaling) as well as rate-of-stimulation cues.

An early experiment by Tong et al. (1983b) demonstrated the independence of percepts evoked by electrode place and pulse rate using a multidimensional scaling experiment (see section 2.2). They used combinations of three electrode positions and three rates of stimulation in their experiment. The resultant stimulus space, shown in Figure 7.5, is clearly a two-dimensional array with electrode place and rate of stimulation being represented along orthogonal dimensions. In a more recent experiment, McKey et al. (2000) showed that, when discriminating stimuli differing in small increments of both place and rate parameters, implantees could optimally combine these two independent perceptual changes to enhance their

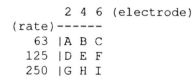

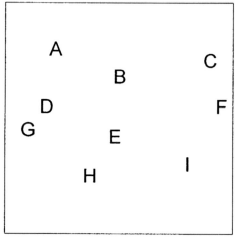

FIGURE 7.5. The results of a multidimensional scaling experiment by Tong et al. (1983b), using stimuli that varied in electrode position and repetition rate of stimulation. The implant used was an early prototype with an array of 10 electrodes labeled in apical to basal direction. The stimuli had multiple biphasic pulses, separated by 1 ms, in the first half of every repetition period. The stimulus space is a two-dimensional array, showing that repetition rate and place were perceived as separate, independent dimensions.

discrimination ability, regardless of whether the two changes produced pitch changes in opposite directions or the same direction.

2.3.2 Perception of Amplitude Modulation and Gap Detection

Two main types of psychophysical experiments have been used to measure how well implantees can detect amplitude variations over time in an electrical signal. This ability is called *temporal resolution. Gap detection tasks* measure the duration of a just-detectable gap in a stimulus. *Amplitude modulation detection tasks* measure the amplitude of a periodic (usually sinusoidal) fluctuation in a signal when the fluctuation is just detectable by the subject.

In an early experiment, Shannon (1989) measured the ability of implantees to detect gaps in sinusoidal or pulsatile stimuli. The duration of just-detectable gaps was 20 to 50 ms for low-level stimuli, with performance improving at high levels to detection of gaps of 2 to 5 ms. The

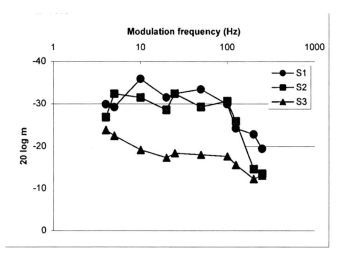

FIGURE 7.6. The ability of three cochlear implantees to detect amplitude modulation of a 1000-Hz carrier pulse train as a function of modulation rate. (Data adapted from Busby et al. 1993.)

similarity of this performance to that of normal-hearing listeners with analogous acoustic stimuli indicated that temporal resolution is a central process that is largely unaffected by sensorineural deafness.

The central process involved in temporal resolution tasks has some similarity to the effect of a low-pass filter. That is, slow changes are easily detected, but fast changes get "smoothed out" and become undetectable. The low-pass characteristic of temporal resolution is best illustrated by the results of amplitude modulation detection experiments (Shannon 1992; Busby et al. 1993). When the just-detectable amplitude of an amplitude modulation is plotted against the frequency of the modulation, the resultant curve is called a *temporal modulation transfer function* (Viemeister 1979). These measurements usually show that when the modulation frequency is increased, the modulation becomes increasingly difficult to detect (that is, the modulation amplitude has to be increased to allow detection). Figure 7.6 shows examples of temporal modulation transfer functions for three implantees. Two of these transfer functions have a typical shape, similar to that seen in acoustic hearing, indicating that, as the frequency of the modulation increases above about 100 Hz, the amplitude of the modulation must be increased in order for the modulation to remain detectable. One subject (S3), who is a prelingually deaf teenager, performed relatively poorly at this task and also did not show a clear low-pass characteristic. It is possible that such differences in ability to perceive amplitude modulations may contribute to the variability of speech perception ability among implantees, since there is much information about speech that is conveyed in the amplitude envelope of a speech signal (Cazals et al. 1991; Fu 2002).

The fact that modulation detection and rate discrimination have a similar low-pass characteristic evokes the question of whether the same mechanism underlies the two perceptual tasks. Certainly, we know that amplitude modulations may produce a pitch percept, both in acoustic and electric hearing (Burns and Viemeister 1976; McKay et al. 1994). Patterson et al. (1978) have addressed this question using amplitude-modulated wide-band noise and normal-hearing listeners. They measured the modulation depth required for subjects to tell whether two fixed modulation frequencies were different in pitch. For modulation frequencies above 70 Hz, this modulation depth was a constant multiple of the modulation depth required to detect the presence of amplitude modulation. They inferred that the ability to detect amplitude modulation was the major factor limiting perception of modulation-rate pitch above 70 Hz. On the other hand, experiments have shown that pitch perception involves processing that is additional to or different from what would be needed to detect fluctuations in the amplitude envelope of a sound. In particular, the central processing of the time intervals in the fine temporal structure of a sound appears to be important for pitch perception. For example, McKay et al. (1995b) and McKay and Carlyon (1999) have shown that, for amplitude modulations of pulsatile electrical stimulation, the underlying intervals in the carrier pulses contribute to the pitch percept as well as the amplitude modulation period. In a more recent experiment using equal-amplitude electric pulse trains with various patterns of interpulse intervals, Carlyon et al. (2002) showed that the resultant pitch was most consistent with the proposition that the central auditory system utilized the time intervals between successive stimulus pulses, rather than an overall periodicity determination, to extract the pitch of a sound. More research is needed for us to fully understand the central processes involved in nonspectral pitch perception.

2.4 Loudness Effects at a Single Electrode Position

Loudness is the perceptual quality that we use to describe the intensity of a sound, along a dimension from soft to loud. Although loudness is monotonically related to intensity for a fixed sound, there is not a 1:1 relationship between intensity and loudness across different sounds. This is because the frequency content (or spectrum) of a sound and its temporal qualities (how its amplitude varies over time) have an effect on the perceived loudness in addition to the overall intensity. Although the physiological correlates of loudness are not completely understood, it is generally accepted that the loudness of a sound is related to the amount of neural excitation it evokes in the auditory nerve. To make a judgment of overall loudness, the central auditory system must combine information about neural excitation coming from different parts of the cochlea, and obtain information about the rate of neural excitation by monitoring such information over time.

In this section we first discuss the effects on loudness of changing the intensity of an electrical stimulus by altering the current amplitude or pulse duration. The discussion is then extended to include the effects on loudness of the temporal characteristics of the stimulus pattern. The effects on loudness of using multiple electrodes concurrently is addressed in section 3.

2.4.1 The Effects of Current Amplitude and Pulse Duration on Loudness

The simplest electrical signals consist of a sinusoidal analog current or a train of biphasic current pulses. A biphasic current pulse consists of a constant-current pulse, followed by a reversed-polarity pulse of the same amplitude and duration. When the current amplitude is increased in analog or pulsatile stimuli, more nerve fibers are activated on each cycle of the stimulation, resulting in an increase in loudness. The increase in neural activity with current amplitude is contributed to by two mechanisms. First, an increased number of neurons are activated by the current. That is, nerve fibers with higher thresholds are recruited by the higher current, leading to a greater total number of neurons activated. Second, for neurons above threshold and not yet firing at their saturation rate, there is an increased probability that each will fire on a particular cycle. For example, a neuron that fired on average every third cycle of the stimulus may increase its firing rate to an average of once every second cycle when the current amplitude is increased.

The effect of current amplitude on loudness for pulsatile stimuli is illustrated in Figure 7.7. The figure shows the result of a loudness estimation task for an implantee using two stimuli. The two stimuli were delivered to the same electrode and had the same rate of stimulation (200 Hz), but differed in their pulse duration. It can be seen that loudness increases monotonically with current for both pulse durations. A similar function of loudness growth is also obtained when increasing the amplitude of sinusoidal stimuli (Shannon 1983a).

The range of currents (dynamic range) that produced a loudness change between threshold and comfortably loud for the subject in Figure 7.7 was 2 to 3 dB. This represents a much steeper loudness growth function than seen for acoustic stimulation in normal hearing (where the typical equivalent dynamic range of hearing is about 80 dB). Although electrical dynamic range varies among subjects and electrodes, and varies with parameters such as pulse duration, rate, and number of activated electrodes, it is always the case that the dynamic range is much smaller than for acoustic hearing. This is probably because the compressive function produced by cochlear mechanics in normal hearing is not present in electrical stimulation (Zeng and Shannon 1994).

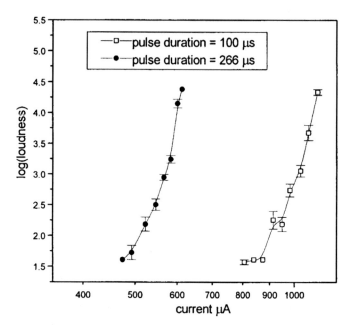

FIGURE 7.7. The result of a loudness estimation experiment in our laboratory with one subject, showing how loudness increases with current amplitude for pulsatile stimuli. The two stimuli in this case have a rate of 200 Hz, and use pulse durations of 100 and 266 μs.

One can also see from Figure 7.7 that the current needed to achieve a particular loudness is lower when the pulse duration is longer (see also Zeng et al. 1998; McKay and McDermott 1999; Chatterjee et al. 2000). To understand the relationship of pulse duration to loudness, it is useful to examine the mechanism by which nerve fibers are activated by an electric current. (This mechanism is described fully in Abbas and Miller, Chapter 5.) For the purposes of our discussion here, there are two important features of the process of electrical stimulation of nerve fibers. First, the total charge that is deposited on the neural membrane by the stimulus current pulse is the important parameter that helps to determine whether or not a particular nerve fiber will initiate a spike. Doubling the duration of the current pulse doubles the total charge delivered to the neural membrane, in the same way as doubling the current amplitude. It may seem, then, that doubling the current or the pulse duration should have an equal effect on neural excitation and therefore on loudness. But this is not the case. In fact, doubling the current causes a greater loudness change than doubling the pulse duration. This is due to a second important feature of neural charging: the neural membrane does not hold all the charge that is deposited on it, but loses some of this charge during the time of the current pulse. The

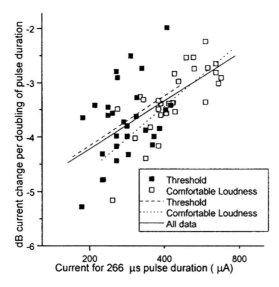

FIGURE 7.8. The results of an experiment by McKay and McDermott (1999) with six implantees and various electrode positions, in which the current at threshold of hearing and the loudness-balanced comfortably loud levels were obtained for stimuli using two different pulse durations (266 μs, and a shorter duration between 50 and 70 μs). The vertical axis shows the current reduction needed to keep the loudness the same when pulse duration was doubled. The horizontal axis shows the current of the stimulus with 266-μs pulse duration for each equally loud pair of stimuli. The two broken lines are the best linear fits for the threshold data and the comfortably loud data, and the solid line is the best fit for the combined data.

neural membrane is often referred to as a "leaky integrator" when this charging property is under discussion. The longer the duration of the current pulse, the more charge can leak away within the duration of the pulse. Thus a current pulse that has half the current and double the pulse duration of another pulse will not be able to cause as great a change in potential of the neural membrane as the other pulse, even though it delivers an equal amount of charge. In other words, shorter duration pulses are more efficient in the task of exciting a neuron than longer duration pulses containing the same total charge. This means that increasing the current by a certain ratio results in more increase in loudness than the same ratio increase in pulse duration.

One way to measure this difference in the effects of current and pulse duration changes on loudness is to ask implant users to match the loudness of two stimuli with different pulse durations by adjusting the current amplitude of one of them. This procedure is called a *loudness balancing task*. Figure 7.8 shows the result of such an experiment with six implantees. The vertical axis shows the current reduction required to match the loudness when the pulse duration is doubled. A halving of the current is equivalent to 6-dB reduction. It can be seen that the current was never reduced by as much as 6 dB when the pulse duration was doubled, but by amounts ranging from 2 to 5.5 dB. It can also be seen that there is a correlation between the absolute current level and the amount of current adjustment that gives the same loudness change as doubling the pulse duration. At lower currents, the current and pulse duration changes are almost equivalent (close to 6 dB current adjustment per doubling of pulse duration), but at higher current levels, doubling pulse duration has less effect on loudness than doubling the current (see also Zeng et al. 1998). The reason for this correlation is still unclear, but it may be due to the fact that higher currents access neurons or parts of neurons that are more distant from the electrodes in the cochlea, and these neurons may have different charging characteristics to the closer neural structures. The overall effect of this correlation is that, if a longer pulse duration is used, the electrical dynamic range (change of current in dB to go from threshold of hearing to an uncomfortable loud sound) is larger than when a shorter pulse duration is used.

In summary, both current amplitude and pulse duration can be used to control the loudness of a sound when using pulsatile electrical stimuli. The way that loudness grows with increasing current or pulse duration is somewhat different, however. In addition, the exact shape of the loudness versus current relationship is still a matter for investigation. It can be seen from Figure 7.7 that, for practical purposes, a linear fit of loudness to current, when both are on a logarithmic scale, describes the relationship reasonably well. This translates to loudness being a power function of current. Some researchers (e.g., Zeng and Shannon 1992) have noted a better fit of the loudness-current relationship with an exponential function. More recently, McKay et al. (2003) proposed a loudness function that was a combination of power and exponential functions. In most implantees all these functions provide a reasonable fit for practical purposes, at least over a restricted range of stimulus parameters. It is likely, however, that variations among the spatial distribution of surviving nerve cells, and variations in the shape of electrical fields due to varying electrode geometry, modes of stimulation, and pathological processes in the cochlea, will lead to individual variations in how loudness grows with current. These issues are still the subject of research.

2.4.2 Intensity Discrimination

The relationship between the amplitude of an electrical or acoustic stimulus and loudness is important for many perceptual tasks. For a steady sound, loudness provides valuable clues about the type of sound source and its distance from the listener. When listening to complex and fluctuating sounds such as speech, the discrimination of different intensities within the signal provides very important cues for identifying and understanding the sounds

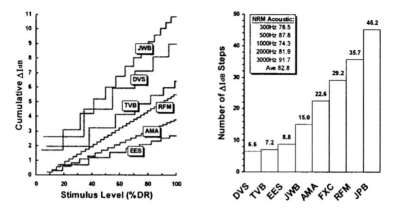

FIGURE 7.9. Individual discriminable steps of current for eight cochlear implantees. The number of discriminable loudness steps for these subjects varied from 7 to 45; however, the number of steps was not correlated with the absolute size of the electrical dynamic range. (From Nelson et al. 1996.)

(Shannon et al., Chapter 8). The way the overall intensity of a signal changes over time (*amplitude envelope cues*) and the relative intensity of components of different frequencies at the same time (*spectral shape*) provide vital cues to how the sound was made, and hence its identity. In normal hearing, the change in acoustic intensity that is just detectable ranges from 1 to 2 dB at low sensation levels to about 0.5 dB at high levels. Considering that the range of intensities from threshold to uncomfortable loudness in normal hearing is generally at least 80 dB, one can estimate the number of discriminable steps of loudness in normal hearing to be at least 50 to 60.

Intensity discrimination experiments with cochlear implantees have shown large individual differences in the size of current steps that produce discriminable loudness changes. Figure 7.9 shows the discriminable current steps within the perceptual dynamic range for eight subjects. These steps were obtained using a three-alternative forced-choice task where the subject had to nominate the loudest of three presented stimuli, where one randomly selected stimulus had a higher current than the other two. It can be immediately seen that although one subject had 45 discriminable steps, which is a similar order of magnitude to the number in normal hearing, other subjects had as few as seven discriminable steps. Thus, there is great variation between people in their ability to detect small level variations in an electrical signal, and this would lead to variation in their ability to use this information to recognize sounds or understand speech. Nelson et al. (1996) noted a correlation among their subjects in that smaller discriminable steps were associated with higher thresholds and smaller dynamic ranges. However, this correlation did not hold for different electrode positions in the same subject.

2.4.3 The Effect of Temporal Parameters on Loudness

Although the control of loudness is achieved in practical situations (such as in the output of a speech processor) by controlling the amount of charge delivered by each current pulse, or the current amplitude of sinusoidal stimulation, the temporal features of an electrical signal (how the current amplitudes vary over time or the relative timing of individual pulses) also affect the loudness.

The next two subsections discuss the effect of frequency or rate of stimulation on loudness and the effect of duration and temporal amplitude envelope fluctuations.

2.4.3.1 The Effect of Frequency, Rate of Stimulation, and Interphase Gap on Loudness

For simple pulse-train stimuli, there are two time intervals (apart from the pulse duration) that affect the loudness perceived by implantees. The one with the most important consequences for loudness coding in speech processors is the repetition period, which gives the frequency or rate of stimulation. The effects of frequency and rate on loudness are discussed below.

The other stimulus time interval that affects loudness is the small gap between the two phases of a biphasic pulse (*interphase gap*). To understand the effect of interphase gap on loudness, it is important to know about the timing of neural action potentials relative to the onset and offset of individual stimulus pulses. Electrophysiological studies have shown that acoustic nerve fibers can fire up to 100 μs after the offset of a stimulus current pulse. For example, Shepherd and Javel (1999) found that the threshold current for production of an action potential increased from its value for monophasic pulses when a reverse phase with interphase gap less than 100 μs was added. When the second phase of the biphasic pulse begins, any charge that has built up on the neural membrane due to the first phase will begin to be removed. If the neuron has not already fired by the start of the second phase, its likelihood of firing steadily reduces during the second phase, thus elevating the threshold current. Shepherd and Javel also found that this elevation in neural threshold was more marked when using narrower pulse durations. This is probably because there is a higher likelihood that the nerve has not yet fired at the end of the first phase when that phase is of shorter duration. Figure 7.10 shows the effect of interphase gap on loudness for five users of the Nucleus C124 implant. It can be seen that greater current is needed to achieve the same loudness when shorter interphase gaps are used. It can also be seen that, consistent with the electrophysiological data, the effects of interphase gap are greater for shorter interphase gaps and reduce as the gap widens to 100 μs. The interphase gaps used in speech processors are generally always less than 100 μs (and in some cases zero) and there is a trend to make both interphase gaps and pulse

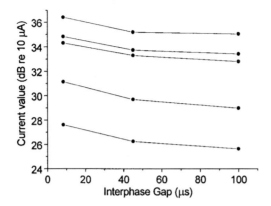

FIGURE 7.10. Equal-loudness curves, for five users of the Nucleus C124 implant, for different interphase gaps. Current must be increased as the interphase gap is reduced in order to maintain the same loudness. The pulse trains used for this experiment had a rate of 1000 Hz and a pulse duration of 26 μs.

durations as short as possible to allow the use of higher rates of stimulation. Thus the elevation of current that these short interphase gaps necessitate leads to important consequences for the power efficiency of speech processors. Since interphase gaps are generally not varied for an individual once they are set in the speech processor map, the effects of interphase gap on loudness are important only at the stage of setting appropriate stimulus parameters for that individual, rather than having importance for how variations in loudness are controlled by the speech processor.

The rate of stimulation (or frequency of sinusoidal stimulation) also affects loudness. For pulsatile stimulation, the current amplitude must be decreased as rate is increased in order to maintain an equally loud sensation (McKay and McDermott 1998; Zeng and Shannon 1999). The fact that loudness increases when the rate increases is not surprising when one supposes that the central auditory system must make an estimate of the overall rate at which neural excitation is occurring in order to judge the loudness of a sound.

To measure the rate at which excitation is occurring, the excitation must be summed or integrated over a small time interval in the central auditory pathways. McKay and McDermott (1998) studied the effects on loudness of the interpulse intervals within a stimulus, and concluded that the loudness data were consistent with such an integration of neural excitation over a time of about 7 ms. This *temporal integration window*, with exponential sides, is identical to the one proposed for a model of intensity discrimination in normal hearing by Moore et al. (1996). The acoustic experiments of Moore et al. required the normal-hearing subjects to detect small short-duration increments in intensity within a continuous sound. In other words, they were measuring the temporal resolution ability of the subjects, in a similar way to gap detection and amplitude modulation detection experiments (see section 2.3.2). If one supposes that loudness judgments are based on the output of such a temporal integration window, then it follows that the duration of this window (an equivalent duration of about 7 ms) determines the limitation for resolving fast changes in intensity in a stimulus, as well as the way rates of stimulation affect loudness. In the former case, only changes in intensity that happen for durations longer than the integration window are noticed, leading to the low-pass characteristic for temporal resolution discussed in section 2.3.2. In the latter case, the rate of stimulation starts to affect the loudness when increases in rate cause more stimulus pulses to fall within the window, that is, for rates above about 100 Hz.

Temporal integration is not the only factor that contributes to the way rate of stimulation affects loudness. Another major factor is the *refractoriness* of nerve fibers (see Abbas and Miller, Chapter 5). After a neuron has produced an action potential, it enters what is called a refractory phase in which it is, at first, unable to be excited again (the *absolute refractory phase*, which lasts for about 1 ms), and then has a finite but reduced probability of being excited by another current pulse (the *relative refractory phase*, which lasts for about 10 ms). Thus, as rates increase above about 100 Hz, each current pulse in a pulse train activates fewer neurons than the same current pulse in a lower rate pulse train. The two features of temporal integration and neural refractoriness, therefore, act in opposite ways on loudness as rate is increased. An increase in rate causes the output of the temporal integration window to increase because there are more stimulus pulses within the window. An increase in rate also causes the amount of neural excitation for *each* stimulus pulse to decrease because of neural refractoriness. The net effect for pulsatile stimulation is generally an increase in loudness as rate increases. The effect of refractoriness on psychophysical data has been modeled by McKay and McDermott (1998) and on physiological data by Bruce et al. (1999).

The effect of frequency on threshold of sine-wave stimuli, compared to the effect of rate on pulsatile stimuli, is shown in Figure 7.11 (see also Zeng and Shannon 1999). It can be seen that the shape of the threshold curve for the sine wave is quite different from those for pulsatile stimuli. When the frequency of the sine wave is increased, the period, or time, between the peaks, is reduced as it is in a pulsatile stimulus, leading to similar effects of temporal integration and neural refractoriness. However, unlike changing the rate in pulsatile stimulation, changing the frequency of sine-wave stimulation also changes the duration of the stimulating time in each period of the stimulus. Doubling the frequency (while keeping current amplitude constant) halves the total amount of charge delivered to the neural membrane in each cycle, because the duration of the stimulus waveform is halved, whereas doubling the rate of a pulsatile stimulus does not change the total charge delivered in each period. In addition, halving the duration of the

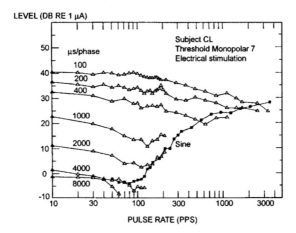

FIGURE 7.11. Threshold current versus pulse rate of pulsatile stimulation and frequency of sinusoidal stimulation for one subject (Shannon 1985). It can be seen that the effect of frequency of sinusoidal stimulation on threshold is greatly different from that of pulse rate in pulsatile stimulation. It can also be seen that much of the difference can be accounted for by the change of effective pulse duration or charge per period as frequency changes in sinusoidal stimulation.

stimulating cycle has effects on the efficiency of the neural-membrane charge-up process, analogous to the effects discussed above for changing the pulse duration in pulsatile stimulation. Thus, the effect of frequency on loudness for sine-wave stimulation is due to a complicated interaction of features of the stimulus waveform, characteristics of how neural membranes are charged, and how the central auditory system processes the auditory nerve impulses.

2.4.3.2 The Effect of Duration and Temporal Envelope on Loudness

Another temporal factor that affects the perceived loudness of any sound is its total duration. For both acoustic and electrical hearing, the loudness of a sound increases as the duration increases, up to about 100 to 200 ms (when the stimulus intensity is kept constant). Donaldson et al. (1997) measured the decrease in threshold for 100-Hz pulse trains as the duration of the pulse train increased from 0.4 to 630 ms and found that the slope of this function varied across electrodes and subjects from 0.06 to 1.94 dB/doubling of duration. These slopes are much shallower than the 2.5 dB/doubling typical of normal hearing, although when expressed as a percentage of the dynamic range the electrical slopes vary both above and below the acoustic ones.

Although the term *temporal integration* is also used to refer to the effect of the duration of a signal on the loudness percept, the time scale over which the duration effect operates (more than 100 ms) is much longer than that of the temporal integration window discussed above (7 ms). This suggests that the two time windows arise from quite different mechanisms.

Earlier models of the duration effect assumed it was the result of stimulus or neural activity summing. However, this led to a paradox, whereby one had to evoke two very different time courses of this integration to account for both intensity resolution effects and duration effects. The shorter time window has been associated with people's ability to discriminate changes in intensity of a signal using tasks such as detection of gaps or stimulus increments or decrements within a signal, as well as the detection of a continuous amplitude modulation of a signal. Research by Donaldson and Viemeister (2000) has suggested that the neural information used by the brain to detect intensity differences between separate signals may also be related to the information used to detect low-frequency (4-Hz) amplitude modulations within a single signal. In all these tasks, the data can be explained by supposing that some physical quantity (the power of the stimulus, or the amount of neural excitation, for example) is summed or integrated over a short time of several milliseconds. It is plausible that this summation is performed by neurons in the central auditory system that have multiple inputs from different acoustic nerve fibers, since the time scale is similar to that characterizing the excitation properties of these central neurons.

In contrast, the long time window involved in the effect of duration on loudness may be the result of higher order processing of the signal over time. For example, Viemeister and Wakefield (1991) have suggested that there is no physical integration of a signal over periods greater than a few milliseconds. They suggested an alternative *multiple-look model* in which the output of the short temporal integration window is "sampled" at a fairly high rate, and these "looks" are stored in memory (which is retained for several hundred milliseconds) and can be accessed and processed selectively. A longer stimulus allows more looks, and so the probability of detecting the stimulus increases (and the threshold decreases) as the duration increases. Donaldson et al. (1997) showed that the duration effect they measured was consistent with the multiple-look theory, and that the variability in the magnitude of the effect across electrode positions and subjects was predictable from (inversely correlated with) the steepness of the psychometric functions describing detection probability versus level. Another way to say this is that, in electrical hearing, the sensation magnitude versus level function is much steeper than in normal hearing, so smaller current adjustments are needed to evoke a fixed change in hearing sensation. It is interesting to note that expressing the dB/doubling data in terms of percentage of the dynamic range did not "normalize" the data so that it resembled the acoustic data, whereas using the psychometric function *at threshold* to predict the slope for individual cases produced a relationship that was consistent with the acoustic case. This is perhaps further evidence that the slope of the loudness growth function (in log-log coordinates) is not constant across the dynamic range (as discussed above).

The way a signal's amplitude fluctuates over time also affects the loudness percept. Zhang and Zeng (1997) measured the difference in loudness between two (analog) signals applied to a cochlear implant electrode. These signals were harmonic complexes that differed only in their phase relationship, so that each had the same root mean square (rms) energy but differed in their temporal envelope (one having a peakier temporal pattern). They found that near threshold, the two signals were equally loud (consistent with rms energy determining loudness), but as the signals were increased in level the peakier waveform became louder than the less peaky one (until at the maximum loudness, the loudness was consistent with the peak determining the loudness rather than the rms level). The authors suggested that their data were consistent with the shorter integration window (a few milliseconds) determining loudness for a dynamic stimulus, and that the level-dependent effect was consistent with the expansive loudness growth function in electrical hearing.[3]

Similar conclusions were drawn by McKay et al. (2001, 2003). They developed a model of electrical loudness that was able to predict the loudness of arbitrary pulsatile stimulus patterns. In their model, neural activity is integrated at each cochlear place (over the short integration window) and then converted via a nonlinear (expansive) function to specific loudness (the loudness contribution at each cochlear place). The total loudness is the sum of the specific loudnesses across electrode positions. These authors developed a practical method of applying this model in which the loudness contribution of each pulse within the integration window is obtained (using a current versus loudness contribution function) and summed to give the overall loudness. Because of the shape of the current-versus-loudness function, the model predicts that, at higher levels, the highest current in the window has an enhanced contribution to the overall loudness compared to the lower currents, consistent with the earlier data of Zhang and Zeng (1997).

3. Perception of Multiple-Electrode Electrical Stimuli

The output of a multiple-electrode cochlear implant sound processor consists of analog or pulsatile stimuli that are distributed over multiple electrode positions, and changing dynamically according to the acoustic input (see Wilson, Chapter 2). As discussed in section 2.2, the goal of multiple-electrode stimulation is to restore to some degree the capability for spectral analysis which is lost in profound hearing impairment. The spectral

---

[3] For more discussion on this topic see Zeng (2003).

shape of a speech sound is a crucially important cue for its identity. For example, to identify a vowel sound, the frequencies at which the first two peaks in intensity in the spectral envelope occur (called the first and second formant frequencies, F1 and F2) must be resolvable. That is, the listener must hear within the overall vowel sound the fact that there are two separate peaks in intensity at the formant frequencies. The capability of multiple-electrode electrical stimuli to enable such *frequency resolution* is discussed in section 3.1. An important factor that may affect this frequency resolution ability in cochlear implantees is the degree of channel interaction. That is, stimulation on a nearby electrode just prior to or simultaneously with a particular electrode affects the neural response elicited by that electrode. This type of channel interaction is discussed in section 3.2. In section 3.3, the perception of temporal information within multiple-electrode stimuli is discussed, and in section 3.4, the effect of multiple-electrode stimuli on loudness is discussed.

3.1 Spectral Shape Perception with Multiple-Electrode Stimuli

The simplest multiple-electrode stimulus that has been studied psychophysically is one in which successive stimulus pulses are alternated between two electrode positions. Tong et al. (1983a) studied the perception of such stimuli in a multidimensional scaling task. They found that the subject perceived two independent perceptual dimensions in the stimuli, one corresponding to each of the two electrode positions. The stimulus space they found, showing the two-dimensional array of stimulus representations, is shown in Figure 7.12. This experiment demonstrated the

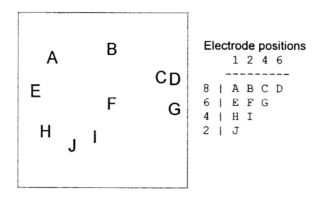

FIGURE 7.12. The stimulus space obtained by Tong et al. (1983a) in a multidimensional scaling experiment with dual-electrode stimuli. The two-dimensional array in the stimulus space shows that the subject perceived two independent dimensions related to the two electrode positions.

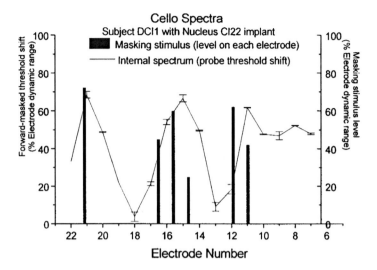

FIGURE 7.13. The forward-masking pattern (or internal spectrum) produced by a multiple-electrode electrical stimulus. The masker is a simulation of the output of a Spectra processor, using the SPEAK strategy, for an acoustic input produced by a cello. (Data from Stainsby 2000.)

feasibility of coding the first two formant frequencies of vowels by the position of two concurrently activated electrodes, and was the basis of an early speech-processing strategy called F0/F1/F2. The fundamental frequency (F0) was coded as the repetition rate of the speech-processor output, providing a perception of voice pitch (see section 2.3.1). A later study by McKay et al. (1996) showed that dual-electrode stimuli retained this two-dimensional perceptual quality even when the electrode positions were adjacent (0.75 mm separation) and the time between the pulses on the two electrodes was reduced to a minimum of 0.62 ms. Thus, at least for stimuli with sequential pulses, the percept remained two-dimensional even though the neural excitation patterns for the two electrodes were presumably substantially overlapping.

Another way of investigating the percepts evoked by multiple-electrode stimuli is to study the forward-masking pattern produced when such stimuli are used as a masking stimulus. Figure 7.13 is an example of such a forward-masking pattern. In this experiment the acoustic signal was produced by a cello playing a musical note (B above middle C). The output of the Spectra processor, using the SPEAK strategy, was examined for this signal input, and a psychophysical masking stimulus was constructed that simulated the electrical output of the processor. This masking stimulus is represented in the graph by the vertical bars, which indicated the electrode positions and level of the six pulses that are generated approximately every 4 ms in the

SPEAK strategy. The line represents the pattern of threshold shift for a probe stimulus (a brief pulse train of duration 20 ms on a single electrode) across different probe electrode positions. It can be seen that the pattern of threshold shifts has a representation of the amplitude peaks in the electrical stimulus itself. The forward-masking pattern for a complex sound is often referred to as the *internal spectrum*, because it can be thought of as a perceptual representation of the external signal spectrum (Moore and Glasberg 1983).

It should be noted that neither of the experimental procedures described above (multidimensional scaling and forward masking) test directly whether implantees can perceptually resolve concurrent electrical stimulation at more than one cochlear position. However, the results of the experiments are consistent with the proposition that multiple-electrode stimulation results in a pattern of neural responses across cochlear position that is analogous to that resulting from basilar membrane mechanics in normal hearing. There is a lot of individual variability among implantees in their ability to perceive spectral shape, as evidenced by variations in vowel category identification ability (see Shannon et al., Chapter 8). This variation may be due to many factors related to the individual listener or the implant design, including the so-called channel interaction to be discussed below.

3.2 Channel Interactions in Multiple-Electrode Stimuli

The term *channel interaction* may refer to three different types of phenomena, all of which lead to multiple-electrode stimuli evoking a percept that is not merely a sum of individual percepts from the individual electrodes (Eddington et al. 1978; Shannon 1983b). First and most basic is the interaction caused by direct summation of simultaneous currents. Second, interactions can occur at the neural stimulus interface, when a preceding stimulus pulse from one electrode can affect the neural response evoked by a subsequent stimulus pulse from a second electrode position. Third, there is interaction at the perceptual level, whereby the presence of stimulation on one electrode can change the way the person perceives the stimulation on another electrode.

Channel interactions at any of these levels are not necessarily problematic. Most interactions at the perceptual level are the result of normal across-channel central auditory processing, and can potentially play a similar role for perception of complex electrical stimuli to analogous processing in normal hearing. The other two, peripheral, interactions (at the stimulus and acoustic-nerve levels) have the potential to degrade and distort the perception of sounds by implantees if allowed to occur in a random or uncontrolled way. However, these same interactions might also be intentionally used in a controlled way to improve the perception of implantees.

When intracochlear currents are simultanteous at two electrode positions (for example, when the stimulation is continuous rather than pulsatile, or when current pulses overlap in time) they can sum directly (see Wilson, Chapter 2). The first and most obvious effect of such current summation is on the loudness of the multiple-electrode stimulus. As we saw in section 2.4.1, small increments or decrements in current can lead to significant loudness changes. Consider two electrodes, both with sinusoidal stimuli. If the two stimuli are in phase, there is an increase in total current at every point where there is nonzero current from both electrodes, compared to the situation when each electrode is activated on its own. On the other hand, if the two stimuli are 180 degrees out of phase, the currents from each electrode have an opposite sign and the total current is reduced at all points where there is nonzero current from both electrodes. Thus the loudness of such a dual-electrode simultaneous stimulus varies according to the phase difference between the two waveforms, even when the current amplitudes are kept constant.

In a study of channel interaction and loudness perception, Shannon (1983b) utilized the effect on loudness summation of phase differences in simultaneous sinusoidal stimuli to investigate the extent of direct current summation across two electrodes in the Ineraid device. Figure 7.14 shows that the in-phase condition (0 or 360 degrees phase shift) led to a significantly louder percept than the out-of-phase condition (180 degrees phase shift). Shannon further showed that, as the separation of the two electrodes was increased, the effect of phase differences on loudness decreased because the spatial region where the two electrodes both made a significant contribution to the current density became smaller. Thus, this is a method that can be used to directly measure the distance from an electrode at which the current density has fallen to an insignificant level.

Apart from the above loudness effects, another important consequence of direct current summation in multiple-electrode stimuli is alteration in the shape of the neural excitation pattern. If we consider the dual-electrode sinusoidal stimulus discussed above, the phase conditions not only differ in loudness, but also produce quite different excitation pattern shapes. The in-phase condition produces more excitation in the area of overlapping currents than the out-of-phase condition, so that the overall excitation pattern for the two electrodes has relatively more excitation central to the two electrode positions in the in-phase case. In contrast, the out-of-phase case suppresses the excitation in the overlapping current region. Thus, the ability of the implantee to perceptually resolve the stimulation on the two separate electrodes may vary with the relative phase of the two sinusoids.

In sound processors that generate independent simultaneous analog current waveforms on a number of electrodes, the phase relationships between the electrodes can change over time in a difficult-to-predict way. Thus, unlike the simple psychophysical dual-electrode stimulus with equal-frequency sinusoids and fixed phase delay, the loudness of the speech-

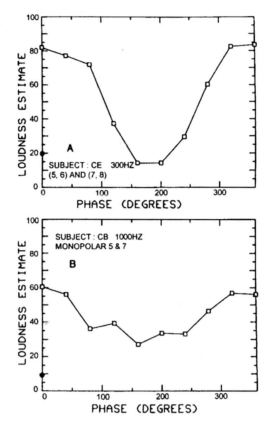

FIGURE 7.14. Results of an experiment by Shannon (1983b), showing the effect on loudness of altering the relative phase of two simultaneous sinusoidal stimuli at different electrode positions. One stimulus was set to a loudness of 50 and the other was set to a value that is shown by the closed symbol on the ordinate.

processor output and the neural excitation shape produced can vary from moment to moment independent of any spectral changes in the acoustic signal. This was the main reason behind the idea of using sequential current pulses rather than simultaneous analog stimulation. In processors that use sequential current pulses, there are no simultaneous currents at more than one electrode position, and thus currents can never directly sum. The newest generation of implantable electrodes that lie closer to the modiolus may theoretically led to a reduction in the amount of simultaneous current summation, because the required currents are lower and the excitation patterns from adjacent electrodes may not overlap to the same extent as with older electrode array designs (see Wilson, Chapter 2).

Current summation in simultaneous stimulation of nearby electrodes can be used in a controlled way, however, to shape the resultant excitation pattern in some desired way. For example, if simultaneous or overlapping current pulses are applied at more than one electrode position, it is possible to sharpen the resultant excitation pattern compared to using a single current pulse on one electrode. An example of this use of simultaneous current is the quadrupolar electrode stimulation mode proposed by Jolly et al. (1996). In a similar way, the controlled use of simultaneous currents might be a feature of future speech-processing strategies to keep the pitch stable for different overall levels, to mimic more accurately the spatiotemporal excitation pattern produced in a normal cochlea, or to shape the neural excitation pattern in some desired way (for example, to sharpen the peaks in excitation). In all these cases, the relative phase or timing of stimulation at adjacent electrode position needs to be finely controlled, as well as the relative current amplitudes.

There is a way in which even sequential current pulses from different electrode positions can interact so that the overall peripheral neural response is not just the sum of the individual neural responses that would have occurred if the single-electrode pulses were delivered in isolation. The mechanism for this interaction is neural refractoriness. The effect of neural refractoriness on the relation between rate of stimulation and loudness in single-electrode stimuli was discussed in section 2.4.3.1. In multiple-electrode stimulation, nerve cells can be activated by currents from more than one electrode position. Thus, the probability that a nerve cell will fire in response to one particular current pulse is affected not only by any preceding pulses from the same electrode within the previous 10 ms (approximately), but also by preceding stimulus pulses originating at other nearby electrode positions. The precise effects of neural refractoriness on the excitation pattern shape produced by multiple-electrode stimuli has not been determined, but it is theoretically likely to be highly dependent on the temporal pattern, order, and relative levels of the individual pulses in such a stimulus.

3.3 Perception of Temporal Information in Multiple-Electrode Stimuli

In section 2.3, we discussed the perception of temporal information (rate of stimulation, amplitude modulation, gap detection) in a single-electrode stimulus. In this section we discuss two complementary questions: (1) Does stimulation on one electrode interfere with the perception of temporal information on another electrode? (2) Can an implantee combine fine temporal information from different electrode positions to perceive an overall temporal pattern? These questions can be reduced to a more basic third question: Is there across-channel processing of temporal information? One main role of central processing of across-channel temporal information is thought to be the perceptual grouping of components of a complex signal, or the perceptual segregation of the components of different sound sources. Thus research studying across-channel processing is of importance in understanding how we can understand one speaker when a second speaker is talking, for example.

In normal hearing, two tones fall into different auditory processing channels if they do not significantly activate a common auditory filter. The current evidence from research with normal-hearing listeners is that cross-channel processing of temporal information probably occurs for some listening conditions and/or tasks and not for others. For example, a phenomenon called *modulation detection interference* has been studied, in which the ability of a listener to detect amplitude modulation in one frequency channel is interfered with by the presence of amplitude modulation in a second channel (Yost and Sheft 1989; Moore et al. 1991). Experiments that have examined the ability of normal-hearing listeners to detect phase differences in components of a complex tone have shown that this ability is enhanced when the components fall within the same auditory filter or channel (Nelson 1994). However, other studies have found that subjects can still discriminate these phase differences above chance levels when the components do not fall in the same auditory filter (Strickland et al. 1989; Yost and Sheft 1989). The detection of temporal gaps in complex tones has also been studied for normal-hearing people in which the gap occurs within or across frequency channels. Generally, subjects find the detection of a gap that occurs between spectrally different sounds (i.e., each side of the gap has a different frequency content) to be more difficult than the detection of a gap in a single sound, suggesting that the within-channel task is mediated by a different or additional process to that used in the across-channel task. Oxenham (2000) measured gap detection ability in stimuli where the markers (each side of the gap) differed in spectral or temporal (F0) or binaural (interaural time and intensity differences) features. He found that gap detection was most affected when the spectral content of the markers differed (meaning that within-channel processing was most important for gap detection). Temporal differences in markers of the same spectral envelope produced no effect (interaural time differences) or a small effect (F0 difference) on gap detection. In the case of F0 differences, they postulated that the temporal fluctuations in the signals made it harder to detect the gap (see also Chatterjee et al. 1998). They concluded that there was no evidence from these data that neurons are spatially tuned to temporal features of signal (e.g., modulation frequency channels).

The evidence for across-channel temporal processing in cochlear implantees is also equivocal. In this context, stimulation on two different electrodes would be considered to be separated into different channels if the two neural excitation patterns did not significantly overlap. Some experiments have shown that the ability to process temporal information that is shared across two electrode positions becomes increasingly difficult as the electrode spatial separation increases. These experiments have been interpreted as evidence that across-electrode temporal processing may occur only when the excitation patterns from the two electrodes overlap (i.e., that implantees are performing only within-channel analysis). In an experiment by McKay and McDermott (1996), two 500-Hz pulse trains, each on a separate electrode, were amplitude-modulated at 100 Hz, and the relative phase of the two pulse trains was varied. Subjects could detect differences in phase delay only when the electrode separation was less than a certain amount (varying from 2.25 to 7 mm for different subjects). In addition, the subjects heard a pitch related to the overall temporal pattern when the electrodes were very close, but heard the pitch of the individual-electrode temporal pattern only when the separation was increased. Similarly, Richardson et al. (1998) investigated the ability of implantees to detect amplitude modulation on one electrode when a concurrent modulated stimulus was applied to a second electrode. They found that the modulated stimulus on the second electrode interfered with detection of modulation on the first electrode only when the electrode separation was less than a subject-dependent critical distance. The interpretation of these experiments, that implantees perform only within-channel temporal analysis in these tasks, and do not perform across-channel analysis, was brought into question by an experiment by Carlyon et al. (2000) in which implantees could detect phase differences between stimuli on very widely separated electrodes, and with a masking stimulus located between the two electrodes. Nevertheless, it is evident that temporal analysis of signals on two electrodes is easiest when the electrodes produce overlapping excitation patterns.

The interpretation of the results of gap detection experiments, in which the two stimulus markers on either side of the gap are at different electrode positions, is not unequivocal either. Hanekom and Shannon (1998) found that the smallest detectable gap increased in duration as the electrode separation was increased. Although a similar interpretation was suggested for these results as for the experiments above (that the discrimination of a gap is dependent on direct neural activation overlap from stimuli on either side of the gap), this interpretation was later questioned by van Wieringen and Wouters (1999). The latter authors measured the effect on gap discrimination of varying stimulus complexity (number of electrodes), electrode separation within and across multichannel pre- and postgap markers, and the rate of stimulation. They also provided training in the gap-detection task, and found that the ability to detect gaps across different electrodes improved markedly with practice. They concluded from their results that the perception of gaps across separate electrode positions depended only mildly on the amount of direct neural interaction, but more on the cognitive ability of the subject to attend to the gap in a stimulus that also included other changes.

In summary, the results of all these experiments show that, as electrode separation increases, the ability of implantees to perceive temporal patterns that are distributed across electrode positions declines. This certainly means that within-channel temporal processing is very important for those tasks that were studied. However, it appears from the data so far collected that across-channel temporal processing may also occur with electrical stimulation for some perception tasks. Further research is needed to investigate the conditions under which this processing is important, and how it can be fruitfully exploited in speech-processing schemes to aid the perception of complex sounds.

3.4 Loudness of Multiple-Electrode Stimuli

In section 2.4 we discussed how the parameters of a single-electrode stimulus affected the loudness of the percept evoked. As well as the stimulus magnitude (current amplitude and/or pulse duration), the temporal pattern of the pulses in pulsatile stimuli or the frequency of continuous stimuli also affects the loudness perceived, due to the effects of central temporal integration and neural refractoriness. The discussion here includes the effects of the spatial distribution of multiple-electrode electrical stimuli on loudness.

Before discussing the above features of loudness summation in electrical stimulation, it is useful to look briefly at what happens in normal hearing when additional frequency components are added to a complex sound. The loudness of complex acoustic signals has generally been interpreted in terms of auditory filter models (Zwicker and Scharf 1965; Moore et al. 1996). First, in these models, the signal is passed through a bank of overlapping bandpass filters, and the outputs of the filters are used to determine the spatial pattern of activation along the cochlea. Second, this pattern is converted to a specific loudness pattern, representing the loudness contribution at each point along the cochlea, via a compressive power function, which models the transformation to neural excitation and to loudness. Third, the specific loudness is integrated across the cochlear place to determine the overall loudness of the sound. These models correctly predict that the loudness of a sound of constant total power with multiple components increases with increasing frequency separation of the components, up to the separation beyond which the excitation patterns of the individual components no longer overlap. This is because closer signal components directly combine within an auditory filter before the excitation pattern is determined and the compressive power function applied, whereas components that do not excite the same filter contribute independently to loudness (that is, after the compressive power function is applied). Thus, for components that fall into non-overlapping auditory filters, the total loudness is the sum of the individual loudnesses produced by each individual component. For other stimuli, where the components are more closely spaced, the overall loudness is less than the sum of the individual loudness contributions.

These acoustic models, however, cannot be applied directly to the electrical case of sequential pulsatile stimulation on multiple electrodes. The

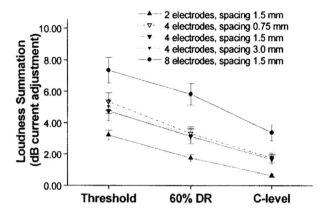

FIGURE 7.15. The amount of loudness summation when multiple electrodes are activated within the same stimulus period. The vertical axis shows the current reduction of the multiple-electrode stimulus required to equalize its loudness to that of each of its single-electrode components. The horizontal axis shows the level in each subject's dynamic range at which the measurements were taken. (Data adapted from McKay et al. 2003.)

pulses are not simultaneous, so stimulus currents are never directly combined before processing by the auditory system, even if the current distributions spatially overlap. Also, the compressive power function that relates the output of the acoustic filters to loudness is not applicable to the electrical case because it arises from the nonlinear nature of basilar membrane mechanics, contributed to by the active process of the outer hair cells. The only part of the acoustic models that we can apply to the electrical case with some degree of confidence (assuming that central auditory processing is the same in the electrical case) is the last stage. That is, the loudness contributions arising from neural activation at different cochlear places are summed to produce the total loudness. Therefore, we can predict that, provided different electrodes activate non-overlapping populations of neurons, the total loudness should be equal to the sum of the loudnesses produced by individual electrodes within the stimulus pattern. However, we know that neural excitation patterns from different electrodes *do* overlap spatially, especially when the electrode positions are close. So the important question to address in electrical stimulation is the effect of the spatial overlap in excitation distributions on loudness.

Figure 7.15 shows the effect on loudness in a multiple-electrode stimulus of adding pulses on additional electrodes while keeping the per-electrode pulse rate constant at 500 Hz. The data are mean loudness summation measurements for six subjects with the Nucleus CI24M implant, using monopolar mode. The multiple-electrode stimuli were constructed by loudness-balancing single-electrode pulse trains to a reference stimulus and then interleaving these pulse trains at two, four, or eight electrode positions. The vertical axis shows the amount of current reduction required in the multiple-electrode stimulus to keep its loudness the same as the reference single-electrode stimulus. The horizontal axis shows the level within each subject's electrical dynamic range at which the measurements were taken. There are three important features that these data show about the loudness of multiple-electrode stimuli. First, increasing the number of electrodes activated in the stimulus period had a large and consistent effect on the loudness. For example, for a stimulus with eight electrodes, each activated at their individual threshold currents, there needs to be more than a 7-dB reduction of current to keep the multiple-electrode stimulus at the threshold of hearing. This current change is the same order of magnitude as a typical dynamic range for a pulse-train stimulus. Second, there is very little effect of electrode separation apparent in the data. The data for the three electrode separations when there were four electrodes activated per period showed that the spacing of the electrodes had a nonsignificant effect compared to the effect of the number of electrodes. Third, the data show quite clearly that the current adjustments needed to overcome the effects of loudness summation are greater when the measurements are performed at lower levels.

In section 3.2 we discussed the interactions that occur at the peripheral neural level as a result of overlap in the neural excitation distributions from two electrode positions. In the absence of direct current or charge summation, the major factor in such interactions is the effect of neural refractoriness. As discussed previously, the magnitude of neural refractory effects in places where the excitation patterns overlap has not been precisely determined. We can say, however, that if neural refractory effects are significant, then there will be a reduction of total excitation compared to the sum of excitation when the two electrode positions are activated separately. Furthermore, this reduction in excitation will be larger when the relative area of overlapping excitation patterns is greater.

The most simplistic model of the relationship between neural excitation and loudness would be that loudness is related to the total amount of neural excitation produced across the cochlea in a fixed short time interval. Using this simple model, the loudness of a dual-electrode stimulus would be predicted to increase with increasing electrode separation and asymptote at a loudness equivalent to the sum of the loudnesses for the two corresponding single-electrode components. An early experiment by Tong and Clark (1986) showed such a pattern of loudness change with electrode separation in two subjects. However, more extensive measurements in later experiments (including that shown in Fig. 7.15) have not supported this initial data (McKay et al. 1995a, 2001). Interestingly, some of the data of McKay et al. (2001, 2003), particularly at low levels, showed that the loudness *increased* at the smallest separation (0.75mm) compared to greater separations. McKay et al. (2001) developed an alternative model of the relationship between neural excitation and loudness to account for the new data (see section 2.4.3.2). The nonlinear transformation in the model between (integrated) neural excitation and specific loudness allows for the possibility (seen in some of the data) of loudness increasing when electrode separation is decreased. This hypothetical model is able to predict more successfully the way loudness of complex stimuli changes with various parameters, but it needs to be tested further on a wider range of stimuli.

In summary, for multiple-electrode stimuli using biphasic sequential current pulses, one of the main stimulus features that affects loudness is the number of current pulses within a fixed stimulus time interval, with the relative positions of the electrodes playing a very minor role. That is, the overall stimulus pulse rate (not the rate per electrode) has a large effect on the loudness. McKay et al. (2003) used these basic observations to implement a simplified procedure for predicting the loudness of arbitrary pulsatile stimulus patterns. However, we are still some distance from a full understanding of how a complex electrical stimulus consisting of varying temporal and spatial patterns of pulses is processed by the central auditory system to produce an overall perception of loudness.

How does the above picture change we considering simultaneous current pulses or continuous stimulation on multiple electrodes? In section 3.2 we considered the potential channel interactions that result from simultaneous currents being present in the cochlea. We also discussed how the instantaneous phase relationships between the current waveforms on the different electrodes have an influence on the loudness perceived. A model that could predict the loudness of a complex stimulus with simultaneous currents faces an additional challenge to the case of sequential current pulses. It would be necessary to devise a method of calculating the interactions of currents at all points along the cochlea, and how these change in real time for a complex stimulus. It is possible that a general loudness model as proposed by McKay et al. (2001) would be applicable to both simultaneous and sequential currents in the cochlea. This has yet to be tested. However, the practical and general application of such a model depends on being able to predict the spatial and temporal patterns of neural excitation produced by an arbitrary electrical stimulus.

4. Hearing with Bilateral Cochlear Implants

In section 4.1 different advantages of binaural listening in normal hearing are briefly summarized.[4] Section 4.2 examines the psychophysical evidence obtained so far for the ability of bilaterally implanted people to obtain

---

[4] For a more in-depth and complete discussion of binaural listening in normal hearing, as well as reviews of the scientific literature, see the excellent chapters on binaural listening in Moore (1997) and Gelfand (1998).

similar advantages from binaural listening. As there are relatively few bilaterally implanted people at this stage, experimental results have been reported mostly for single subjects or very small subject groups. Also the range of experiments that have been completed is limited. Thus, there are some questions that remain to be answered definitively. Early results have also indicated that subjects may differ greatly in their ability to make use of binaural cues. This variation may be partly due to central auditory pathway changes following deafness (Leake and Rebscher, Chapter 4; Hartmann and Kral, Chapter 6). Also of importance is the issue of whether current speech processors and implants can accurately provide all the information about signal differences at the two ears, and the corresponding pattern of peripheral neural inputs that the listener needs for performing binaural listening tasks.

4.1 Advantages of Binaural Hearing

There are four main mechanisms whereby normal-hearing listeners gain advantage from using two ears instead of one: binaural loudness summation, the head-shadow effect, localization ability, and binaural release from masking. In this section each is briefly described.

4.1.1 Binaural Loudness Summation

Binaural loudness summation leads to a sound heard with both ears being perceived as louder than the same sound heard with only one ear. The threshold of hearing for tones (with normal hearing) is lower by about 3 dB when listening binaurally, compared to listening monaurally (provided the monaural thresholds are equal). At higher levels, a tone heard binaurally is twice the loudness of the tone heard monaurally (Marks 1978). Generally this means that about 6 to 10 dB reduction in level of the tone is required in the binaural case to make the tone equal in loudness to the monaural case. Thus, at least for suprathreshold hearing, there appears to be approximately perfect binaural loudness summation. That is, the loudness of the binaurally heard sound is the sum of the loudness produced in the two ears when heard monaurally.

4.1.2 The Head-Shadow Effect

The ability to hear signals located on either side of the head is enhanced when listening binaurally. The head-shadow effect refers to the fact that high-frequency sounds (above about 1 kHz) do not bend around the head. Thus, there is a "shadow" on the side of the head opposite to the sound, where the sound level is significantly attenuated (up to 20 dB for frequencies above 4 kHz). Low-frequency sounds can bend around the head without significant attenuation. For example, the attenuation of sounds with frequencies below 1 kHz is generally less than 2 to 5 dB, and is negligible at

200 Hz. If we had only one ear, our ability to hear a high-frequency sound located on the opposite side of the head would be significantly reduced. Binaural hearing allows us to detect such a sound with the ear located more favorably with respect to the sound source, and thus eliminates this potential disadvantage of monaural hearing.

The head-shadow effect can also be used to advantage when listening to a signal in background noise, and the signal and noise sources are in different locations. In this case, we can position our head so that, for one of our ears, the noise is attenuated by the head-shadow effect and the signal is not, thus improving the signal-to-noise ratio at that ear, and hence improving perception of the signal. Of course, the opposite is true for the ear facing the noise (the signal-to-noise ratio is lowered by the head-shadow effect). The fact that we can attend to the favorable ear and disregard the unfavorable ear means that we can often improve our signal detection (or speech perception) when using two ears instead of one. Note that this advantage is present only if the signal and noise sources are located at significantly different angles in the horizontal plane and the noise has high-frequency content.

4.1.3 Sound Localization

There are both monaural and binaural cues to sound localization in normal hearing. The monaural cues are related to how reflections of sound from the head and external ear change the spectral shape of the sound in a way specific to the direction from which the sound has arrived at the ear. These cues are particularly useful for localization of sounds in the vertical plane. Binaural cues for sound localization are related to the fact that our two ears are separated by a fixed horizontal distance (i.e., the width of our head), and these binaural cues [interaural time difference (ITD) and interaural intensity difference (IID)] are therefore useful for localizing sounds in the horizontal plane. Both IIDs and ITDs have been shown to be coded by auditory pathway neurons in the brainstem (see Hartmann and Kral, Chapter 6).

The central auditory system can extract the ITD by comparing the phase of signals arriving at each ear or computing the delay in the amplitude envelope of a fluctuating signal. For a pure tone above 700 Hz, the phase difference between the ears gives ambiguous cues for ITD, as there is an unknown number of stimulus periods between the two signals. In a complex sound, however, the comparison of phase information across frequency channels can reduce this ambiguity. For fluctuating and noncontinuous sounds, ongoing amplitude envelope variations provide ITD cues. The smallest detectable ITD ranges from 6 to 60 μs depending on the frequency content and amplitude envelope of the signal. Thus, perception of ITDs and discrimination of small differences in ITD are very powerful and precise cues for sound localization.

In general, the smallest difference in interaural intensity (IID) that can be detected by normal-hearing listeners is between 0.5 and 2.5 dB, depending on the frequency of the sound and the reference intensity (Yost and Dye 1988). IIDs are less useful for sound localization for frequencies below about 1000 Hz (because the head-shadow effect is reduced for low frequencies). Thus the relative importance of IID and ITD cues differs according to the frequency content of the sound being localized, with ITDs being important for low frequencies (and complex sounds) and IIDs being important for high frequencies.

Using independently controlled signals delivered via headphones, it is possible to investigate the separate effects of IID and ITD on the perceived location of a sound. In these experiments the sound is perceived by the subjects to be located inside their head, but *lateralized* toward one ear or the other ear. Some researchers (e.g., Harris 1960), suggest that ITDs and IIDs can be "traded off" against each other. Others (e.g., Jeffress and McFadden 1971) have found that stimuli that have "nonconsistent" ITDs and IIDs can produce two lateralized images: a time image that is related to the cycle-by-cycle ITDs and is unaffected by the size of the IID, and an intensity image that is related to both the ITDs and IIDs.

4.1.4 Binaural Release from Masking

Timing differences in a signal at the two ears also underlie the phenomenon of binaural release from masking. This refers to the fact that the ability to hear a signal in the presence of background noise is enhanced when listening with two ears compared to when listening with one ear, provided that the signal and noise are coming from different directions. At first this may seem to be explained by the effects of the head-shadow effect, as described in 4.1.2. That is, the actual signal-to-noise ratio is enhanced in the ear facing the signal source, making the signal easier to detect in that ear. However, the binaural release from masking effect is an additional phenomenon dependent on central auditory processing in the brainstem, and is not related to the actual signal-to-noise ratio changes that occur at the ear due to the head-shadow effect.

Binaural release from masking is best illustrated in experiments in which subjects listen to monaural or binaural stimuli under headphones and the threshold of a signal in noise is measured. In this case, the phenomenon is called *masking level difference* (MLD). For example, the threshold of a tone in noise is lowered by 13 to 15 dB compared to the monaural case, when the tone and noise are presented binaurally with either the tone or the noise 180 degrees out of phase between the ears (Hirsh 1948).

For signals confined to frequencies above 500 Hz, the MLD effect is smaller (for example, 3-dB reduction in signal masked threshold for frequencies above 2 kHz). It is interesting to note that, although speech detectability in noise can be improved by about 13 dB using similar binaural stimuli, the resultant advantage for speech intelligibility at suprathreshold levels in much smaller. For example, Levitt and Rabiner (1967) found that the signal level for 50% intelligibility was improved by only 3 to 6 dB. It should also be noted that the psychophysical experiments conducted under headphones generally test the condition where the difference in interaural phase relationship for the signal and noise is maximized. In natural listening conditions, this would rarely be the case, and so the advantages of binaural release from masking are generally not as high as measured by the psychophysical experiments. Nevertheless, there are many listening situations where only a small change in intelligibility can lead to a significant improvement in communication success.

It is interesting to ask whether the central auditory processing of interaural temporal cues is the same for sound localization as it is for binaural release from masking. There is evidence that the central auditory system processes temporal information in different ways for these two perceptual tasks, especially for complex signals. The ITDs which are common across frequency components of a complex sound can be perceptually integrated to improve the localization of a sound source. In contrast, the common ITDs are not used by the auditory system to improve the perception of a signal in noise (Gabriel and Colburn 1981; Durlach et al. 1986). Instead, binaural release from masking is more accurately predicted by models in which the signal is detected by means of detection of frequency regions in which the noise is de-correlated between the ears due to the presence of the phase-shifted signal (Culling and Summerfield 1995).

4.2 Binaural Listening with Two Cochlear Implants

4.2.1 Binaural Loudness Summation and Head-Shadow Effects with Bilateral Cochlear Implants

Binaural loudness summation is present for electrical hearing in the same way as for acoustic hearing. For example, van Hoesel and Clark (1997) have shown that, for two bilaterally implanted subjects, the loudness for two single-electrode pulse trains presented bilaterally was about double what it was for each stimulus presented monaurally. The currents of the single-electrode stimuli were adjusted to equalize the loudness contribution from each ear, and the electrode positions in each ear were either matched or unmatched in pitch. The principle that loudness contributions add for the two ears has more recently been confirmed for multiple-electrical stimulation (van Hoesel, personal communication).

Cochlear implantees have also been shown to be able to take advantage of the head-shadow effect when listening to speech signals in noise (van Hoesel and Tyler 2003). That is, when the signal and noise are in different locations, the implantee can attend successfully to the ear that has the better signal-to-noise ratio. This effect is discussed more fully in Wilson, Chapter 2.

4.2.2 Localization with Bilateral Cochlear Implants

Psychophysical experiments have been conducted with bilateral implantees to measure how well they can detect IIDs and ITDs in bilateral electrical stimuli. In these studies, the subject is presented with a pulse train or amplitude modulated pulse train binaurally, with varying relative intensity (or current), and varying interaural time delay. These experiments are analogous to the experiments with normal-hearing listeners using headphones. The subject is asked to indicate whether they heard a single (fused) sound, and if so whether it was heard centrally or lateralized to one side or the other. Alternatively, they may be asked to compare the perceived location of a sound with that of a reference sound.

Early studies completed with two bilaterally implanted subjects (van Hoesel et al. 1993; van Hoesel and Clark 1995, 1997) showed that they could detect very small IIDs but their sensitivity to ITDs (thresholds of more than 500 µs) was at least an order of magnitude worse than that shown by normal-hearing listeners. The threshold ITD for these subjects remained poor even in the conditions where the electrode places were matched using x-ray or pitch-matching techniques. A later report on a single subject (Lawson et al. 1998) indicated a threshold for ITD of 150 µs or less. More recently, van Hoesel and Tyler (2003) have measured ability to detect ITDs of between 100 and 200 µs in five bilaterally implanted subjects. However, in no case has the threshold ITD for a cochlear implantee been as small as those seen in normal hearing (which can be less than 10 µs). In one subject studied by Long et al. (1998), an ITD of 300 µs significantly altered the location of the fused binaural image. However, in keeping with the other results (e.g., van Hoesel and Tyler 2003), the IID variations had a more significant effect on the location of the sound image than the ITD variations. The single subject studied by Lawson et al. (1998) was only required to give a two-alternative lateralization response (left or right) and was not able to achieve 100% correct even for large ITDs, whereas her score quickly rose to 100% for small IIDs. In summary, these psychophysical experiments have shown that both IIDs and ITDs can alter the perceived location of a binaurally fused sound, but that implantees do not show the same sensitivity to ITDs as seen in normal-hearing listeners.

Other experiments have tested whether bilaterally implanted subjects can correctly localize external sounds when they are using two speech processors. All five subjects tested by van Hoesel and Tyler (2003) showed a significantly improved ability to localize external sounds in the horizontal plane compared to the situation where they were using only one implant. An example of the improvement seen in localization ability for one of their

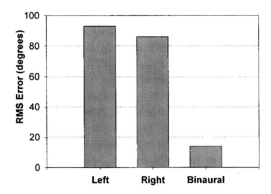

FIGURE 7.16. Improved identification of sound direction with two implants (Nucleus C124) compared to just one. The figure shows root mean square (RMS) errors for one subject who was asked to identify the correct sound source with signals presented randomly from an array of 11 loudspeakers. Loudspeakers in an anechoic chamber were spaced at 18-degree intervals in the frontal hemisphere. Signals were pink noise bursts at 60 dB sound pressure level. (Adapted from van Hoesel and Tyler 2003.)

subjects is shown in Figure 7.16. The typical error of the subjects in the localization task was only 10 degrees when using bilateral implants. This accuracy is better than that predicted by the ITD sensitivity of the same subjects, leading to the conclusion that they were using their better perception of IIDs to perform the task. There is so far only variable evidence that implantees are using ITDs in localization tasks in addition to IIDs.

In summary, investigations with bilaterally implanted subjects have shown convincing and consistent evidence of the ability of implantees to perceive IIDs, and variable ability to perceive ITDs. Experiments investigating lateralization have shown that implantees can fuse (pitch-matched) binaural stimuli into a single sound image that has a location related to IIDs and, to a lesser degree, ITDs. Improved localization of external sound sources has been shown in the laboratory, and is reported by the subjects in everyday situations.

4.2.3 Binaural Release from Masking with Bilateral Cochlear Implants

Psychophysical unmasking experiments analogous to those conducted on normal-hearing listeners under headphones have not yet been carried out with cochlear implantees. Thus it has not yet been specifically tested whether implantees can make use of different phase relationships for signal and noise in order to improve their detection of a signal, or improve their perceptual segregation of two speakers, when all other stimulus parameters (such as amplitudes) are held constant. Experiments have concentrated instead on the issue of whether speech understanding of subjects in background noise can be improved using binaural cochlear implants.

In these experiments, speech and noise stimuli are presented in the free field, and the electrical stimuli are generally the output of bilaterally fitted speech processors. The interpretation of the results of such experiments is a little tricky, as care has to be taken to consider the effects of binaural loudness summation and head-shadow effects, as well as the signal processing done by the speech processor, when deciding whether there is a benefit attributable to binaural release from masking. However, there is considerable subject variability in this type of test, and the results would have to considered equivocal at this stage. van Hoesel and Tyler (2003) found that, averaged across four subjects, the speech reception threshold in noise was improved in binaural listening by less than 2 dB ($p = .04$), compared to the best monaural score (from the ear facing away from the noise). They used a novel bilateral processing scheme (PDT, for peak derived timing) in which the fine-timing cues are preserved in each ear. This processing scheme was developed because of a concern that the output of current commercially available speech processors may not produce a temporal and spatial pattern of neural excitation in each ear that would allow the central binaural processing of de-correlations between the ears. More research is needed at the psychophysical level to ascertain the electrical stimulus patterns that can produce an unequivocal binaural release from masking effect.

5. Summary and Future Research

We have discussed the basic psychophysical percepts that underlie the ability of normal-hearing people to detect, discriminate, scale, and identify sounds in the environment. The restoration of these hearing abilities to profoundly deaf individuals is the goal of cochlear implants. As we have seen, cochlear implantees in general can perform these hearing tasks, but with variable success. A major goal of future research is to gain a better understanding of this individual variability so that implants and sound-processing schemes can be optimized to maximize the hearing ability of all implantees.

There are several areas in which implantees currently cannot perform auditory tasks that normal-hearing people take for granted. One of these is the perception of musical pitch. The ability to easily extract the pitch from a complex sound underlies our enjoyment of musical performance. To hear the pitch of such a sound (such as the cello note illustrated in Fig. 7.13), it is necessary for the listener either to resolve some of the individual harmonics in the sound, or to perceive the fundamental frequency (F0) of the sound as amplitude fluctuations. For example, the cello note illustrated in Figure 7.13 has a fundamental frequency of 494 Hz. Thus the individual harmonics in the note are spaced at intervals of 494 Hz. This fundamental frequency is too high to be reliably detected or perceived by implantees as an amplitude-modulation pitch (see section 2.3.2). Also, the spacing of the harmonics cannot be perceived from the spatial pattern since the allocation of frequency ranges to electrode positions is fixed and the frequency spacing is quite broad. This problem also has important consequences for understanding tonal languages, where the speaker's fundamental frequency modulations provide important cues for speech understanding (for a review, see Zeng 1995).

The difficulty that an implantee has in perceiving the structure of a complex sound is compounded further when there is more than one sound present, for example more than one instrument is playing, or someone is singing with musical accompaniment, or two people are talking. As well as perceiving the harmonic structure of each sound present, listeners must be able to sort out which harmonic components belong to which sound source so that they can perceptually separate the different sounds. For example, if two people are speaking at the same time, a normal-hearing listener can hear each voice as a separate entity and hence can follow the conversation with one of the speakers. As another example, a person may be singing a song accompanied by a musical instrument (or several instruments). A normal-hearing person is able to follow the tune and words of the singer, and hear the separate musical accompaniment as a pleasing harmonic addition. In contrast, a cochlear implantee finds both of these situations very difficult because the additional sounds cannot be separated from the sound they most want to hear. These difficult perceptual tasks pose a challenge for future cochlear implant psychophysical research.

Cochlear implants have provided a variably successful means of partially restoring hearing abilities to profoundly deaf people. Understanding this variability and devising ways to compensate individually for differences among impalntees is a major focus of current psychophysical research. There are, unfortunately, some aspects of normal hearing that are not currently restored by these devices, and that will be the focus of future psychophysical research.

References

Bruce IC, Irlicht LS, White MW, O'Leary SJ, et al. (1999) A stochastic model of the electrically stimulated auditory nerve: pulse-train response. IEEE Trans Biomed Eng 46:617–629.

Burns EM, Viemeister NF (1976) Nonspectral pitch. J Acoust Soc Am 60:863–869.

Busby PA, Tong YC, Clark GM (1993) The perception of temporal modulations by cochlear implant patients. J Acoust Soc Am 94:124–131.

Carlyon RP, Geurts L, Wouters J (2000) Detection of small across-channel timing differences by cochlear implantees. Hear Res 141:140–154.

Carlyon RP, van Wieringen A, Long CJ, Deeks JM, Wouters J (2002) Temporal pitch mechanisms in acoustic and electric hearing. J Acoust Soc Am 112:621–633.

Cazals Y, Pelizzone M, Kasper A, Montandon P (1991) Indication of a relation between speech perception and temporal resolution for cochlear impalntees. Ann Otol Rhinol Laryngol 100:893–895.

Chatterjee M, Shannon RV (1998) Forward masked excitation patterns in multi-electrode electrical stimulation. J Acoust Soc Am 103:2565–2572.

Chatterjee M, Fu QJ, Shannon RV (1998) Within-channel gap detection using dissimilar markers in cochlear implant listeners. J Acoust Soc Am 103:2515–2519.

Chatterjee M, Fu Q-J, Shannon RV (2000) Effects of phase duration and electrode separation on loudness growth in cochlear implant listeners. J Acoust Soc Am 107:1637–1644.

Cohen LT, Busby PA, Clark GM (1996a) Cochlear implant place psychophysics. 2. Comparison of forward masking and pitch estimation data. Audiol Neurootol 1:278–292.

Cohen LT, Busby PA, Whitford LA, Clark GM (1996b) Cochlear implant place psychophysics 1. Pitch estimation with deeply inserted electrodes. Audiol Neurootol 1:265–277.

Collins LM, Zwolan T, Wakefield GH (1997) Comparison of electrode discrimination, pitch ranking, and pitch scaling data in postlingually deafened adult cochlear implant subjects. J Acoust Soc Am 101:440–455.

Culling JF, Summerfield Q (1995) Perceptual separation of concurrent speech sounds: absence of across-frequency grouping by common interaural delay. J Acoust Soc Am 98:785–797.

Donaldson GS, Viemeister NF (2000) Intensity discrimination and detection of amplitude modulation in electric hearing. J Acoust Soc Am 108:760–763.

Donaldson GS, Viemeister NF, Nelson DA (1997) Psychometric functions and temporal integration in electric hearing. J Acoust Soc Am 101:3706–3721.

Durlach NI, Gabriel KJ, Colburn HS, Trahiotis C (1986) Interaural correlation discrimination: II. Relation to binaural unmasking. J Acoust Soc Am 79:1548–1557.

Eddington DK, Dobelle WH, Brackmann DE, Mladejovsky MG, Parkin JL (1978) Auditory prostheses research with multiple channel intracochlear stimulation in man. Ann Otol Rhinol Laryngol 87:5–39.

Fu QJ (2002) Temporal processing and speech recognition in cochlear implant users. NeuroReport 13:1635–1639.

Gabriel KJ, Colburn HS (1981) Interaural correlation discrimination: I. Bandwidth and level dependence. J Acoust Soc Am 69:1394–1401.

Gelfand SA (1998) Hearing. An Introduction to Psychological and Physiological Acoustics. New York: Marcel Dekker.

Hanekom JJ, Shannon RV (1998) Gap detection as a measure of electrode interaction in cochlear implants. J Acoust Soc Am 104:2372–2384.

Harris GG (1960) Binaural interactions of impulsive stimuli and pure tones. J Acoust Soc Am 32:685–692.

Henshall KR, McKay CM (2001) Optimizing electrode and filter selection in cochlear implant speech processor maps. J Am Acad Audiol 12:478–489.

Hirsh IJ (1948) The influence of interaural phase on interaural summation and inhibition. J Acoust Soc Am 20:536–544.

Jeffress LA, McFadden D (1971) Differences of interaural phase and level in detection and lateralization. J Acoust Soc Am 49:1169–1179.

Jolly CN, Spelman FA, Clopton BM (1996) Quadrupolar stimulation for cochlear prostheses: modeling and experimental data. IEEE Trans Biomed Eng 43:857–865.

Lawson DT, Wilson BS, Zerbi M, van den Honert C, et al. (1998) Bilateral cochlear implants controlled by a single speech processor. Am J Otol 19:758–761.

Levitt H, Rabiner LR (1967) Binaural release from masking for speech and gain in intelligibility. J Acoust Soc Am 42:601–608.

Long CJ, Eddinton DK, Colburn HS, Rabinowitz WM, Whearty ME, Kadel-Garcia N (1998) Speech processors for auditory prostheses. Quarterly report of NIH contract N01-DC-6-2100, pp. 1–13.

Marks LE (1978) Binaural summation of the loudness of pure tones. J Acoust Soc Am 64:107–113.

McDermott HJ, McKay CM (1997) Musical pitch perception with electrical stimulation of the cochlea. J Acoust Soc Am 101:1622–1631.

McKay CM, Carlyon RP (1999) Dual temporal pitch percepts from acoustic and electric amplitude-modulated pulse trains. J Acoust Soc Am 105:347–357.

McKay CM, McDermott HJ (1996) The perception of temporal patterns for electrical stimulation presented at one or two intracochlear sites. J Acoust Soc Am 100:1081–1092.

McKay CM, McDermott HJ (1998) Loudness perception with pulsatile electrical stimulation: the effect of interpulse intervals. J Acoust Soc Am 104:1061–1074.

McKay CM, McDermott HJ (1999) The perceptual effects of current pulse duration in electrical stimulation of the auditory nerve. J Acoust Soc Am 106:998–1009.

McKay CM, McDermott HJ, Clark GM (1994) Pitch percepts associated with amplitude-modulated current pulse trains in cochlear implantees. J Acoust Soc Am 96:2664–2673.

McKay CM, McDermott HJ, Clark GM (1995a) Loudness summation for two channels of stimulation in cochlear implants: effects of spatial and temporal separation. Ann Otol Rhinol Laryngol Suppl 166:230–233.

McKay CM, McDermott HJ, Clark GM (1995b) Pitch matching of amplitude-modulated current pulse trains by cochlear implantees: the effect of modulation depth. J Acoust Soc Am 97:1777–1785.

McKay CM, McDermott HJ, Clark GM (1996) The perceptual dimensions of single-electrode and nonsimultaneous dual-electrode stimuli in cochlear implantees. J Acoust Soc Am 99:1079–1090.

McKay CM, McDermott HJ, Carlyon RP (2000) Place and temporal cues in pitch perception: Are they truly independent? 1:25–30.

McKay CM, Remine MD, McDermott HJ (2001) Loudness summation for pulsatile electrical stimulation of the cochlea: Effects of rate, electrode separation, level, and mode of stimulation. J Acoust Soc Am 110:1514–1524.

McKay CM, Henshall KR, Farrell RJ, McDermott HJ (2003) A practical method of predicting the loudness of complex electrical stimuli. J Acoust Soc Am 113:2054–2063.

Moore BCJ (1997) An Introduction to the Psychology of Hearing. London: Academic Press.

Moore BCJ, Glasberg BR (1983) Masking patterns for synthetic vowels in simultaneous and forward masking. J Acoust Soc Am 73:906–917.

Moore BCJ, Glasberg BR, Gaunt T, Child T (1991) Across-channel masking of changes in modulation depth for amplitude- and frequency-modulated signals. Q J Exp Psychol 43A:327–347.

Moore BCJ, Peters RW, Glasberg BR (1996) Detection of decrements and increments in sinusoids at high overall levels. J Acoust Soc Am 99:3669–3677.

Nelson DA (1994) Level-dependent critical bandwidth for phase discrimination. J Acoust Soc Am 95:1514–1524.

Nelson DA, Van Tasell DJ, Schroder AC, Soli S, Levine S (1995) Electrode ranking of "place pitch" and speech recognition in electrical hearing. J Acoust Soc Am 98:1987–1999.

Nelson DA, Schmitz JL, Donaldson GS, Viemeister NF, Javel E (1996) Intensity discrimination as a function of stimulus level with electric stimulation. J Acoust Soc Am 100:2393–2414.

Oxenham AJ (2000) Influence of spatial and temporal coding on auditory gap detection. J Acoust Soc Am 107:2215–2223.

Patterson RD, Johnson-Davies D, Milroy R (1978) Amplitude-modulated noise: the detection of modulation versus the detection of modulation rate. J Acoust Soc Am 63:1904–1911.

Pijl S, Schwarz DW (1995) Melody recognition and musical interval perception by deaf subjects stimulated with electrical pulse trains through single cochlear implant electrodes. J Acoust Soc Am 98:886–895.

Richardson LM, Busby PA, Clark GM (1998) Modulation detection interference in cochlear implant subjects. J Acoust Soc Am 104:442–452.

Schiffman SS, Reynolds LM, Young FW (1981) Introduction to Multidimensional Scaling: Theory, Methods and Applications. New York: Academic Press.

Shannon RV (1983a) Multichannel electrical stimulation of the auditory nerve in man. I. Basic Psychophysics. Hear Res 11:157–189.

Shannon RV (1983b) Multichannel electrical stimulation of the auditory nerve in man. II. Channel interaction. Hear Res 12:1–16.

Shannon RV (1985) Threshold and loudness functions for pulsatile stimulation of cochlear implants. Hear Res 18:135–143.

Shannon RV (1989) Detection of gaps in sinusoids and pulse trains by patients with cochlear implants. J Acoust Soc Am 85:2587–2592.

Shannon RV (1992) Temporal modulation transfer functions in patients with cochlear implants. J Acoust Soc Am 91:2156–2164.

Shepherd RK, Javel E (1999) Electrical stimulation of the auditory nerve: II. Effect of stimulus waveshape on single fibre response properties. Hear Res 130:171–188.

Stainsby TH (2000) The perception of musical sounds with cochlear implants. PhD Thesis, the University of Melbourne.

Strickland EA, Viemeister NF, Fantini DA, Garrison MA (1989) Within- versus cross-channel mechanisms in detection of envelope phase disparity. J Acoust Soc Am 86:2160–2166.

Tong YC, Clark GM (1986) Loudness summation, masking, and temporal interaction for sensations produced by electric stimulation of two sites in the human cochlea. J Acoust Soc Am 79:1958–1966.

Tong YC, Dowell RC, Blamey PJ, Clark GM (1983a) Two-component hearing sensations produced by two-electrode stimulation in the cochlea of a deaf patient. Science 219:993–994.

Tong YC, Blamey PJ, Dowell RC, Clark GM (1983b) Psychophysical studies evaluating the feasibility of a speech processing strategy for a multiple-channel cochlear implant. J Acoust Soc Am 74:73–80.

Townshend B, Cotter N, Van Compernolle D, White RL (1987) Pitch perception by cochlear implant subjects. J Acoust Soc Am 82:106–115.

van Hoesel RJ, Clark GM (1995) Fusion and lateralization study with two binaural cochlear implant patients. Ann Otol Rhinol Laryngol Suppl 166:233–235.

van Hoesel RJ, Clark GM (1997) Psychophysical studies with two binaural cochlear implant subjects. J Acoust Soc Am 102:495–507.

van Hoesel RJ, Tyler RS (2003) Speech perception, localization, and lateralization with bilateral cochlear implants. J Acoust Soc Am 113:1617–1630.

van Hoesel RJ, Tong YC, Hollow RD, Clark GM (1993) Psychophysical and speech perception studies: a case report on a binaural cochlear implant subject. J Acoust Soc Am 94:3178–3189.

van Wieringen A, Wouters J (1999) Gap detection in single- and multiple-channel stimuli by Laura cochlear implantees. J Acoust Soc Am 106:1925–1939.

Viemeister NF (1979) Temporal modulation transfer functions based upon modulation thresholds. J Acoust Soc Am 66:1364–1380.

Viemeister NF, Wakefield GH (1991) Temporal integration and multiple looks. J Acoust Soc Am 90:858–865.

Yost WA, Dye RH Jr (1988) Discrimination of interaural differences of level as a function of frequency. J Acoust Soc Am 83:1846–1851.

Yost WA, Sheft S (1989) Across-critical-band processing of amplitude-modulated tones. J Acoust Soc Am 85:848–857.

Yost WA, Popper AN, Fay RR, eds. (1993) Human Psychophysics, SHAR vol. 3. New York: Springer-Verlag.

Zeng FG (1995) Cochlear implants in China. Audiology 34:61–75.

Zeng FG (2004) Compression in cochlear implants. In: Bacon SP, Popper AN, Fay RR, eds. Compression: From Cochlea to Cochlear Implants. SHAR series. New York: Springer-Verlag pp. 184–220.

Zeng FG, Shannon RV (1992) Loudness balance between electric and acoustic stimulation. Hear Res 60:231–235.

Zeng FG, Shannon RV (1994) Loudness-coding mechanisms inferred from electric stimulation of the human auditory system. Science 264:564–566.

Zeng F-G, Shannon RV (1999) Psychophysical laws revealed by electric hearing. NeuroReport 10:1931–1935.

Zeng F-G, Galvin JJr, Zhang C (1998) Encoding loudness by electric stimulation of the auditory nerve. NeuroReport 9:1845–1848.

Zhang C, Zeng FG (1997) Loudness of dynamic stimuli in acoustic and electric hearing. J Acoust Soc Am 102:2925–2934.

Zwicker E, Scharf B (1965) A model of loudness summation. Psych Rev 72:3–26.

8
Speech Perception with Cochlear Implants

ROBERT V. SHANNON, QIAN-JIE FU, JOHN GALVIN and
LENDRA FRIESEN

1. Introduction

Cochlear implants, besides restoring hearing sensation to otherwise deaf individuals, provide an excellent tool with which to investigate how the human central nervous system (CNS) processes complex patterns of sensory information. Throughout the lifetime of normal-hearing persons, the auditory CNS has been continually trained to extract meaningful speech (and other meaningful sounds) from a constant barrage of auditory sensory information. The CNS establishes networks to process auditory sensory information; for complex pattern recognition tasks, these networks can take as long as 10 to 12 years to fully develop (see Hartmann and Kral, Chapter 6). Once these networks are fully mature, auditory pattern recognition is highly robust to degradations in the sensory signal, as revealed by decades of speech perception research. For example, military cryptologists in the 1940s searched for a type of signal degradation that would render speech unintelligible during transmission (but could be decoded at the receiving end by reversing the degradation, thereby restoring intelligibility). To their amazement, even severe alterations to the speech signal did not destroy its intelligibility. One of the most well-known examples is the work of Licklider and Pollack (1948), who eliminated all amplitude information of the speech signal by means of "infinite clipping" (the signal waveform was simply absent or present, according to an amplitude threshold). Although the speech signal was effectively reduced to zeros and ones, the speech remained highly intelligible.

When people become severely hearing impaired or deaf, their central pattern recognition mechanisms do not receive enough sensory information to understand speech. But as Licklider's experiment showed, considerable information can be deleted from the speech signal with only minor deleterious effects. In designing prosthetic devices for deaf or hearing-impaired listeners, we need to understand the relative importance of different acoustic features found in speech. Because present cochlear implant technology cannot preserve all the acoustic features found in speech, those

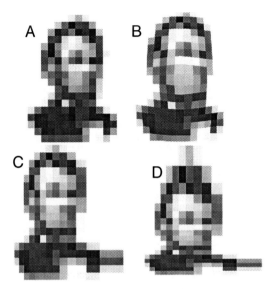

FIGURE 8.1. Visual example of pattern recognition with reduced resolution. Familiar picture of Abraham Lincoln is recognizable even when highly pixelated (A), but not when distorted by bulging (B), logarithmic skewing on one axis (C), or log-log skewing on both axes (D).

features that are critically important for speech recognition must be preserved, while those that are less important may be sacrificed. The question is: Which features are most important?

Consider the following example to clarify the interaction between the quality of information provided by the sensory system and the brain's pattern recognition. Figure 8.1A shows a familiar picture that has been "pixelized"; the visual sensory information in the picture has been reduced to a coarse grid of shaded squares. Yet most people will recognize that this is a picture of a man, and many people (at least in the United States) will even be able to identify the man in the picture—Abraham Lincoln. We are able to identify Lincoln because this is not just any picture of the president—it is the most familiar picture of Lincoln: the one on the $5 bill. If the same degree of pixelization were applied to a less familiar picture (even an unfamiliar picture of Lincoln), most people would not be able to identify the person in the picture. An unfamiliar picture would require much more sensory resolution (smaller pixels) before it was recognizable.

Auditory and visual pattern recognition seem to share some general properties. Speech recognition is a highly trained and well-practiced skill based on auditory pattern recognition, particularly for familiar talkers in the listener's native language. Recognizing familiar speech may be analogous to the recognizing a familiar picture of Lincoln—a coarse representation may be enough. However, if the talker is unfamiliar, or speaking with a heavy accent, or talking in a noisy environment, additional sensory detail may be required to recognize the words, just as a less familiar picture would require additional visual detail.

With a prosthetic device like the cochlear implant, the sensory pattern provided to the brain may be coarse as well as distorted, due to idiosyncrasies in individuals' surviving neural populations, the number and location implanted electrodes, and/or speech processing of the acoustic signal. Thus, for many cochlear implant patients, the brain must overcome both the reduced sensory resolution of and distortion to the auditory pattern. A visual analogy is provided in Figure 8.1B–D. Even if people were able to recognize the image of Lincoln when it is pixilated (Fig. 8.1A), it is unlikely that they would be able to recognize a pixelized Lincoln that had been distorted by bulging in the middle (Fig. 8.1B), logarithmically warping along the horizontal axis (log-Lincoln, Fig. 8.1C), or logarithmically warping along both horizontal and vertical axes (log-log Lincoln, Fig. 8.1D). For some types of distortion, even better sensory resolution (smaller pixels) may not be sufficient to restore recognition. To design better cochlear implant speech processors, it is important to understand the factors and parameters that are most important for auditory pattern recognition.

2. Signal Processing for Cochlear Implants

Noise-band vocoders (Shannon et al. 1995) have been used to "pixelize" the auditory spectrum in a manner that is analogous to the pixelized image of Lincoln. This type of processing is also similar to that performed by cochlear implant speech processors. Many of the experiments described in the following sections were performed with normal-hearing subjects listening to noise-band vocoders that simulated features of cochlear implant signal processing.

Figure 8.2 shows a block diagram of a noise-band vocoder and Figure 8.3 shows examples of speech processed with a four-channel noise-band vocoder. First, the acoustic signal is spectrally analyzed by four bandpass filters (typically having 12 to 18 dB/octave filter slopes). Next, the time-amplitude envelope is extracted from each band by half-wave rectification, followed by low-pass filtering (typically, −6 to −12 dB/octave at 160 Hz). The extracted envelope from each channel is then used to modulate wide-band (white or pink) noise; each noise band is then bandpass filtered with the filters used to spectrally analyze the original acoustic signal. The result is a series of spectrally contiguous noise bands, each of which is modulated in time by the envelope of its respective acoustic spectral region. Figure 8.3A shows the spectrogram of the unprocessed phrase "shoo-cat"; Figure 8.3B shows the spectrogram of "shoo-cat" after processing by a four-channel noise-band vocoder. Note that the gross spectral distribution of energy and temporal envelope and preserved, but the fine structure in both the spec-

Noise-Band Processor (4 bands)
Shannon et al., Science, 1995

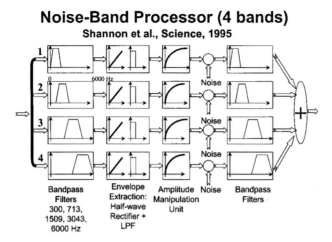

FIGURE 8.2. Block diagram of a noise-band vocoder. Signals are bandpass filtered, the envelope is extracted from each channel by rectification and low-pass filtering, and the resulting envelope signal is used to modulated a noise band.

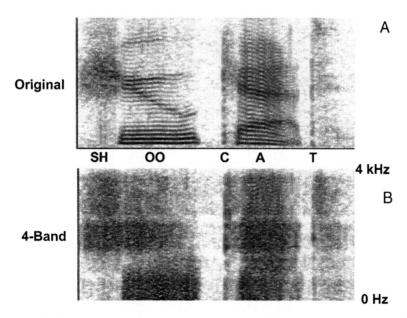

FIGURE 8.3. Spectrograms of a sentence comparing the original (upper or A, B), with a noise-band vocoded version of the same sentence in which the spectral resolution is reduced to four (lower) modulated noise bands.

tral and temporal domains is removed by this processing. Despite the dramatic reduction in spectral and temporal detail, four-channel noise-vocoded speech remains highly intelligible (Shannon et al. 1995). Note the similarity between the pixelized picture of Lincoln shown in Figure 8.1A and the noise-band vocoded speech shown in Figure 8.3B: visual information is quantized in terms of shape and contrast across the receptor array (which corresponds to visual space), while auditory information is quantized in terms of frequency (which corresponds to distance across the auditory receptor array).

Cochlear implant speech-processing attempts to replace the function of the cochlea that is relevant for speech understanding. Acoustic signals are analyzed into different frequency bands and the speech information from each band is presented to an electrode along the scala tympani that represents the corresponding frequency region. In the 1970s, most cochlear implants either presented an amplitude-compressed version of the acoustic signal directly to the electrodes, or attempted to extract important speech features from the acoustic signal and present the extracted information to the electrodes in coded form (see Wilson, Chapter 2; Niparko, Chapter 3). The disadvantage of analog electrical stimulation was that electrical fields from adjacent electrodes would add and subtract, depending on the instantaneous waveform presented to each electrode. This type of electrical field interaction was so problematic with early implant designs that new strategies were designed in which the electrodes were stimulated nonsimultaneously, with electrical pulses interleaved in time. These nonsimultaneous stimulation strategies evolved into the commonly used continuous interleaved sampling (CIS) strategy (Wilson et al., 1991); CIS processing is quite similar to the noise-band vocoder processing shown in Figure 8.2. However, in CIS processing for implants, an amplitude mapping function is utilized after the acoustic envelope extraction to map signal amplitudes from the acoustic into electrical domains. In addition, in CIS processing for implants, the carrier signal is a biphasic electrical pulse train (rather than a noise band). The output of each CIS processing channel is a modulated electrical pulse train, which is delivered to an electrode placed in the cochlea.

In later-generation implant devices, increased processing speeds and stimulation rates allowed feature extraction processors to provide more speech information. These feature-extraction techniques evolved to be the presently used spectral peak (SPEAK) and advanced combined encoder (ACE) strategies. These signal-processing strategies are discussed in more detail in Chapter 2. In this chapter we describe the types of acoustic cues necessary for speech recognition and how those cues are coded for cochlear implant listeners.

Many psychophysical capabilities of implant patients have been characterized in research laboratory settings (see McKay, Chapter 7). However, in a clinical setting, there is usually not enough time to measure all the psychophysical abilities of each patient. Thus, it is important to know which speech-processor parameters are most critical for individual patients' optimal use of the implant, and obtain psychophysical measures for those parameters in the limited time available to clinicians (Allen 1994). If speech recognition is relatively robust to a poorly set parameter, then the clinician need not spend time optimizing that parameter carefully for each individual patient. Rather, clinical time should be spent measuring those parameters that vary significantly across patients and that have a significant effect on speech recognition. In the following sections, we describe the results of speech-processor manipulations to quantify the effects of various parameters on speech recognition in both normal-hearing (NH) and cochlear implant (CI) listeners: amplitude, temporal, spectral, and binaural cues.

3. Amplitude Cues

Licklider and Pollak's (1948) aforementioned "infinite clipping" studies suggest that amplitude cues are of little importance for speech recognition; speech remained perceptible despite the complete removal of amplitude information. However, amplitude cues may be more important for speech recognition by hearing-impaired listeners (Turner et al. 1995) and cochlear implant users (Fu and Shannon 1998; Zeng and Galvin 1999; Zeng et al. 2002). Hearing-impaired (HI) and implant listeners have greatly reduced dynamic ranges and distorted loudness growth functions, compared to those of NH listeners. If the prosthetic device (hearing aid or implant) does not map loudness correctly from the acoustic to the prosthetic domain, the relative loudness relations within speech sounds will be distorted, and possibly result in reduced recognition (Plomp 1988; Freyman et al. 1989, 1991; Drullman 1995). Normal-hearing listeners may be more resistant to amplitude distortions because the full spectral processing mechanisms remain intact. A correct amplitude mapping may be more important for HI listeners who have reduced spectral selectivity (Van Tasell et al. 1987, 1992), or in implant listeners whose spectral selectivity is limited by the number of implanted electrodes (Zeng and Galvin 1999; Loizou et al. 2000a). Let us review some studies that have measured speech recognition as a function of parametric distortions to the amplitude mapping functions in CI listeners, and in NH subjects listening to implant simulations.

3.1 Loudness Mapping Function

Most contemporary commercial cochlear implant speech processors use a logarithmic function to map acoustic sound pressure to electrical current (in microamperes). Fu and Shannon (1988) investigated the effect of manip-

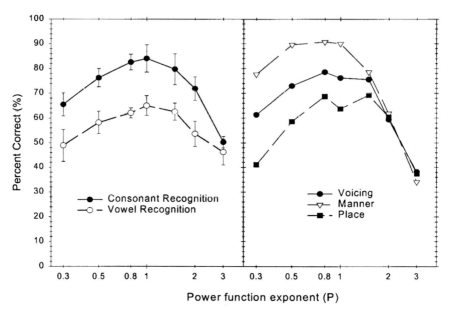

FIGURE 8.4. Effect of amplitude mapping on consonant and vowel recognition by normal-hearing (NH) listeners ($n = 7$). (From Fu and Shannon 1998.)

ulations of this logarithmic amplitude mapping function on speech recognition. For all experimental speech stimuli, amplitudes were measured for the acoustic envelope extracted from each band of a multiband speech processor, and a probability histogram was compiled for the frequency of occurrence of each amplitude value. The acoustic maximum (>99th percentile) was mapped to the electrically stimulated maximum comfortable loudness, and the acoustic minimum (<1st percentile) was mapped to electrically stimulated threshold. Within these two end points, the acoustic amplitude levels (A) were mapped to electrical levels (E) by a power law, $E = A^p$. Varying the value of the exponent p changes the shape of the acoustic-to-electrical mapping function. Assuming that an optimal value of p best preserves the normal loudness relationships within speech sounds (Zeng and Shannon 1992, 1994, 1999; Zeng and Galvin 1999; Zeng et al. 1998), values of p larger or smaller than this optimal value will distort the loudness relations within speech. Figures 8.4 and 8.5 show the results of varying the exponent p of the amplitude mapping function on consonant and vowel recognition (multitalker sets of 16 medial consonants, a/C/a, and 12 medial vowels, h/V/d). Figure 8.4 shows average vowel and consonant recognition for seven NH subjects listening with a four-channel noise-band processor, and Figure 8.5 shows similar data for three CI subjects listening with custom four-channel CIS cochlear implant processors.

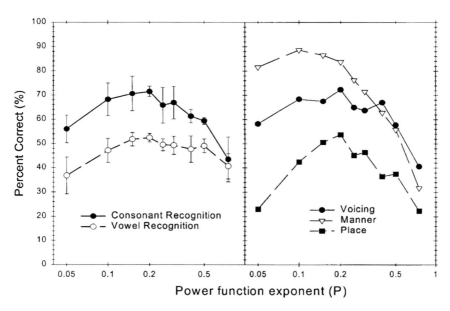

Figure 8.5. Effect of amplitude mapping on consonant and vowel recognition by cochlear implant (CI) listeners ($n = 3$). (From Fu and Shannon 1998.)

Functions for both NH and CI listeners show a broad, shallow peak in performance. NH listeners performed best when the exponent $p = 1.0$ (resulting in a linear amplitude mapping between the acoustic and simulated electrical domains). CI listeners performed best when the exponent $p = 0.2$ (slightly more compressive than logarithmic amplitude mapping used in their regular speech processors). Loudness growth functions were also measured in these implant listeners using magnitude estimation; the resulting estimates were well fit by a power function with an exponent of 2.72 (Fu and Shannon 1998). Because NH listeners' loudness growth functions are generally well fit by a power function with an exponent of 0.6, the predicted cross-modal loudness match between acoustic and electrical amplitude should have an exponent of the ratio between these two exponents ($0.6/2.72 = 0.22$). This value agrees quite well with the peak in the performance function of Figure 8.5, indicating that best performance is achieved when the loudness function is mapped properly from acoustic to electrical amplitudes. However, note that performance drops only 10 to 15 percentage points even when the amplitude functions are mapped poorly with an exponent that is twice or half of the optimal value. Overall, these results show that while the best performance is achieved when loudness is properly mapped, the acoustic-to-electrical amplitude mapping is not a critical parameter for speech perception by CI users (at least in quiet testing conditions). Misestimating the loudness

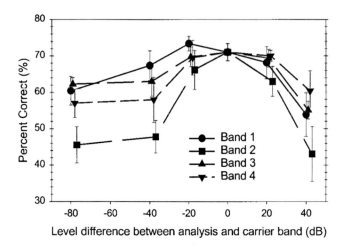

FIGURE 8.6. Effect of one misadjusted band in four-channel noise-band processor on vowel recognition ($n = 4$). (From Fu 1997.)

exponent by as much as a factor of 2 has only a relatively small effect on performance.

Note that even the amount of information transmitted via voicing, manner, and place cues (Miller and Nicely 1955) was similar for implant and acoustic listeners (right panels of Figures 8.4 and 8.5), as a function of the amplitude mapping exponent p. In fact, the amount of information received via voicing and manner cues was almost identical for NH and CI listeners. The primary difference between NH and CI listeners' overall level of performance was due to differences in the amount of information received from place cues (right panels).

3.2 Errors in Loudness Mapping on One Electrode

Fu (1997) measured the effect of local errors in amplitude mapping, i.e., distortions to the acoustic-to-electrical mapping on only one electrode in the array. NH subjects listened to a four-channel noise-band acoustic simulation of a cochlear implant speech processor. The gain of one of the four channels' output was amplified/attenuated to simulate an erroneously measured dynamic range for one of four electrodes in a CI patient. The results (Figures 8.6 and 8.7) show that vowel recognition was remarkably robust to such an erroneously measured dynamic range. Speech recognition was affected only when one channel's gain was amplified/attenuated more than 20 dB, relative to the other three channels (Fig. 8.6). This effect was reduced as the number of electrodes was increased, as shown in Figure 8.7. With the

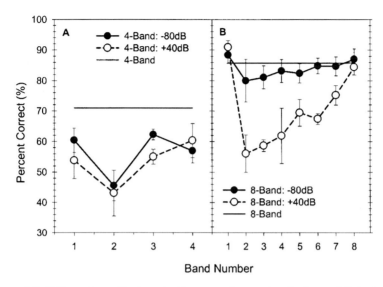

FIGURE 8.7. Effect of amplitude misadjustment in one band of a multiband noise processor ($n = 4$). (From Fu 1997.)

four-channel processor (Fig. 8.7, left panel), attenuation of band 2 had the largest effect, most likely because this band contained the envelope from the second formant region of speech. In contrast, turning bands 1, 3, or 4 off (−80 dB) caused performance to drop by only 10 percentage points. With the eight-channel processor (Fig. 8.7, right panel), performance dropped by less than 10 percentage points when any single band was turned off (−80 dB). However, amplifying (+40 dB) any one of bands 2 through 7 significantly degraded speech recognition; performance dropped as much as 30 percentage points when band 2 was amplified by 40 dB, presumably because of masking effects. Thus, with increased spectral resolution (i.e., more electrodes/channels), underestimating any single electrode's dynamic range does not affect performance as much as overestimating an electrode's dynamic range.

3.3 Custom Loudness Functions for Each Electrode

Cosendai and Pelizzone (2001) compared speech recognition between processors that implemented a standard logarithmic loudness function on all electrodes and one that implemented a customized loudness function on each electrode. Their results showed a modest improvement (10–15 percentage points) in sentence recognition with the custom loudness functions.

3.4 Peak Clipping and Center Clipping

If the parameters of the amplitude mapping function are not set properly in a cochlear implant, the amplitude envelopes of speech could be truncated, resulting in either peak clipping or center clipping. For example, underestimating the stimulation thresholds of electrodes would result in low-amplitude portions of the acoustic envelope being presented at sub-audible levels (i.e., center clipping). An overly compressive mapping would result in high-amplitude portions of the acoustic envelope being presented at or near the maximum comfort level of electrical stimulation (i.e., peak clipping).

Drullman (1995) studied the effects of peak and center clipping on the speech perception of 60 NH listeners. Center clipping was found to be slightly more detrimental than equivalent amounts of peak clipping, but neither truncation significantly affected speech recognition until the clipping was severe (>50% amplitude range).

Shannon et al. (2001a) studied the effects of peak and center clipping in conditions of reduced spectral resolution with NH listeners using acoustic noise-band simulations of cochlear implant speech processors. The acoustic speech envelope amplitudes were measured for each frequency band of the four-channel processors. The amplitude mapping functions were set to clip the input amplitudes at the 25th, 50th, or 75th percentile of the total amplitude envelope distribution. Figure 8.8 shows recognition results for multitalker vowels (10 male, 10 female talkers) and consonants (15 talkers) by seven NH subjects. The left panel shows the results when the acoustic input was peak clipped. The middle panel shows results when the peak-clipped acoustic input was then expanded across the entire output dynamic range. The right panel shows the results when the acoustic input was center clipped. Note that 25% of the amplitude distribution could be peak and/or center clipped with only minimal effect on vowel or consonant recognition. Similar results were reported in cochlear implant users (Zeng and Galvin 1999; Zeng et al. 2002).

3.5 Amplitude Quantization

Zeng and Galvin (1999) reported the results of experiments that reduced the number of amplitude steps in Nucleus-22 implant fitted with the SPEAK processing strategy. Even when the amplitudes for each electrode were limited to just two levels (on or off), sentence recognition was not noticeably affected. This result may be partially due to the nature of the SPEAK processing strategy, in which only the six highest amplitude frequency bands are represented for each 4-ms interval. The SPEAK processing strategy implicitly performs center clipping on the acoustic input

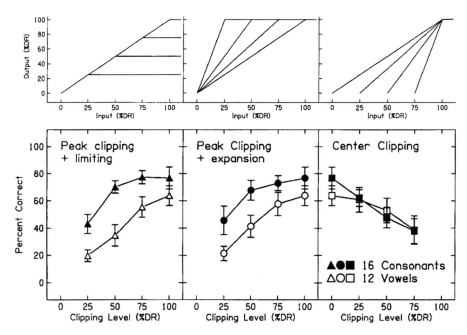

FIGURE 8.8. Effect of center and peak clipping on vowel and consonant recognition. (From Shannon et al. 2001a.)

because only the highest energy peaks in the spectrum are selected for stimulation.

Loizou et al. (2000a) tested NH subjects listening to acoustic simulations of a six-channel cochlear implant speech processor, in which the output amplitude range was compressed to have the same intensity resolution commonly found in CI patients [20–30 just-noticeably-different (jnd) steps in terms of intensity]. Within this restricted intensity range, performance was similar between NH subjects listening to the implant simulation and better-performing CI listeners fitted with a six-electrode CIS processor. Loizou et al. (1999, 2000b) also measured the effects of amplitude compression/quantization and limited spectral resolution in NH and CI listeners. They found that more amplitude resolution was necessary when spectral cues were highly limited. But when six or more channels of spectral information were available, speech recognition was highly resistant to amplitude compression and quantization. Only four to eight discrete amplitude steps were necessary to achieve good speech recognition, as long as a minimum of four to six channels of spectral information were provided. Even fewer amplitude steps are required if more spectral channels are available, as originally demonstrated by Licklider and Pollack's (1948) "infinite clipping" experiments.

4. Temporal Factors

Rosen (1992) and Plomp (1983) have characterized the temporal information in speech to fall into three categories—envelope, periodicity, and fine
structure—corresponding to temporal fluctuations 1 to 50 Hz, 50 to 500 Hz,
and 500 to 5 kHz, respectively. The capability of implant listeners to use
the temporal information found in each of these categories is reviewed
below.

4.1 Temporal Envelope Cues (1–50 Hz)

Steeneken and Houtgast (1980) measured the spectrum of the temporal
envelope in running speech. They found a bell-shaped curve (when plotted
in terms of log frequency) with a peak at about 4 Hz, corresponding to
the average syllable rate observed in speech. Relatively few temporal
fluctuations in speech were slower than 1 Hz or faster than 50 Hz. According to the temporal modulation transfer functions (TMTFs) measured
in implant listeners (Shannon 1992), CI users are able follow temporal
fluctuations in this range. Overall, CI users' perception of envelope
fluctuations in this range is relatively normal as long as the loudness is
mapped properly from acoustic to electrical domains; center or peak clipping would necessarily reduce the degree of amplitude fluctuations. Thus,
temporal envelope cues are largely dependent on amplitude coding, as
discussed in the previous section. Several experiments have shown that
temporal envelope information below 20 Hz is most important for
speech recognition. Classic work using vocoded speech has shown that
good-quality speech could be reconstructed using only the temporal
envelope information below 20 Hz from each frequency band. Shannon
et al. (1995), testing NH subjects listening to an implant simulation,
systematically reduced the low-pass cutoff frequency on the envelope
filter for each frequency band; after the temporal envelope was extracted
from each frequency band, the envelope was "smoothed" by applying a low-
pass filter before modulating carrier-band noise. Even for processors having
only four spectral channels, performance was unchanged as the cutoff frequency was reduced from 500 to 50 Hz; only a small reduction in performance was observed when the cutoff frequency was further reduced to
16 Hz. Drullman et al. (1994a,b) measured speech recognition in NH listeners when either slow or fast envelope modulations were parametrically
removed from speech. Speech recognition was unchanged as long as the
envelope fluctuations below 16 Hz were preserved. Further reductions in
envelope frequencies below 16 Hz resulted in a significant decrement, primarily for consonant recognition.

4.2 Periodicity Cues (50–500 Hz)

Temporal fluctuations found between 50 and 500 Hz provide periodicity information. NH listeners are able to perceive temporal fluctuations in this range purely in the temporal domain, i.e., without spectral analysis (Viemeister 1979; Burns and Viemeister 1976, 1981; Bacon and Viemeister 1985). CI patients are also able to perceive and discriminate temporal information in this range (Shannon 1983a, 1992; Shannon and Otto 1990; Zeng 2002), with some patients able to perceive and discriminate temporal information at rates as high as 1000 Hz (Wilson and colleagues, unpublished results). This range of temporal information is critically important for CI listeners because the intracochlear electrodes are generally not inserted deep enough in the cochlea to reach tonotopic regions below 1000 Hz. Because speech information in the periodicity range cannot be delivered by a cochlear implant to the correct tonotopic location, it may be essential that periodicity information be delivered and perceived temporally. Periodicity cues up to 300 Hz might provide information about voice pitch and intonation contours, but would not include first-formant frequency information. If CI users were able to receive periodicity information up to 1000 Hz, first-formant distinctions might be possible using only temporal cues. Individual differences in the ability to make use of periodicity cues may account for some differences in performance between implant patients (Fu 2002).

4.3 Temporal Fine Structure: Time or Place?

While CI patients generally cannot detect temporal fluctuations faster than 300 to 500 Hz, some patients may be able to detect temporal fluctuations up to 800 to 1000 Hz. The ability to perceive temporal fine structure may be a key difference in performance between good and poor implant users. The frequency region between 500 and 1500 Hz is critical for speech perception, and CI users may not perform well if they are unable to discriminate information in this frequency range, either temporally or spectrally. CI listeners who cannot access temporal information at frequencies higher than 500 Hz can only access information in this frequency range from the tonotopic place of the electrodes. It is likely that for all CI listeners, the temporal information above 1500 Hz will have to be represented spatially by the tonotopic place of activation rather than by temporal coding.

4.4 Pulse Rate Per Electrode

Modern cochlear implants generally use high pulse rates for electrical stimulation. High rates are used to convey the temporal properties of speech and to put the electrically stimulated auditory nerve into a mode of neural firing that is more like the normal acoustically stimulated nerve. As discussed in previous sections, CI listeners can access temporal information up to 300 to 500 Hz; stimulation rates greater than 1000 Hz should be high enough to accurately represent this information.

In addition, high stimulation rates allow for more stochastic firing patterns in electrically stimulated nerves than low stimulation rates (Rubinstein et al. 1999). It has been demonstrated that the refractory properties of the electrically stimulated auditory nerve (as inferred from intracochlear CAP recordings) are strongly affected by pulse rate (Wilson 1995; Wilson et al. 1997). At low pulse rates, the neural firing is highly synchronized to the pulse rate. At medium rates, the nerve refractory time and the stimulation rate can exhibit complex interactions and produce response patterns that alternate or "beat" between the pulse rate and the neural firing rate. At high pulse rates, the probabilistic recovery from adaptation desynchronizes the neural response from the stimulation rate. The high stimulation rates may be necessary to avoid any "aliasing" or "beating" effects between the nerve response and the stimulation rate.

While these neural phenomena are quite clear and well established, it is unclear whether they play any role in speech recognition. Significant improvements in speech recognition have been shown with stimulation rates as high as 4000 pulses/sec/electrode (ppse) (Brill et al. 1997). However, other studies found no change in performance for stimulation rates between 200 and 2500 ppse (Vandali et al. 2000; Fu and Shannon 2000). Thus, although high stimulation rates undoubtedly improve the stochastic firing properties of auditory neurons, it remains unclear whether these stochastic response properties are important for higher order feature extraction and pattern recognition in the nervous system.

4.5 Individual Differences in Temporal Processing

As suggested in previous sections, differences in temporal processing may account for differences in performance between good and poor implant users. Studies have revealed differences in temporal processing between good and poor implant CI users (Chatterjee and Shannon 1998; Fu 2002).

Chatterjee and Shannon (1998) measured recovery from forward masking functions for eight CI users. Three patients with relatively poor speech recognition skills showed a significantly different recovery time constant than patients possessing good speech recognition skills. Paradoxically, the poorer-performing patients exhibited much faster recovery from masking—approximately twice as fast as the time constant found in the better-performing patients. Assumedly, faster recovery might provide some advantage in speech recognition, allowing the system to better respond to rapidly changing stimuli. However, such fast recovery might also indicate that these patients primarily have high spontaneous-rate (SR) neurons remaining (Relkin and Doucet 1991), which may not be good for suprathreshold pattern recognition (Zeng et al. 1991). In normal acoustic environments, these high-SR neurons are mostly saturated at normal conversational speech levels, and therefore may not be useful for pattern recognition.

Fu (2002) measured the detection of a 100-Hz modulation rate as a function of the sensation level. The average modulation detection threshold across the entire perceptual dynamic range was highly correlated with speech recognition for nine CI listeners (see also Cazals et al. 1994). These results suggest that CI listeners' psychophysically measured temporal resolution is related to their speech recognition abilities. Understanding this relation will undoubtedly contribute to future speech-processor designs that strive to maximize CI users' utilization of temporal cues.

5. Spectral (Tonotopic) Cues

As mentioned in earlier in this chapter, complex pattern recognition may not always require the full resolution of the sensory system. But how much resolution is needed for understanding speech? How much resolution is needed if the tonotopic pattern is slightly distorted? A cochlear implant can create a complex spectral pattern if each electrode stimulates an independent tonotopic neural region. If electrodes are not completely independent and stimulate overlapping neural regions, the tonotopic resolution of the spectral pattern will become "blurred." In this section we report the results of experiments designed to measure the effects of electrical stimulation parameters on tonotopic selectivity as well as experiments designed to assess the effect of tonotopic selectivity on speech recognition.

5.1 Electrode Configuration

All electrical stimulation requires a pair of electrodes: a current source and a current sink (ground). A pair of electrodes that are closely spaced is typically called a bipolar pair. If one electrode of the pair is located outside the cochlea, the stimulation is called "monopolar" (see Wilson, Chapter 2). Because one of the electrodes is located outside the cochlea (the "ground" electrode in placed in the temporalis muscle), monopolar stimulation produces a broad current field and low thresholds. Many commercial implant devices presently use monopolar stimulation to achieve low power consumption in the speech processors, allowing behind-the-ear processors. Theoretically, monopolar stimulation should produce an even broader activation of auditory nerve fibers than widely spaced bipolar pairs, resulting in almost complete loss of tonotopic selectivity (van den Honert and Stypulkowski 1984, 1987).

The trade-off between sharp tonotopic resolution and stimulation level has yet to be fully explored. The data show no clear advantage with either monopolar or bipolar stimulation, for either electrode discrimination or speech recognition measures (Pfingst et al. 2001; Zwolan et al. 1996). Surprisingly, several patients performed better with monopolar stimulation. This paradoxical finding may be partially explained by psychophysical measures that show little difference in tonotopic selectivity between bipolar and monopolar stimulation modes, when equated for loudness (Hanekom and Shannon 1998; Chatterjee et al. 2001).

Fu and Shannon (1999a) measured speech recognition in cochlear implant patients as a function of the electrode configuration and as a function of the frequency assignments to electrodes. They found that the apical-most member of each bipolar pair was the most important in terms of determining the characteristic pitch. When frequency information is assigned to bipolar electrodes, the tonotopic location that should be matched is the location of the apical member of the pair.

5.2 Pulse Phase Duration and Tonotopic Selectivity

Threshold and loudness with electrical stimulation are related to the amount of electrical charge in each stimulating pulse. Thus, a pulse with a long phase duration requires a lower current amplitude to produce the same perceptual level as a pulse with a short phase duration. Low current amplitudes may allow better tonotopic selectivity because the effective current field covers a smaller tonotopic region. However, for interleaved multielectrode stimulation, long phase durations necessarily require slower overall pulse rates. So short pulses allow fast stimulation rates but may cause more electrode interaction than long pulses.

Studies of the trade-off between pulse phase duration and overall pulse rate have produced mixed results (Wilson et al. 2000). Some patients appeared to achieve better performance with longer pulses at a slower overall rate, while others appeared to do best with the fastest possible rates, which require very short pulses. Loizou et al. (2000a) found that some patients achieved better performance with longer pulses at a slower overall rate, suggesting that the optimal performance in a given patient might be achieved with a custom combination of pulse rate and pulse duration. At the present time it is not clear if short pulses produce more electrode interaction than longer pulses, or if there is any effect on speech recognition.

5.3 Tonotopic Selectivity and Electrode Interaction

The normal cochlea exhibits exquisite frequency selectivity both physiologically and perceptually. It is estimated that the normal ear is capable of processing 30 to 50 independent channels of frequency information.

In contrast, cochlear implants represent this spectral information by the relatively small number of implanted electrodes. If the output of those electrodes do not stimulate independent neural regions, the electrode interaction can further reduce the number of effective spectral "channels" of information. How many channels are necessary for speech understanding and how much "crosstalk" or interaction is tolerable between spectral channels?

Experiments with normal-hearing listeners indicate that spectral selectivity is not highly critical for speech recognition (e.g., Boothroyd et al. 1996; Dubno and Dorman 1987; ter Keurs et al. 1992, 1993). Boothroyd et al. (1996) "smeared" the spectral representation of speech by multiplying the speech waveform with low-pass noise. They found that a smearing bandwidth of 8000 Hz was necessary to reduce recognition to the level observed with no spectral cues. ter Keurs et al. (1992, 1993) smeared the spectral representation of speech using a fast Fourier transform (FFT) overlap-and-add method. They found no significant decrease in performance until the spectrum was smeared by more than one critical band. Speech recognition was reduced only moderately even when the spectrum was smeared over an octave. Dubno and Dorman (1987) selectively smeared the spectral representation of the first and second vowel formants and found that speech discrimination was remarkably robust to spectral smearing. These studies suggest that a sharp spectral representation is not required for good speech recognition. However, while the spectral cues were smeared in these studies, all other speech cues (amplitude, temporal, etc.) remained intact. Similar results were found in acoustic cochlear implant simulations by altering the slope of the noise carrier bands (Shannon et al. 1998; Fu and Shannon 1999b).

Fu (1997) simulated the effects of electrode interaction in implants by altering the slope of the carrier bands in a noise-band processor. Spectral information was first quantized into four or six bands to simulate a four- or six-electrode implant. Electrode interaction was simulated by changing the slope of the carrier noise band filters from steep (100 dB/octave) to shallow (6 dB/octave). This manipulation simulated both the quantizing of the spectral information and the spectral smearing that would be produced by electrode interactions. The results (Fig. 8.9, left panel) again demonstrate remarkable resistance to spectral smearing, even in the presence of spectral quantization. The right panel of Figure 8.9 shows that for both the four- and six-channel processors, vowel recognition was largely unchanged when only one band of noise was broadened (simulating a local spectral smearing). Combining results from CI subjects, NH subjects listening to comparable implant simulations, and NH subjects listening to conditions of spectral smearing, sharp spectral resolution is not critical for speech recognition, at least for quiet testing conditions.

Actual measurements of interaction across electrodes in a cochlear implant are complex. When electrical signals are presented simultaneously

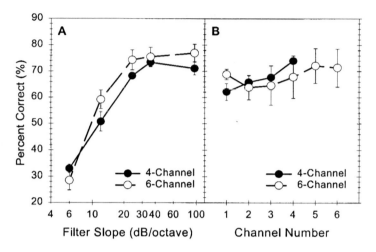

FIGURE 8.9. Effect of changing slope of noise carrier bands ($n = 4$). (From Fu 1997.)

to two electrodes, the current fields can add and subtract directly. Since the dynamic range of perception with cochlear implants is so small, two electrical fields adding can produce an uncomfortable loud sensation even if each electrode alone only produces a soft or inaudible sound. But electrodes can also interact even if their stimulation occurs sequentially, as in the CIS processor strategy. If a pulse is presented on one electrode and neurons in that region are activated, then those neurons are not available to fire in response to a pulse on another electrode. To the extent that the neural populations excited by two electrodes are overlapping, then even nonsimultaneously presented stimulation can show interaction.

Figure 8.10 presents data from electrode interaction measures using five different techniques: electrode discrimination, gap detection (Hanekom and Shannon 1998), forward masking (Shannon 1983b; Lim et al. 1989; Chatterjee and Shannon 1998), loudness summation (Shannon 1985), and simultaneous interleaved masking. In this plot a masker was placed on a central electrode in a Nucleus-22 array (electrode pair 10, 12) and interaction was measured for a signal presented on all other electrodes, both apical and basal to the masker electrodes. The ordinate on each plot has been plotted in a direction to make each curve resemble a "tuning curve," and each panel has been arbitrarily scaled to make the depth of each curve similar for comparison. Each method measures a different aspect of the perceptual interaction across electrodes. Figure 8.10 shows results from two implant listeners—one with relatively good speech recognition (N4) and one with relatively poor speech recognition (N3). Although the measures are slightly different from each other, all measures show that N4 had

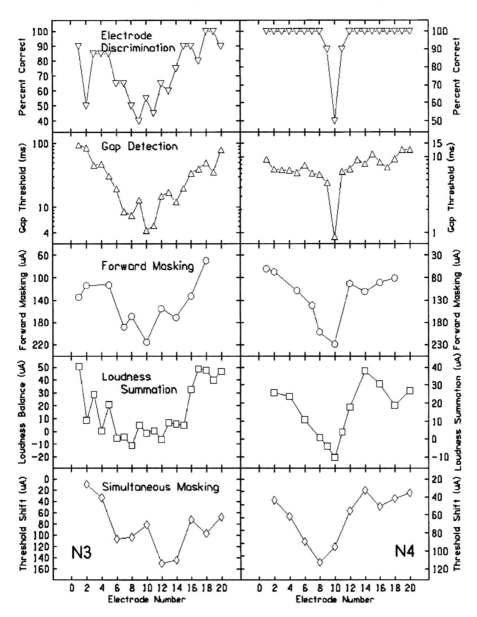

FIGURE 8.10. Five measures of electrode interaction for two implant listeners. A comfortable masker was presented to electrode pair (10, 12) and the indicated measure was collected from more apical and more basal electrode pairs. Individual panels are all oriented to resemble tuning curves and arbitrarily scaled to show comparable tuning depth.

sharper tonotopic selectivity (less electrode interaction) than N3. Gap detection and electrode discrimination measures appear to be sensitive to any interaction between electrodes, showing the most sharply "tuned" patterns of excitation. Other measures show more broadly tuned patterns of interaction. It is not clear exactly how channel interaction is related to speech recognition. Acoustic experiments indicate that considerable spectral smearing can be tolerated in quiet listening conditions, but excessive spectral smearing (channel interaction) probably reduces the number of effective spectral channels.

5.5 Effect of Number of Electrodes

How many electrodes (or channels) are necessary for good speech recognition? The cost and complexity of implant devices increases as the number of electrodes increases (see Niparko, Chapter 3). Is there a point at which an increased number of electrodes provides no further benefit? Can CI listeners utilize all the spectral information presented on their electrodes? Are more electrodes beneficial when listening in noise? This section reviews studies of speech recognition as a function of the number of spectral channels (or number of electrodes).

Historically, speech recognition and quality have been measured as a function of the number of frequency bands used to reconstruct a transmitted speech signal. For example, Hill et al. (1968) analyzed speech into different numbers of frequency bands and then used the envelope from each band to modulate a sinusoid at the center frequency of that analysis band. They found that six to eight modulated sinusoids were adequate for good speech recognition.

Shannon et al. (1995) measured consonant, vowel, and sentence performance in NH subjects listening to a noise-band simulation of a cochlear implant having one to four channels. Performance increased dramatically between one to four channels, with four channels providing a high level of performance (>90%) for all three recognition measures.

Dorman et al. (1997b; Dorman and Loizou 1998) measured speech recognition for vowels, consonants, and sentences in NH listeners as a function of the number of frequency bands used in the representation. They used both sinusoidal carriers (as in Hill et al. 1968) and noise bands (as in Shannon et al. 1995) and found similar results for the two carriers: performance increased as the number of bands increased, up to about six bands. In summary, work in NH listeners suggests that only about four to six bands of frequency information are necessary to allow high levels of speech recognition, at least in quiet listening conditions.

Speech recognition with reduced number of spectral channels has also been measured directly in CI patients. Lawson et al. (1993, 1996) measured consonant recognition in seven CI patients (one Ineraid and six Nucleus-

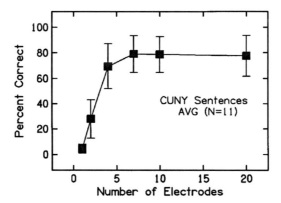

FIGURE 8.11. Sentence recognition as a function of the number of electrodes. (Data replotted from Fishman et al. 1997.)

22, all with percutaneous plug devices). Between one and six channels of CIS processing was implemented in the Ineraid patient, and two and 12 channels of CIS processing in the Nucleus-22 patients. In the Ineraid patient, consonant recognition improved as the number of channels was increased from one to six; in the Nucleus-22 patients, consonant recognition improved up to four channels, after which performance asymptoted.

Fishman et al. (1997) varied the number of electrodes in a SPEAK processor in 11 Nucleus-22 implant patients. Figure 8.11 shows Fishman et al.'s average data for sentence recognition. Performance increased from one to four electrodes for all speech measures, and reached asymptotic levels with seven electrodes. The authors concluded that even patients with excellent tonotopic selectivity were not using the information from all 20 electrodes, and that (in quiet) there was no difference in performance between 7, 10, and 20 electrodes.

Overall, there is remarkable consistency across all studies. Historical vocoder studies, recent implant simulations, and implant results all show that about six channels of spectral information is adequate for good speech recognition in quiet. Performance improves rapidly as the number of channels are increased from one to six, and little or no improvement is observed as the number of channels (noise bands, electrodes) are increased beyond six.

5.6 Effects of Noise

Although six channels of spectral information may be adequate for good speech recognition in quiet listening conditions, additional spectral chan-

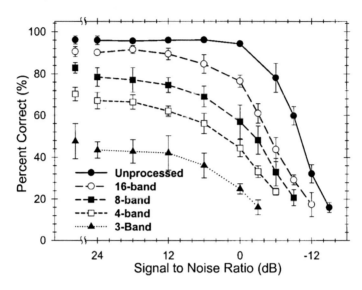

FIGURE 8.12. Vowel recognition as a function of signal-to-noise ratio; the number of spectral bands as the parameter. (From Fu et al. 1998.)

nels may be necessary in noisy conditions. Fu et al. (1998b) measured speech recognition in the presence of noise with three CI listeners and four NH subjects listening to a noise-band simulation of an implant processor. Figure 8.12 shows multitalker medial vowel recognition by NH subjects as a function of noise level for three, four, eight, and 16 channels, as well as for the original unprocessed speech. The number of channels affected performance even in quiet: performance with 16 channels was significantly lower than that with the original speech. CI listeners' performance in noise was comparable to that of NH subjects listening to implant simulations having the same number of channels (Eddington et al. 1997; Fu et al. 1998b).

Friesen et al. (2001) measured phoneme, word, and sentence recognition in noise for 10 Nucleus-22 listeners, nine Clarion listeners, and five NH listeners (listening to a simulation of an implant CIS processor). For all groups, speech recognition was measured as a function of the number of electrodes (or noise bands) at various signal-to-noise ratios (SNRs). Figure 8.13 shows the results for vowel recognition. Note that CI listeners' performance improved as the number of electrodes was increased up to about seven, with no significant improvement observed beyond seven to 10 electrodes. In contrast, NH listeners' performance continued to improve as the number of noise bands was increased to the maximum of 20 channels. This result suggests that the CI listeners were not able to make use of all the spectral cues presented to the multiple electrodes. It had been hypothesized

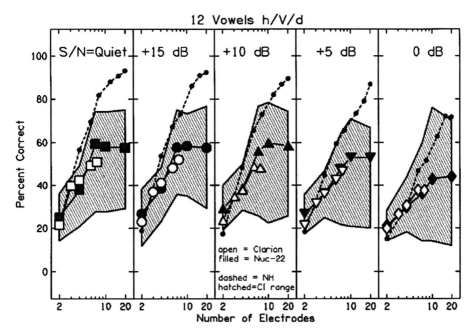

FIGURE 8.13. Medial vowel recognition as a function of the number of electrodes for 10 Nucleus-22 listeners (filled symbols) and nine Clarion listeners (open symbols). The dashed line shows normal-hearing listeners' results with noise-band processors. From left to right, the panels show results for decreasing signal-to-noise ratios. (From Friesen et al. 2001.)

that CI listeners might be able to make use of more channels of spectral information in challenging listening situations (such as noisy or reverberant environments). However, at all SNRs, implant performance reached asymptotic levels at seven to 10 electrodes. The hatched area in Figure 8.13 shows the range of performance across all 19 CI listeners. Note that the top edge of the hatched area is in line with NH subjects' results, suggesting that the best implant listeners were able to utilize all of the spectral information presented, at least up to seven channels.

5.7 Frequency to Electrode Mapping

Spectral information is normally represented at relatively fixed locations in the cochlea; i.e., the 1000-Hz information is always represented predominantly at the 1000-Hz tonotopic location. Thus, central pattern recognition mechanisms are trained over a lifetime of experience to depend on the stability of the cochlear tonotopic representation. However, in a cochlear implant, spectral information is often presented to electrodes that are more basally situated relative to the normal tonotopic locations (because of the limited insertion depth of the implanted electrodes). Thus, for CI listeners, the tonotopic representation of speech patterns may be shifted. In the following subsection, we review the research on the effects of alterations to the frequency-to-place mapping of speech information in the cochlea.

5.7.1 Tonotopic Shift

There is a long history of research into the effects of frequency shifting on speech recognition (e.g., Blesser 1972, Braida et al. 1979). Researchers in the 1940s and 1950s noted that speech remained recognizable when the playback of a tape recorder was sped up or slowed down, thereby changing both the frequency content and duration of the speech signal (Daniloff et al. 1968). Another example is that of deep-sea divers, who breathe a mixture of oxygen and helium that causes voices to sound higher pitched because of the faster speed of sound in the oxygen/helium mixture than in air. While divers' voices sound high pitched and squeaky, they remain intelligible (Mendel et al. 1995).

In the 1960s and 1970s, there was considerable interest in shifting speech frequencies down into the regions of residual hearing for patients with steeply sloping high-frequency hearing loss. Daniloff et al. (1968) reported results of frequency-shifted speech, with and without duration compensation. Normal-hearing listeners were able to tolerate larger frequency shifts when word duration shifted along with the frequency, suggesting that a central mechanism had already performed some normalization on the frequency-shifted speech duration. However, in implants (or any device for the hearing impaired), temporal information is represented in real-time and would not be linked to the frequency shift. Figure 8.14 shows frequency-shifted speech recognition results from three studies in which temporal distortions were corrected such that word and phoneme durations were normal even though frequencies were shifted. Results for the three studies (Tiffany and Bennett 1961; Daniloff et al. 1968; Nagafuchi 1976) are remarkably consistent, showing that listeners can only tolerate a frequency shift of about 35% before performance significantly worsens. When the speech frequencies were shifted by 60%, speech recognition was reduced to about 20% correct.

In cochlear implants, there is both a tonotopic shift of frequency information and a reduction in spectral resolution (due to the limited number of electrodes). Several investigators (Dorman et al. 1997a; Fu 1997; Shannon et al. 1998) have evaluated the effect of a tonotopic shift combined with reduced spectral resolution, using a noise-band simulation of an implant with NH listeners. Speech was first filtered into four or 16 bands and the envelope from each band was extracted; these envelopes were then used to

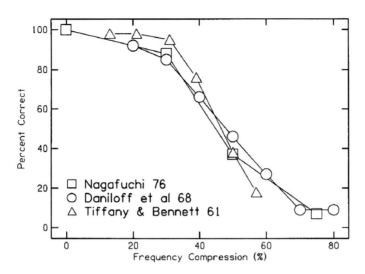

FIGURE 8.14. Comparison of results from three studies that measured speech recognition as a function of frequency compression, when temporal distortion was corrected.

modulate bands of noise. The analysis bands were either matched to the noise carrier bands in terms of cochlear location and extent, or were shifted (in millimeters) relative to the noise carrier bands using Greenwood's (1990) formula. Figure 8.15 shows Fu's results for vowel recognition. When both the analysis and the carrier bands were shifted together, performance was only mildly affected (as shown in the curves with square symbols and dotted lines). The mild decline in performance when both the analysis and the carrier bands were shifted basally was due to the loss of low-frequency speech information. In the implant simulation condition, the carrier bands were fixed to simulate an electrode location where the most apical electrode was located 22 mm from the base while the analysis bands were shifted in terms of cochlear location in millimeters. Performance decreased markedly as the analysis filters were shifted in the apical direction (which caused an upward frequency shift in the speech signal). The effect of tonotopic mismatch was even more pronounced for the 16-band processors than for the four-band processors. For both the 16- and four-band processors, vowel recognition was reduced to chance level when the analysis bands were shifted by 6 mm relative to the carrier bands. The effect of tonotopic mismatch is made clear when comparing the performance of matched and mismatched processors. For the matched processors, performance was only mildly affected as the matched analysis and carrier bands were shifted toward the cochlea base (removing increasing amounts of low-frequency

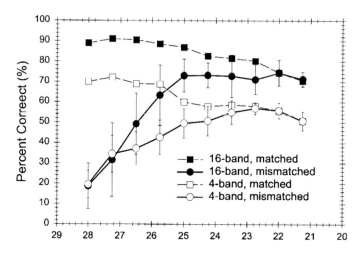

Analysis filerband based on tonotopic location (mm from base)

FIGURE 8.15. Effect of a shift in envelope frequency information relative to the noise carrier bands representing that information (from Fu 1997). The abscissa shows the tonotopic location (in millimeters) of the most apical edge of the frequency analysis bands.

speech information). For the mismatched processors, when the analysis filters were shifted by 6mm (relative to the carrier bands), performance dropped to chance level, even though the apically shifted analysis bands contained more low-frequency speech information. Thus, frequency information must be matched to its normal acoustic tonotopic location or speech recognition will suffer.

Figure 8.16 compares results of Nagafuchi's (1976) frequency-shifted vowel recognition (full-spectrum speech) with the normalized results of Fu and Shannon (1999b), who used noise-band implant simulations (spectrally quantized speech). Note the similarity of all sets of data, demonstrating that speech recognition falls off sharply as a function of spectral shift in either the apical or basal direction, whether the shift is of whole speech, mildly spectrally quantized speech (16 bands), or severely spectrally quantized speech (four bands). These results indicate that a spectral shift can severely limit speech recognition, regardless of the degree of spectral resolution. A shift of an octave (about 4.5mm) can reduce vowel recognition by more than 60 percentage points. For spectral shifts beyond 35% (about 2mm), the reduction in performance is 10 percentage points per millimeter of shift. For cochlear implants, the analysis bands in the speech processor must be matched to the actual tonotopic location of the electrodes in the scala tympani of individual patients. Some of the variability in performance across patients may be due to the different electrode inser-

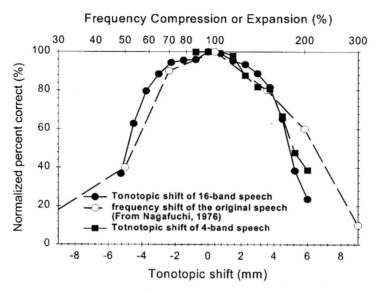

FIGURE 8.16. Comparison of vowel recognition for frequency expansion and compression for original speech and for spectrally quantized speech.

tion depths, resulting in differing degrees of spectral mismatch across patients.

Listeners may be able to adapt to spectrally shifted speech patterns and eventually compensate for deficits caused by a spectral mismatch. For example, Rosen et al. (1999) demonstrated that NH listeners were able to quickly accommodate to frequency-shifted speech; after 3 hours of training with a 6-mm basal shift, performance improved from chance level to 30% correct. Fu et al. (2002) measured CI listeners' accommodation to a 3-mm apical shift; speech recognition was measured over a 3-month period during which the implant patients wore the shifted speech processor all waking hours. Fu et al.'s results suggested that there might be two adaptation periods: a short adaptation period that results in only a partial accommodation, and a longer adaptation period during which a full accommodation may or may not be possible. CI subjects were not able to achieve their baseline performance measured with their everyday processors even after 3 months of daily use with the apically shifted processors. The rapid recovery period (similar to that observed by Rosen et al.) may represent an adjustment to novel speech patterns. The longer time period may be measured in years (rather than months, as shown by Fu et al.), as an implant listener fully accommodates electrically stimulated speech patterns. The degree of spectral shift may also determine the time course needed to fully adapt to frequency-shifted speech; if the mismatch is not severe, adaptation may occur much more quickly. Intensive training and long-term experience with spectrally shifted speech might also allow adult CI users to regain speech recognition at the "unshifted" performance levels.

5.7.2 Effects of Tonotopic Warping

An additional problem encountered in implant patients is that of nonuniform neural survival. Some pathologies may cause uniform nerve survival along the extent of the cochlea, while others may cause a more selective nerve loss in some tonotopic regions (see Leake and Rebscher, Chapter 4). The effects of uneven nerve survival are not clearly known in CI users, but have been simulated in NH subjects listening to noise-band speech processors (Shannon et al. 1998, 2001b). If there is a local nerve loss close to a stimulating electrode, then the current level on that electrode must be increased until the current spreads and activates surviving nerves in distant tonotopic locations. However, these activated nerves will then be stimulated with speech information from the wrong tonotopic location. Assuming that the remaining neural populations were to receive tonotopically matched spectral information, there would be a "warping" of the distribution of spectral information (Skinner et al. 1995; Shannon et al. 1998, 2001b). The spectral information from the tonotopic region where nerves were missing would be "compressed" into a smaller region.

Fu (1997) simulated tonotopic warping of spectral patterns in NH listeners by mismatching the analysis and carrier filter bands in a noise-band processor. The overall spectral extent (in millimeters along the cochlea) was always the same for both the analysis and carrier bands, but the distribution of the filters within this range was different for the analysis filters and noise carrier bands. Figure 8.17 shows the effect of this mismatch on vowel recognition. The analysis filters were either distributed linearly (partition 0), logarithmically (partition 6), or somewhere between linear and log (partition 3); the carrier bands were systematically varied from a linear to a logarithmic distribution. Performance was always best when the analysis and carrier partitions were matched, and decreased rapidly as the difference between the analysis and carrier partitions increased.

5.7.3 Effects of a "Hole" in the Tonotopic Representation

Some types of hearing loss can produce a localized loss of hair cells and may produce a localized loss of auditory neurons as well, sometimes termed "dead regions" (Moore and Glasberg 1997; Moore et al. 2000; Moore 2001; Moore and Alcantara 2001). When stimulating a dead region, thresholds will be elevated at the frequencies corresponding to the dead region. To compensate, gain is usually increased for that frequency region. In a hearing aid the amplification would be increased in that frequency region, and in a cochlear implant the electrical amplitude would be increased for the electrode in that cochlear location. Because the sensory cells and/or neurons

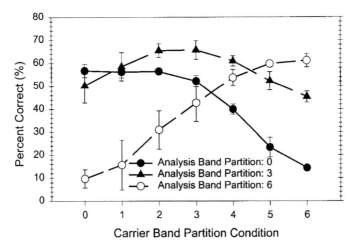

FIGURE 8.17. Effect of mismatched distribution of the analysis and carrier bands on vowel recognition in a four-channel noise-band processor. Carrier band partition condition 0 represents a linear division of the total frequency range, while condition 6 represents a logarithmic division. (From Fu 1997.)

are dead, no activation can actually occur within the dead region. Thus, when the amplitude is increased, the stimulation occurs instead at the edges of the dead region (or "hole"). In effect, this causes a local warping in the frequency-to-place mapping. The speech information from the frequency region of a hole is transmitted to the brain by neurons that normally respond to a different frequency region.

How does a hole in the tonotopic representation affect speech understanding? Lippmann (1996) demonstrated good speech recognition even when the entire midfrequency region from 800 to 3000 Hz was removed. Warren et al. (1995) found that excellent speech recognition was possible when listening through only a few narrow spectral bands (slits). Clearly, large regions of spectral information can be dropped without disrupting speech recognition, due to the spectral redundancy of speech.

Shannon et al. (2001b) measured the effect of such holes in the speech spectrum with CI listeners and in NH subjects listening to a 20-channel noise-band simulation. Holes were simulated by simply deleting several contiguous carrier bands (NH listeners) or by turning off electrodes in the apical, middle, or basal cochlear region. Other experimental conditions reassigned the information lost in the hole to nearby electrodes in an attempt to preserve the speech information from that region. Figure 8.18 shows the results for vowel recognition, normalized to baseline performance (no hole). In general, the NH and CI results were similar, in that both types of listeners could tolerate a "hole" of up to 3 mm with relatively

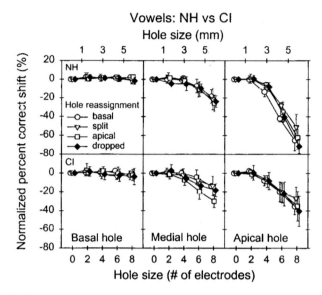

FIGURE 8.18. Relative decrease in vowel recognition for normal-hearing (top panels) and cochlear implant listeners (lower panels) as a function of the size of a "hole" in the tonotopic representation. (From Shannon et al. 2001b.)

little decrease in recognition. Holes larger than 3 mm in the apical region produced large decrements in vowel recognition, while holes as large as 6 mm in the basal region produced almost no deficit. While it appears that NH listeners were more adversely affected than CI listeners by holes in the apical region, differences in baseline performance account for differences in the normalized scores (CI listeners had significantly lower baseline scores). With a 6-mm apical hole, both NH and CI listeners' vowel recognition dropped to chance level. Unfortunately, none of the remapping manipulations succeeded in "rescuing" the information from the hole region; performance was no different when the information was simply dropped. However, remapping information to the edges of the hole (as may often occur with CI patients) did not produce a further decrement in performance. Thus, warping the distribution of spectral information around a hole did not produce a further deterioration in vowel recognition beyond the simple loss of spectral information from the hole region.

5.8 Summary—Spectral Information

Speech can be recognized with great accuracy even with relatively coarse representations of spectral information, in good listening conditions. Four spectral channels are sufficient for good sentence recognition in quiet. However, in poor listening conditions, such as in background noise, higher resolution spectral information is necessary for good speech recognition. CI listeners appear to utilize only four to seven channels of spectral information no matter how many electrodes are available in their implant. It remains unclear why they are unable to use all of the spectral information that is presented. For postlingually deafened CI patients, it may be that the transformation of acoustic frequency to cochlear place is too dissimilar in the implant device from what they were previously accustomed to with acoustic hearing. When the frequency-to-place mapping is overly distorted, listeners may not be able to fully utilize all of the spectral channels that are available. If the warping of frequency-to-place mapping is severe, listeners may lose the ability to utilize spectral channels altogether.

6. Binaural Cochlear Implants

One of the biggest complaints of CI patients is the difficulty they have understanding speech in noisy listening conditions. It has long been known that binaural hearing provides a large advantage for separating signals from background noise, and researchers have begun to investigate the advantages of binaural cochlear implants for speech recognition in noise. Van Hoesel et al. (1993; van Hoesel and Clark 1995, 1997) measured sentence recognition in quiet and in noise by CI users who had bilateral implants. Subjects were able to utilize interaural level differences (ILDs) to lateralize sounds; however, their ability to lateralize using interaural temporal differences (ITDs) was much poorer than that of NH listeners. Nonetheless, a small advantage was observed for the binaural processors in noise over either monaural processor. Many patients have since received binaural implants, and several research groups have studied their effectiveness.

Sound source localization with binaural implants was measured in one CI patient (Long et al. 1998; Long 2000). The time delay, loudness difference, and electrode location were varied between the two implants. In this patient, sounds were localized toward the side that was stimulated first or was louder, similar to localization in normal hearing. This patient was able to utilize ITDs as short as 150 μs to lateralize sounds—a value that is comparable to that of NH listeners. However, these effects were observed only if the test electrodes from the two sides were matched in pitch, suggesting that the electrodes were stimulating the same tonotopic region in the two cochleae. If the stimulating electrodes from the two implants elicited distinctly different pitch percepts, then localization was poor for all time or loudness differences between the two devices. This corresponds well to NH listeners, who can localize sounds only if they are presented to the same cochlear region of both ears, within a few millimeters (Nuetzel and Hafter 1981). In general, signals presented to two ears must stimulate tonotopic regions within a critical band of each other (1 to 2 mm) for the central system to make full use of the ITD for localization. This result indicates that it is highly important to match the two implant devices in pitch (which presumably means cochlear location) to achieve the benefit of binaural processing.

Lawson et al. (1998, 2001) measured binaural speech recognition in quiet and in noise in several CI listeners with binaural implants and found that, while some patients achieved better speech recognition in noise with two implants, others received no benefit over a single implant. Mixed results were also observed for binaural implant users' ability to localize sounds, with some patients able to localize stimuli with ITDs as low as 50 μs, while others were unable to localize at any ITD.

At the present time it is unclear how many patients might benefit from binaural cochlear implants. Even for patients who may ultimately benefit from two devices, the fitting process of binaural implants will be difficult, requiring not only the normal fitting of each individual implant, but also specialized fitting procedures to match the two devices in terms of cochlear place. Amplitude compression schemes commonly used in implant speech processors may have to be linked to produce the correct level difference between devices. The frequency-to-place mapping in each implant would also have to compensate for differences in electrode location between the two implants. Some patients may not have sufficient residual auditory capacity in the central nervous system to make use of binaural cues from two implants, even if they are properly matched in terms of loudness and pitch. Presently, it is difficult to estimate the proportion of patients who might benefit from binaural cochlear implants or to devise a method that might be used to predict this population prior to surgery.

7. Combined Acoustic and Electrical Stimulation

As cochlear implants have provided increasingly better speech recognition to patients over the years, patients selection criteria have been relaxed to allow for implantation of patients who have some residual hearing. Many cases have been reported in which the implant did not damage the residual acoustic hearing of the implanted ear (von Ilberg et al. 1999). Recent clinical trials have investigated whether a cochlear implant could be useful for patients who have only low-frequency residual hearing. Such patients generally do poorly with hearing aids, presumably because they have no remaining hair cells in the basal end of the cochlea. In these patients, a cochlear implant inserted into only the basal end of the cochlea might provide some benefit without destroying the residual hearing at the apical end of the cochlea. The residual acoustic low-frequency hearing could be combined with electric stimulation of the higher frequency regions in the cochlea. Can patients effectively combine acoustic and electric stimulation in the same ear?

The initial clinical trials have so far shown mixed results (Lawson et al. 2001; Turner and Gantz 2001). Some patients (under some conditions) were able to achieve better speech recognition with combined acoustic and electric hearing than with either type of stimulation alone, particularly in noise. However, other patients showed no improvement, and some even showed some deterioration in their residual acoustic hearing after implantation. The results of Turner and Gantz (2001) suggest that acoustic and electrical stimulation should be contiguous for the combined stimulation to be beneficial. If there is a large gap in the tonotopic space that is not stimulated, or if the frequency information conveyed by the electrical stimulation is not appropriate for the stimulated cochlear location, patients may receive little or no benefit.

8. Auditory Brainstem Implants

Neurofibromatosis type 2 (NF2) is a genetic disorder that produces bilateral benign tumors on the vestibular branch of the auditory nerves. These tumors can be removed surgically but the procedure often severs the eighth nerve, leaving these patients deaf and unable to benefit from a cochlear implant. Because these patients do not have a functioning auditory nerve, prosthetic stimulation of the auditory system must target the next stage of neural processing—the cochlear nucleus (CN) complex in the brainstem. The auditory brainstem implant (ABI) was developed for this patient population by the House Ear Institute in collaboration with the Huntington Medical Research Institutes in Pasadena, California, and the Cochlear Corporation. The ABI is similar in design and function to a cochlear implant, but the electrode is placed on the cochlear nucleus rather than in the cochlea. The first ABI was a single-channel device pioneered by William House and William Hitselberger in 1979 (Edgerton et al. 1984; Eisenberg et al. 1987). The modest success of that first device led to the development of multichannel ABI devices (Brackmann et al. 1993; Shannon et al. 1993).

Presently, ABI devices are produced by all major cochlear implant companies (availability in the U.S. is restricted to the device made by Cochlear Corp.). The Cochlear Corp. device has 21 electrodes on a silicone pad that is placed into the lateral recess of the fourth ventricle, which lies adjacent to the cochlear nucleus. Basic psychophysical performance is similar between ABI and CI patients (Shannon and Otto 1990; Zeng and Shannon 1992, 1994), but speech recognition performance is considerably poorer with the ABI (Otto et al. 1998, 2002). Although the ABI has 21 electrodes, patients do not seem able to use the multiple channels of spectral information. Most multichannel ABI patients perform at levels similar to those of single-channel CI users. A few ABI patients (less than 10%) are able to understand a limited number of words in sentences— performance comparable to that with a two- or three-channel cochlear implant. It is possible that the difference in performance between CI and ABI is due to the lack of consistent pitch mapping in the ABI.

In the surface electrode ABI, electrically stimulated pitch percepts do not always increase in a regular fashion from one end of the electrode array to the other. For some ABI patients, pitch percepts increase from lateral to medial electrodes, while others show the reverse pattern; some patients show a disordered pattern of pitch percepts associated with electrodes, while others perceive little change in pitch across the entire electrode array (Otto et al. 1998). The variable relation between pitch and electrode location in the present ABI device is most likely due to the fact that the tonotopic dimension of the human CN is oriented orthogonal to the surface, while the electrode array lies parallel to the surface. To obtain better stimulation along the tonotopic axis of the CN, a new ABI with penetrating microelectrodes has been developed (McCreery et al. 1998). The penetrating microelectrodes will be of different lengths so that they will target different pitch regions of the CN. The first application of this penetrating electrode auditory brainstem implant (PABI) in humans is slated for 2003.

The PABI will present an interesting test of prosthetic design. If poor speech recognition performance by ABI patients is due to poor access to the CN tonotopic organization, then the PABI should correct this problem and PABI patients' performance should be at levels more equivalent to that of CI patients. On the other hand, if poor ABI speech recognition is due to the lack of critical intrinsic neural mechanisms bypassed by the ABI device, there would be no improvement in PABI patients' performance because the PABI device would also bypass these critical neural mechanisms. If the PABI produces CI-like levels of speech recognition, it may be possible to stimulate even higher auditory brainstem nuclei, such as the inferior colliculus, or even the auditory cortex, which might provide easier surgical access than the CN.

9. Implications for Implant Speech-Processor Design

Two factors emerge as major challenges for cochlear implant research and design: CI patients' inability to utilize all channels of the spectral information provided by the implant, and patient outcome variability. Figure 8.9 shows that the best CI users were able to recognize speech at a similar level to NH subjects listening to a comparable number of channels. However, this similarity only holds for less than eight electrodes/channels. Beyond eight channels/electrodes, NH listeners continued to improve while CI listeners did not. To improve patients' speech recognition, it is critical to understand why CI listeners seem to be limited to eight spectral channels. Next-generation implants that have more electrodes may not improve performance until the cause of this limitation can be determined.

Figure 8.9 also showed that while some CI listeners were able to perform as well as NH subjects when listening to a comparable number of spectral channels, some were not. Understanding the causes of the large range in performance among CI patients is another priority for future speech-processor design. Some CI listeners may simply have too much damage to their auditory nerve from the pathology that caused their deafness, or from its sequelae, and so may never be able to take full advantage of cochlear implant technology. For these patients, an appropriate speech-processor fitting protocol might include an assessment of their auditory perceptual capabilities, and tailor the parameters of the speech processor to best exploit those capabilities. For other poorly performing CI patients, the outlook may be much more optimistic. If poor performance is the result of speech processing that distorts the mapping of acoustic frequency information to cochlear place, then simple modifications to the signal processing may dramatically improve speech understanding. As shown earlier, experiments with NH listeners showed that a 5-mm mismatch in the frequency-to-place mapping or a severe warping in the frequency-to-place mapping resulted in a total inability to use spectral information (Shannon et al. 1998; Fu and Shannon 1999b). It is possible that the poor performance of some CI listeners is due to such spectral distortions; restoring the normal acoustic frequency-to-place mapping might result in improved speech recognition. Clinical fitting procedures must be developed to assess the optimal frequency-to-electrode mapping in individual CI patients.

Cochlear implants provide a surprisingly high level of functional hearing to deaf patients, and the sophistication of the implant hardware continues to improve rapidly. Some gains in implant performance over the past 20 years have been directly due to improvements in basic implant technology, such as faster speech processors and more electrodes. However, technological advances may be nearing the point of diminishing returns, given the high costs involved and limited additional benefits they may provide. The next phase of improvement in cochlear implant performance may come not from further development of the implant hardware, but from understanding how implant speech processors may be more effectively programmed and customized for individual patients, so that the capabilities that are already available may be fully utilized.

10. Summary

The studies reviewed in this chapter demonstrate that amplitude mapping (Figs. 8.4 to 8.8) and envelope filter frequency are not critical for speech recognition, at least in quiet listening conditions. Reduced temporal and amplitude cues can cause some reduction in speech recognition, but the effects are relatively small, even in cases where the spectral resolution is severely limited. When spectral cues are available, even severely reduced temporal and/or amplitude cues have almost no effect on speech recognition. However, there may be differences in temporal processing that may account for some differences in implant patient performance. In terms of spectral resolution, relatively little improvement in speech recognition is observed in implant performance as the number of electrodes is increased beyond seven or eight (Figs. 8.11 and 8.13), even for implant patients who have excellent electrode selectivity. Some as-yet-unknown factor appears to limit implant listeners' ability to utilize all the spectral information that is provided by the implant. The two factors that appear to be most significant for implant speech recognition are the tonotopic match between acoustic speech information and the location of stimulating electrodes (Figs. 8.14 to 8.16), and the match between the relative bandwidth of the frequency analysis bands and the extent of neural activation for each electrode (Fig. 8.17).

References

Allen JB (1994) How do humans process and recognize speech? IEEE Trans Speech Audio Proc 2:567–577.

Bacon SP, Viemeister NF (1985) Temporal modulation transfer functions in normal-hearing and hearing-impaired listeners. Audiology 24:117–134.

Blesser B (1972) Speech perception under conditions of spectral transformation: I. Phonetic characteristics. J Speech Hear Res 15:5–41.

Boothroyd A, Mulhearn B, Gong J, Ostroff J (1996) Effects of spectral smearing on phoneme and word recognition. J Acoust Soc Am 100:1807–1818.

Brackmann DE, Hitselberger WE, Nelson RA, Moore JK, et al. (1993) Auditory brainstem implant. I: issues in surgical implantation. Otolaryngol Head Neck Surg 108:624–634.

Braida LD, Durlach NI, Lippmann RP, Hicks BL, Rabinowitz WM, Reed CM (1979) Hearing aids—a review of past research on linear amplification, amplitude compression, and frequency lowering. ASHA Monograph 19:1–114.

Brill SM, Gstottner W, Helms J, von Ilberg C, et al. (1997) Optimization of channel number and stimulation rate for the fast continuous interleaved sampling strategy in the COMBI 40+. Am J Otol 18(6 suppl):S104–106.

Burns EM, Viemeister NF (1976) Nonspectral pitch. J Acoust Soc Am 60:863–869.

Burns EM, Viemeister NF (1981) Played-again SAM: further observations on the pitch of amplitude-modulated noise. J Acoust Soc Am 70:1655–1660.

Cazals Y, Pelizzone M, Saudan O, Boex C (1994) Low-pass filtering in amplitude modulation detection associated with vowel and consonant identification in subjects with cochlear implants. J Acoust Soc Am 96(4):2048–2054.

Chatterjee M, Shannon RV (1998) Forward masked excitation patterns in multi-electrode cochlear implants. J Acoust Soc Am 103(5):2565–2572.

Chatterjee M, Shannon RV, Galvin JJ, Fu Q-J (2001) Spread of excitation and its influence on auditory perception with cochlear implants. Physiological and psychological bases of auditory function. In: Houtsma AJM, Kohlrausch A, Prijs VF, Schoonhoven R, eds. Proceedings of the 12th International Symposium on Hearing. Maastricht, NL: Sharker Publishing BV, pp. 403–410.

Cosendai G, Pelizzone M (2001) Effects of the acoustical dynamic range on speech recognition with cochlear implants. Audiology 40:272–281.

Daniloff RG, Shiner TH, Zemlin WR (1968) Intelligibility of vowels altered in duration and frequency. J Acoust Soc Am 44:700–707.

Dorman MF, Loizou PC (1998) Identification of consonants and vowels by cochlear implant patients using a 6-channel continuous interleaved sampling processor and by normal-hearing subjects using simulations processors with two to nine channels. Ear Hear 19:162–166.

Dorman MF, Loizou PC, Rainey D (1997a) Simulating the effect of cochlear-implant electrode insertion depth on speech understanding. J Acoust Soc Am 102:2993–2996.

Dorman MF, Loizou PC, Rainey D (1997b) Speech intelligibility as a function of the number of channels of stimulation for signal processors using sine-wave and noise-band outputs. J Acoust Soc Am 102:2403–2411.

Drullman R (1995) Temporal envelope and fine structure cues for speech intelligibility. J Acoust Soc Am 97:585–592.

Drullman R, Festen JM, Plomp R (1994a) Effect of temporal envelope smearing on speech perception. J Acoust Soc Am 95:1053–1064.

Drullman R, Festen JM, Plomp R (1994b) Effect of reducing slow temporal modulations on speech perception. J Acoust Soc Am 95:2670–2680.

Dubno JR, Dorman MF (1987) Effects of spectral flattening on vowel identification. J Acoust Soc Am 82:1503–1511.

Eddington DK, Rabinowitz WR, Tierney J, Noel V, Whearty M (1997) Speech processors for auditory prostheses. 8th quarterly progress report, NIH contract N01-DC-6-2100.

Edgerton BJ, House WF, Hitselberger W (1984) Hearing by cochlear nucleus stimulation in humans. Ann Otol Rhinol Otolaryngol 91(suppl):117–124.

Eisenberg LS, Maltan AA, Portillo F, Mobley JP, House WF (1987) Electrical stimulation of the auditory brainstem structure in deafened adults. J Rehabil Res Dev 24:9–22.

Fishman K, Shannon RV, Slattery WH (1997) Speech recognition as a function of the number of electrodes used in the SPEAK cochlear implant speech processor. J Speech Hear Res 40:1201–1215.

Freyman RL, Nerbonne GP (1989) The importance of consonant-vowel intensity ratio in the intelligibility of voiceless consonants. J Speech Hear Res 32:524–535.

Freyman RL, Nerbonne GP, Cote HC (1991) Effect of consonant-vowel ratio modification on amplitude envelope cues for consonant recognition. J Speech Hear Res 34:415–426.

Friesen L, Shannon RV, Baskent D, Wang X (2001) Speech recognition in noise as a function of the number of spectral channels: comparison of acoustic hearing and cochlear implants. J Acoust Soc Am 110:1150–1163.

Fu Q-J (1997) Speech perception in acoustic and electric hearing. Ph.D. dissertation, University of Southern California, Los Angeles, CA.

Fu Q-J (2002) Temporal processing and speech recognition in cochlear implant users. NeuroReport 13:1–5.

Fu Q-J, Shannon RV (1998) Effects of amplitude nonlinearities on speech recognition by cochlear implant users and normal-hearing listeners. J Acoust Soc Am 104:2570–2577.

Fu Q-J, Shannon RV (1999a) Effects of electrode configuration and frequency allocation on vowel recognition with the Nucleus-22 cochlear implant. Ear Hear 20(4):332–344.

Fu Q-J, Shannon RV (1999b) Recognition of spectrally degraded and frequency-shifted vowels in acoustic and electric hearing. J Acoust Soc Am 105:1889–1900.

Fu Q-J, Shannon RV (2000) Effect of stimulation rate on phoneme recognition in cochlear implants. J Acoust Soc Am 107(1):589–597.

Fu Q-J, Shannon RV, Wang X (1998b) Effects of noise and number of channels on vowel and consonant recognition: acoustic and electric hearing. J Acoust Soc Am 104:3586–3596.

Fu Q-J, Shannon RV, Galvin JJ III (2002) Perceptual learning following changes in the frequency-to-electrode assignment with the Nucleus-22 cochlear implant. J Acoust Soc Am 112:1664–1674.

Greenwood DD (1990) A cochlear frequency-position function for several species—29 years later. J Acoust Soc Am 87:2592–2605.

Hanekom JJ, Shannon RV (1998) Gap detection as a measure of electrode interaction in cochlear implants. J Acoust Soc Am 104(4):2372–2384.

Hill FJ, McRae LP, McClellan RP (1968) Speech recognition as a function of channel capacity in a discrete set of channels. J Acoust Soc Am 44:13–18.

Lawson D, Wilson B, Finley C (1993) New processing strategies for multichannel cochlear prostheses. In: Allum JA, Allum-Mecklenburg DJ, Harris FP, Probst R, eds. Natural and Artificial Control of Hearing and Balance. Progress in Brain Research, vol. 97. Amsterdam: Elsevier, pp. 313–321.

Lawson DT, Wilson BS, Zerbi M, Finley CC (1996) Speech processors for auditory prostheses. Third quarterly progress report, NIH contract N01-DC-5-2103.

Lawson DT, Wilson BS, Zerbi M, van den Honert C, et al. (1998) Bilateral cochlear implants controlled by a single speech processor. Am J Otol 19(6):758–761.

Lawson DT, Brill S, Wolford R, Wilson BS, Schatzer R (2001) Speech processors for auditory prostheses. 9th quarterly progress report, NIH contract N01-DC-8-2105.

Licklider JCR, Pollack I (1948) Effects of differentiation, integration, and infinite peak clipping on the intelligibility of speech. J Acoust Soc Am 20:42–51.

Lim HH, Tong YC, Clark GM (1989) Forward masking patterns produced by intracochlear stimulation of one and two electrode pairs in the human cochlea. J Acoust Soc Am 86:971–980.

Lippmann RP (1996) Accurate consonant perception without mid-frequency energy. IEEE Trans Speech Audio Proc 4:66–69.

Loizou PC, Dorman MF, Tu Z (1999) On the number of channels needed to understand speech. J Acoust Soc Am 106(4):2097–2103.

Loizou PC, Poroy O, Dorman M (2000a) The effect of parametric variations of cochlear implant processors on speech understanding. J Acoust Soc Am 108(2):790–802.

Loizou PC, Dorman M, Poroy O, Spahr T (2000b) Speech recognition by normal-hearing and cochlear implant listeners as a function of intensity resolution. J Acoust Soc Am 108:2377–2387.

Long CJ (2000) Bilateral cochlear implants: basic psychophysics. Ph.D. dissertation, Massachusetts Institute of Technology.

Long CJ, Eddington DE, Colburn HS, Rabinowitz WM, Whearty ME, Kadel-Garcia N (1998) Speech processors for auditory prostheses. 11th quarterly progress report, NIH contract N01-DC-6-2100.

McCreery DG, Shannon RV, Moore JK, Chatterjee M, Agnew WF (1998) Accessing the tonotopic organization of the ventral cochlear nucleus by intranuclear microstimulation. IEEE Trans Rehabil Eng 6:391–399.

Mendel LL, Hamill BW, Crepeau LJ, Fallon E (1995) Speech intelligibility assessment in a helium environment. J Acoust Soc Am 97:628–636.

Miller G, Nicely P (1995) An analysis of perceptual confusions among some English consonants. J Acoust Soc Am 27:338–352.

Moore BCJ (2001) Dead regions in the cochlea: diagnosis, perceptual consequences, and implications for the fitting of hearing aids. Trends Amplification 5:1–34.

Moore BCJ, Alcántara JI (2001) The use of psychophysical tuning curves to explore dead regions in the cochlea. Ear Hear 22:268–278.

Moore BCJ, Glasberg BR (1997) A model of loudness perception applied to cochlear hearing loss. Aud Neurosci 3:289–311.

Moore BCJ, Huss M, Vickers DA, Glasberg BR, Alcántara JI (2000) A test for the diagnosis of dead regions in the cochlea. Br J Audiol 34:205–224.

Nagafuchi M (1976) Intelligibility of distorted speech sounds shifted in frequency and time in normal children. Audiology 15:326–337.

Nuetzel JM, Hafter ER (1981) Discrimination of interaural delays in complex waveforms: spectral effects. J Acoust Soc Am 69:1112–1118.

Otto SR, Shannon RV, Brackmann DE, Hitselberger WE, Staller S, Menapace C (1998) The multichannel auditory brainstem implant: performance in 26 patients. Otolaryngol Head Neck Surg 118:291–303.

Otto SA, Brackmann DE, Hitselberger WE, Shannon RV, Kuchta J (2002) The multichannel auditory brainstem implant update: performance in 60 patients. J Neurosurg 96:1063–1071.

Pfingst BE, Franck KH, Xu L, Bauer EM, Zwolan TA (2001) Effects of electrode configuration and place of stimulation on speech perception with cochlear implants. J Assoc Res Otolaryngol 2:87–103.

Plomp R (1983) The role of modulation in hearing. In: Klinke R, Hartmann R, eds. Hearing—Physiological Bases and Psychophysics. Berlin: Springer-Verlag, pp. 270–276.

Plomp R (1988) The negative effect of amplitude compression in multichannel hearing aids in light of the modulation-transfer function. J Acoust Soc Am 83:2322–2327.

Relkin EM, Doucet JR (1991) Recovery from prior stimulation. I. Relationship to spontaneous firing rates of primary auditory neurons. Hear Res 55:215–222.

Rosen S (1992) Temporal information in speech and its relevance for cochlear implants. Philos Trans R Soc Lond Ser B Biol Sci 336:367.

Rosen S, Faulkner A, Wilkinson L (1999) Adaptation by normal listeners to upward spectral shifts of speech: implications for cochlear implants. J Acoust Soc Am 106:3629–3636.

Rubinstein JT, Wilson BS, Finley CC, Abbas PJ (1999) Pseudospontaneous activity: stochastic independence of auditory nerve fibers with electrical stimulation. Hear Res 127:108–118.

Shannon RV (1983a) Multichannel electrical stimulation of the auditory nerve in man: I. Basic psychophysics. Hear Res 11:157–189.

Shannon RV (1983b) Multichannel electrical stimulation of the auditory nerve in man: II. Channel interaction. Hear Res 12:1–16.

Shannon RV (1985) Loudness summation as a measure of channel interaction in a multichannel cochlear implant. In: Schindler RA, Merzenich MM, eds. Cochlear Implants. New York: Raven Press, pp. 323–334.

Shannon RV (1992) Temporal modulation transfer functions in patients with cochlear implants. J Acoust Soc Am 91:1974–1982.

Shannon RV, Otto SR (1990) Psychophysical measures from electrical stimulation of the human cochlear nucleus. Hear Res 47:159–168.

Shannon RV, Fayad J, Moore J, Lo W, et al. (1993) Auditory brainstem implant: II. Postsurgical issues and performance. Otolaryngol Head Neck Surg 108:634–642.

Shannon RV, Zeng F-G, Kamath V, Wygonski J, Ekelid M (1995) Speech recognition with primarily temporal cues. Science 270:303–304.

Shannon RV, Zeng F-G, Wygonski J (1998) Speech recognition with altered spectral distribution of envelope cues. J Acoust Soc Am 104:2467–2476.

Shannon RV, Fu Q-J, Wang X, Galvin J, Wygonski J (2001a) Critical cues for auditory pattern recognition in speech: implications for cochlear implant speech processor design. In: Breebaart DJ, Houtsma AJM, Kohlrausch A, Prijs VF, Schoonhoven R, eds. Physiological and Psychological Bases of Auditory Function: Proceedings of the 12th International Symposium on Hearing. Maastricht, NL: Shaker Publishing BV, pp. 500–508.

Shannon RV, Galvin JJ, Baskent D (2001b) Holes in hearing. J Assoc Res Otolaryngol 3:185–199.

Skinner MW, Holden LK, Holden TA (1995) Effect of frequency boundary assignment on speech recognition with the SPEAK speech coding strategy. Ann Otol Rhinol Laryngol 104(suppl 166):307–311.

Steeneken HJM, Houtgast T (1980) A physical method for measuring speech-transmission quality. J Acoust Soc Am 67:318–326.

ter Keurs M, Festen JM, Plomp R (1992) Effect of spectral envelope smearing on speech reception. I. J Acoust Soc Am 91:2872–2880.

ter Keurs M, Festen JM, Plomp R (1993) Effect of spectral envelope smearing on speech reception. II. J Acoust Soc Am 93:1547–1552.

Tiffany WR, Bennett DA (1961) Intelligibility of slow-played speech. J Speech Hearing Res 4:248–258.

Turner C, Gantz B (2001) Combining acoustic and electric hearing for patients with high frequency hearing loss. Abstracts of the 2001 Conference on Implantable Auditory Prostheses, Asilomar, CA, p. 33.

Turner CW, Souza PE, Forget LN (1995) Use of temporal envelope cues in speech recognition by normal and hearing-impaired listeners. J Acoust Soc Am 97(4):2568–2576.

Vandali AE, Whitford LA, Plant KL, Clark GM (2000) Speech perception as a function of electrical stimulation rate: using the Nucleus 24 cochlear implant system. Ear Hear 21:608–624.

Van den Honert C, Stypulkowski PH (1984) Physiological properties of the electrically stimulated auditory nerve. II. Single fiber recordings. Hear Res 14:225–243.

Van den Honert C, Stypulkowski PH (1987) Single fiber mapping of spatial excitation patterns in the electrically stimulated auditory nerve. Hear Res 29:195–206.

van Hoesel RJM, Clark GM (1995) Evaluation of a portable two-microphone adaptive beamforming speech processor with cochlear implant patients. J Acoust Soc Am 97:2498–2503.

van Hoesel RJM, Clark GM (1997) Psychophysical studies with two binaural cochlear implant subjects. J Acoust Soc Am 102:495–507.

van Hoesel RJM, Tong YC, Hollow RD, Clark GM (1993) Psychophysical and speech perception studies: a case report on a binaural cochlear implant subject. J Acoust Soc Am 94:3178–3189.

Van Tasell DJ, Soli SD, Kirby VM, Widin GP (1987) Speech waveform envelope cues for consonant recognition. J Acoust Soc Am 82:1152–1161.

Van Tasell DJ, Greenfield DG, Logemann JJ, Nelson DA (1992) Temporal cues for consonant recognition: training, talker generalization, and use in evaluation of cochlear implants. J Acoust Soc Am 92:1247–1257.

Viemeister NF (1979) Temporal modulation transfer functions based upon modulation thresholds. J Acoust Soc Am 66:1364–1380.

von Ilberg C, Kiefer J, Tillein J, Pfenningdorff T, et al. (1999) Electric-acoustic stimulation of the auditory system. New technology for severe hearing loss. ORL J Otorhinolaryngol Relat Spec 61:(6)334–340.

Warren RM, Reiner KR, Bashford JA Jr, Brubaker BS (1995) Spectral redundancy: intelligibility of sentences heard through narrow spectral slits. Percept Psychophys 57(2):175–182.

Wilson BS (1997) The future of cochlear implants. Br J Audiol 31:205–225.

Wilson BS, Finley CC, Lawson DT, Wolford RD, Eddington DK, Rabinowitz WM (1991) New levels of speech recognition with cochlear implants. Nature 352:236–238.

Wilson BS, Lawson DT, Zerbi M, Finley CC, Wolford RD (1995) New processing strategies in cochlear implantation. Am J Otol 16:669–675.

Wilson BS, Finley CC, Lawson D, Zerbi M (1997) Temporal representations with cochlea implants. Am J Otol 18(6 suppl):S30–S34.

Wilson BS, Wolford RD, Lawson DT (2000) Speech processors for auditory prostheses. 7th quarterly progress report. NIH contract N01-DC-8-2105.

Zeng F-G (2002) Temporal pitch in electric hearing. Hear Res 174(1–2):101–106.

Zeng F-G, Galvin J (1999) Amplitude mapping and phoneme recognition in cochlear implant listeners. Ear Hear 20:60–74.

Zeng F-G, Shannon RV (1992) Loudness balance between electric and acoustic stimulation. Hear Res 60:231–235.

Zeng F-G, Shannon RV (1994) Loudness coding mechanisms inferred from electric stimulation of the human auditory system. Science 264:564–566.

Zeng F-G, Shannon RV (1999) Psychophysical laws revealed by electric hearing. NeuroReport 10(9):1–5.

Zeng F-G, Turner CW, Relkin EM (1991) Recovery from prior stimulation. II: effects upon intensity discrimination. Hear Res 55(2)223–230.

Zeng F-G, Galvin J, Zhang C-Y (1998) Encoding loudness by electric stimulation of the auditory nerve. NeuroReport 9(8):1845–1848.

Zeng F-G, Grant G, Niparko J, Galvin J, et al. (2002) Speech dynamic range and its effect on cochlear implant performance. J Acoust Soc Am 111(1 pt 1):377–386.

Zwolan TA, Kileny PR, Ashbaugh C, Telian SA (1996) Patient performance with the Cochlear Corporation "20 + 2" implant: bipolar vs monopolar activation. Am J Otol 17:717–723.

9
Learning, Memory, and Cognitive Processes in Deaf Children Following Cochlear Implantation

David B. Pisoni and Miranda Cleary

1. Introduction

A cochlear implant is a sensory aid that uses a surgically implanted electrode array threaded within the scala tympani of the cochlea to present an electrical representation of sound directly to the auditory nerve of individuals who have a severe to profound bilateral sensorineural hearing loss. Cochlear implants work and they work reasonably well in many profoundly deaf adults and children. For these patients, a cochlear implant (CI) is a form of intervention, an alternative way of providing access to sound using electrical stimulation of the auditory system (see Niparko, Chapter 3). For postlingually deafened adults, a cochlear implant serves primarily as a sensory aid to restore lost hearing and regain contact with the world of sound as they knew it before the onset of their deafness. For the prelingually deaf child, in contrast, the electrical stimulation provided by a cochlear implant represents the introduction of a new sensory modality and an additional way to acquire knowledge about sound, sound sources, and the correlations between objects and events in their environment. Perhaps the most important benefit of a cochlear implant, however, is that it provides the prelingually deaf child with access to information about speech and spoken language. Because speech is a multimodal event, the electrical stimulation provided by the cochlear implant also provides the child with a rich source of new information about the cross-modal relations between the auditory and visual correlates of speech that reflect the common underlying articulatory gestures of the talker. Finally, a cochlear implant provides the deaf child with direct auditory feedback about the consequences of his/her own vocal articulation and motor control in speech production, which affects the development of speech and language acquisition after implantation.

Despite the success of cochlear implants in many deaf patients, enormous individual differences have been reported in adults and children on a wide range of outcome measures (see Shannon et al., Chapter 8). This finding is observed in all research centers around the world. Some patients do extremely well with their cochlear implant while others derive only minimal benefits after receiving their implant. Understanding the reasons for the large individual differences following cochlear implantation is one of the most important and challenging research problems in the field today (NIH Consensus Conference 1995). It is not immediately obvious why some patients do well while others struggle and derive only small benefits after receiving a cochlear implant. Many factors may be responsible for these differences in outcome, and numerous complex interactions among these factors should be explored.

Our initial interest in studying individual differences in children following cochlear implantation came from several reports in the literature demonstrating that a small number of deaf children displayed exceptionally good performance with their cochlear implants. They acquired spoken language quickly and easily and seemed to be on a developmental trajectory that paralleled children with normal hearing. These children are often called "stars" and until recently their exceptionally good performance appeared to be an anomaly to many clinicians and researchers.

The finding that some children with cochlear implants display exceptionally good performance can be taken, at first glance, as an "existence proof" for the efficacy of cochlear implants: cochlear implants work with some deaf children and they facilitate the processes of speech perception and language development. The major problem right now, however, is that cochlear implants do not work well with *all* children, and some children derive only minimal benefits from their implants. Why does this occur? What sensory, perceptual, cognitive, and environmental factors are responsible for the differences in performance among deaf children with cochlear implants?

Our theoretical motivation for studying individual differences following cochlear implantation is based on an extensive body of research in the field of cognitive psychology over the last 25 years on "expertise" and "expert systems" theory (Ericsson and Pennington 1993). Many important new insights have come from studying expert chess players, radiologists, and other individuals who have highly developed skills in specific knowledge domains like computer programming, spectrogram reading, and even chicken-sexing (Biederman and Shiffar 1987).

The rationale underlying our approach is quite straightforward. If we can learn more about the exceptionally good users of cochlear implants and the reasons why they do so well, perhaps we can use this information to develop new intervention techniques with children who are not benefiting from their implants. Knowledge and understanding of the exceptionally good users, the "stars," might also be useful for developing new preimplant predictors of performance, modifying current criteria for candidacy, and creating better methods of assessing performance and measuring outcome and benefit after implantation. Thus, many important clinical benefits could result from research on individual differences in these particular children.

The large individual differences following cochlear implantation also have implications for several important theoretical issues dealing with neural plasticity and development. Deaf children who receive cochlear implants have been deprived of sound input for some length of time after birth, and their nervous systems have continued to develop in the absence of normal sensory stimulation during the critical period for language learning. These children represent a unique clinical population to study because they can provide new information about the effects of early auditory deprivation on cognitive and linguistic development. What happens to the nervous system and brain of these children as a result of deafness and lack of auditory stimulation over this period of time? Can the atypical pattern of development in these children be modified or reversed after the introduction of sound?

At the present, we know that a small number of demographic factors are strongly associated with a variety of speech perception and language development outcome measures in these children. However, the investigation of higher level perceptual, cognitive, and linguistic factors has not received much attention until recently. One reason for the lack of knowledge about cognitive processes in these children is that most of the clinical research on cochlear implants over the last 10 to 15 years has been carried out by audiologists and speech-language pathologists who have been concerned mainly with questions of device efficacy and assessment of outcome. Their primary interests have been focused on demonstrating that cochlear implants work and provide benefit to deaf patients. Historically, these researchers have had relatively little interest in variation and individual differences in performance. Research on treatment efficacy requires well-defined assessment measures of outcome performance that are familiar to surgeons and clinicians who work with deaf patients. In contrast, research on variability and individual differences in performance deals with a fundamentally different problem, the clinical effectiveness of cochlear implants, that is, explaining why cochlear implants do not work well in all patients who receive them.

Interest in individual differences following cochlear implantation has also become a high priority of the federal government, which funds basic and clinical research on hearing and deafness. In 1995 the National Institutes of Health published a "Consensus Statement on Cochlear Implants in Adults and Children" to provide clinicians with an up-to-date summary of the benefits and limitations of cochlear implants (NIH 1995). The NIH panel concluded that while cochlear implants improve communication abilities in most postlingually deafened adults with severe to profound hearing loss, the outcomes of implantation are much more variable in children, especially prelingually deafened children. Among other findings related to the efficacy and effectiveness of cochlear implants, the panel specifically focused on the wide variation in outcome measures in implant users and recommended that additional basic and clinical research be carried out on individual differences in both adults and children. The NIH panel also suggested that new methods and tools should be developed to study how cochlear implants activate the central auditory system.

An examination of the literature on the effectiveness of cochlear implants in prelingually deaf children suggests that central auditory, cognitive, and linguistic factors may be responsible for some of the variability and individual differences observed in traditional outcome measures (Pisoni 2000; Pisoni et al. 2000). Although the NIH consensus statement on cochlear implants mentioned central auditory factors, the report was not very specific about precisely what these factors might be, or what role higher level cognitive processes might play in the outcome measures.

This chapter summarizes findings that suggest that the observed individual differences in outcome following implantation are related to central cognitive factors associated with perception, attention, memory, learning, and language processing. These new findings are encouraging and suggest additional directions for future research on the effects of early sensory experience and language development in deaf children following cochlear implantation. New process measures of performance have revealed the contribution of working memory, verbal rehearsal, and coding processes to several outcome measures of speech and language development (e.g., measures of open-set word recognition and language tasks requiring the use of phonological processing skills). These recent findings also suggest that the particular form of learning that occurs following cochlear implantation may be "domain-specific" and may be related to the temporal processing of sound sequences and coding speech signals into phonological representations in working memory. These phonological representations form the basic building blocks of spoken language processing that are used in word recognition, comprehension, and speech production (Levelt 1989; Luce and Pisoni 1998; Nittrouer 2002).

2. Effectiveness of Cochlear Implants: Five Key Findings

Five key findings have been consistently reported in the literature on cochlear implants in deaf children (Pisoni et al. 2000). These findings suggest that the investigation of central auditory factors may provide new insights into the enormous variability in outcome and benefits observed following cochlear implantation. This section briefly reviews the five key findings that serve as the starting point for the research presented in this chapter. Then we summarize the major assumptions of the information processing approach to cognition that has guided the development of new process measures of performance.

2.1 Individual Differences in Outcome and Benefit

Large individual differences in outcome and benefit following cochlear implantation are well documented in the clinical literature (Hodges et al. 1999; Pisoni 2000; Blamey et al. 2001; Sarant et al. 2001; Geers et al. 2003a,b). However, all current outcome measures of performance are the final end product of a large number of complex sensory, perceptual, cognitive, and linguistic processes that contribute to the observed variation among cochlear implant users. Until our recent studies on working memory in pediatric implant users, no research had focused on the info processing skills of these children or examined the underlying psychological and cognitive processes used to perceive and produce spoken language. Understanding these central cognitive processes has provided some new insights into the basis of individual differences in outcome and may help in developing new intervention techniques in the future that can be used with children who are deriving only minimal benefits from their implants. Four other findings have also been consistently reported in the literature on cochlear implants in children. These findings place several additional constraints on the problem of individual differences following cochlear implantation.

2.2 Age at Onset of Deafness, Age at Implantation, and Duration of Deafness

Age of implantation and duration of deafness have both been found to affect a wide range of outcome measures. Children who receive an implant at an early age do consistently better on all of the clinical outcome measures than children who are implanted at older ages after longer periods of deprivation (Fryauf-Bertschy et al. 1997; Kirk et al. 2002). Moreover, children who have been deprived of auditory stimulation for shorter periods of time also do much better on a variety of outcome measures than children who have been deaf for longer periods of time. Both findings—age of implantation and duration of deafness—demonstrate the role of sensitive periods in development and the close links between neural development and behavior (Konishi and Nottebohm 1969; Konishi 1985; Marler and Peters 1988; Ball and Hulse 1998), especially the sensory and cognitive processes underlying hearing, speech, and language development (Yoshinaga-Itano and Apuzzo 1998; Yoshinaga-Itano et al. 1998).

The interactions among the factors of age at onset of deafness, duration of deafness, and age at implantation have been observed not only in the behavioral clinical performance measures such as word recognition scores, but also, more recently, in electrophysiological and neuroimaging measures of cortical processing in response to sensory stimuli. For example, in children, cortical evoked responses display certain landmarks soon after the temporal onset of a stimulus that are widely agreed to indicate sound detection. Unimplanted deaf children with no prior auditory experience have been found to display certain features in their cortical evoked responses that resemble the cortical evoked responses of much younger normal-hearing infants prior to normal auditory experience (Sharma et al. 2002a,b). In implanted children, these evoked response patterns develop as a function of experience, such that the latency of the sound detection landmark approaches what might be typical for a normal-hearing child with a similar number of years of hearing experience (Ponton et al. 1996a,b; Sharma et al. 2002a,b). More recently, it has been found that the detection latencies of children deafened within the first year of life but implanted before the age of 3.5 years with several years of implant use, appear to be age-appropriate relative to those of normal-hearing children (Sharma et al. 2002b). This research suggests that with early implantation, the development of these cortical response latencies becomes age-appropriate and not proportional to length of auditory deprivation prior to implantation as observed in children implanted later in life. Ponton and Eggermont (2001) have proposed that without auditory experience, a certain class of neurons in the auditory cortex do not reach their usual mature form, thus affecting the shape of certain landmarks in the evoked cortical response, which, in turn, may indicate a slowed auditory detection response.

Echoing results previously obtained in studies of sensory system development in nonhuman vertebrates (Sur et al. 1990; Sur and Leamey 2001), recent functional neuroimaging studies using positron emission tomography (PET) have found, both in pediatric and adult cochlear implant patients, that auditory deprivation may result in the "recruitment" or "colonization" of cortical areas normally used for auditory processing, by other sensory systems, such as vision (Giraud et al. 2001a,b; Lee et al. 2001). This recruitment appears to be more difficult to reverse in patients with longer periods of auditory deprivation. In some cases, when this recruitment corresponds to increased skill in tasks such as visual lip-reading, some practical advantages accrue; on the other hand, in auditory-only listening situations, the cross-modal response of such cortical areas appears to have little practical advantage.

The findings summarized in this section largely reflect the effects of age at onset of deafness, duration of deafness, and age at cochlear implantation. However, these studies also begin to describe at neural and electrophysiological levels, some of the changes in neural processing function that occur as a result of early sensory and linguistic experience.

2.3 Effects of Early Experience

The nature of the early sensory and linguistic experience after implantation has been found to affect performance on a wide range of outcome measures. Depending on a variety of factors including the recommendations of clinicians and the social services available in the family's community, the parents of a child who receives a cochlear implant usually face a choice between placing the child in an "oral" environment that requires the child to rely primarily on spoken language or in a "total" (or simultaneous) communication environment that encourages the child to use a communication system consisting of manual signs in combination with spoken language. Subcategories of communication styles can be found within each of these larger groupings. For example, in some oral communication environments reliance on lip-reading is actively discouraged, whereas in other oral environments lip-reading plays a prominent role in facilitating communication (Ling 1993). Total communication (TC) environments can also vary in their form; in some approaches there is a strong emphasis on having a strict correspondence between the morphological and syntactic structure of what is signed and what is spoken. In other TC approaches, signs are used only to better convey certain aspects of the spoken signal that are typically the most difficult for hearing-impaired listeners to perceive. For the purposes of the present discussion, however, the dichotomy between oral and TC environments is retained because it is this distinction that has proven to be important in predicting individual differences in clinical outcome measures among children with cochlear implants.

Although both modes of communication, oral and TC, incorporate listening, recent research has demonstrated that children in oral-only communication environments generally do better on standardized tests of speech perception, spoken word recognition, and auditory language processing than implanted children in TC programs (Kirk et al. 2000). The differences in performance between these two groups of children as a function of communication mode are seen most clearly in receptive and expressive language tasks that make use of automatized phonological processing skills such as open-set word recognition, spoken language comprehension, and measures of speech production, especially measures of a child's speech intelligibility and expressive language development (Hodges et al. 1999; Cullington et al. 2000; Kirk et al. 2000; Svirsky et al. 2000b).

2.4 Lack of Behavioral Preimplant Predictors of Outcome

In recent years, a small number of additional factors have emerged as potentially important in helping to account for individual differences in success with a cochlear implant. Although the impact for prelingually deafened, early-implanted pediatric cochlear implant users is as yet unknown, in postlingually deafened adults, the positioning and depth of insertion of the electrode array inside the cochlea has been shown to reliably predict some portion of the variability among adult patients in word recognition scores attained postimplantation (Skinner et al. 2002). Furthermore, although obtaining detailed knowledge about the distribution of auditory nerve survival as a function of (normal) characteristic frequency and associated location along the basilar membrane is difficult to ascertain, a number of studies suggest that individual differences in nerve survival within the spiral ganglion or even more speculatively, at the level of the brainstem or thalamus, also contribute to the observed large individual differences in performance with a cochlear implant. Additionally, with some recent revisions in the candidacy requirements for pediatric cochlear implantation, differences in residual hearing have emerged as a potential source of individual differences in outcome performance; greater amounts of residual hearing measured prior to implantation have generally been found to be associated with better outcome performance in children (Zwolan et al. 1997; Gantz et al. 2000).

These recent developments notwithstanding, substantial variability in outcome performance still remains. Especially for the clinical population of prelingually deafened children who receive cochlear implants, effective predictor measures of future performance that can be collected quickly and noninvasively prior to implantation have yet to be identified. Very few reliable behavioral preimplant predictors of success with a cochlear implant are available for young children (Tait et al. 2000). The lack of reliable preimplant predictors in children is an important finding because it suggests the operation of complex interactions among the newly acquired sensory and perceptual capabilities of a child after a period of sensory deprivation, the properties of the language-learning environment, and the various interactions with parents and caregivers that the child is exposed to early on after receiving a cochlear implant. More importantly, however, the absence of reliable preimplant predictors of outcome also makes it difficult to identify in a timely manner those children who may benefit from specific interventions to improve their speech and language processing skills.

2.5 Emergence of Abilities After Implantation

When all of the outcome and demographic measures are considered together, the current evidence suggests that the underlying sensory and perceptual abilities for speech and language emerge after implantation and that performance with a cochlear implant improves with use over time. Because the outcome and benefit of cochlear implantation cannot be predicted reliably from current preimplant behavioral measures, improvement in performance observed after implantation is assumed to be due to learning and memory processes that are related in complex ways to maturational changes in neural and perceptual development and exposure to the target language in the child's immediate environment.

2.6 Significance and Implications

Taken together, the effects of traditional demographic variables on outcome measures after cochlear implantation suggest several general conclusions about how cochlear implants operate to facilitate the acquisition and development of spoken language. These findings also point to several underlying factors that may account for individual differences on various outcome measures. As noted earlier, although some proportion of the total variance in outcome performance is clearly due to peripheral factors related to audibility and the initial sensory encoding of the speech signal into information-bearing sensory channels in the auditory nerve, additional sources of variance may also come from more central cognitive factors. These additional sources of variance have to do with information processing operations and cognitive demands—that is, how the child uses the initial sensory input that is obtained from the cochlear implant and how the language-learning environment modulates, shapes, and facilitates this learning process. Investigation of the encoding, rehearsal, storage, and retrieval of information may provide some new insights into the underlying basis of the large individual differences in outcome measures of speech and language development.

To gain a better understanding of what deaf children are learning from the information provided by their cochlear implants and how they use sound input, we have adopted a theoretical perspective that looks more closely at the content and flow of information within the nervous system and how it changes over time following implantation. Our research on cochlear implants in children has focused on the underlying cognitive and linguistic processes that mediate speech perception and production (Pisoni 2000).

Little, if any, of the previous research on cochlear implantation has investigated what the children learn after they receive their implants, how they go about the process of acquiring spoken language, or how they develop receptive and expressive language skills. Until recently, there have been very few attempts to study the process of language development of children with CIs and compare their linguistic knowledge and performance with that of normal-hearing children or with other hearing-impaired children who use hearing aids (Robbins and Kirk 1996; Miyamoto et al. 1997; Svirsky et al. 2000a).

These are important new research directions that go beyond the basic questions of clinical assessment, device efficacy, and measuring benefit with traditional audiologically based outcome measures; they are fundamental problems in speech and hearing sciences that deal with the effectiveness of cochlear implants outside the restricted conditions of the hearing clinic or the research laboratory. The exclusive reliance on assessment-based clinical research and prediction of outcome measures following cochlear implantation has changed over the last few years. Several articles have already reported new findings on some of these issues (Kirk et al. 1997; Pisoni et al. 1997; Zwolan et al. 1997; Robbins et al. 1998; Bergeson and Pisoni in press), and other studies are currently under way at a number of research centers around the world.

3. Information Processing Approach to Cognition

To pursue these new research questions and to move beyond the study of demographics and issues surrounding clinical assessment and prediction of outcome measures, it has become necessary to look to other allied disciplines for guidance. New experimental methods and behavioral techniques are available to study individual differences and the emergence of fundamental underlying cognitive and neural processes and how these change over time after implantation. Many useful experimental procedures already have been developed by cognitive and developmental scientists to study perception, attention, learning, and memory in children within the framework of human information processing (Neisser 1967; Haber 1969; Lachman et al. 1979; Kail 1984; Siegler 1998). More importantly, this theoretical approach also has provided a variety of powerful conceptual tools for thinking about the structures and processes involved in cognitive activity and the underlying psychological phenomena (Reitman 1965; Lindsay and Norman 1977; Ashcraft 1998, 2002).

3.1 Learning, Memory, and Cognition Viewed as Information Processing

Cognitive psychology is the study of the basic phenomena of intelligent behavior, usually understood to include the activities of perception, memory, learning, comprehension, reasoning, and decision making. Most experimental research in cognitive psychology over the past 40 years has been structured around the information-processing framework, the basic assumptions of which can be conveyed by contrasting it with another theoretical approach that prevailed in the United States from around 1920 to the mid-1960s, namely behaviorism. Behaviorism proposed that intelligent behavior could be understood by simply studying the outwardly observable relationships between properties of the stimulus and properties of the organism's response; explanation was couched in terms of rules relating input to output, divorced from any particular theory regarding the intermediate processing that presumably took place within the organism.

In contrast, the basic assumption of the information-processing approach is that an understanding of the intermediate processing stages is fundamental to characterizing the complexities of human behavior. The information-processing approach to cognition is generally concerned with arriving at an accurate description, at the functional level, of how the human nervous system receives sensory information and operates upon this information to achieve some higher level goal. Using an information-processing framework, researchers typically seek evidence to support or disconfirm the existence of distinct stages in information processing, each accomplishing a particular function. An information-processing framework does not have as its foremost concern a description of exactly how these functions are physically instantiated in the human nervous system; however, in practice it recognizes the need for the functional description to be compatible with what is known about the physical implementation of these processes in the brain (Posner 1989; Gazzaniga 2000).

The existence of different stages in processing is typically demonstrated by providing evidence that some transformation of information takes place. Indeed, one of the primary concerns of the information-processing framework is to describe the nature of these transformations and how information is represented at each stage of processing. Researchers working in an information-processing framework are often also concerned with determining the capabilities and limitations of each processing stage; key issues include how quickly each stage operates, and what limitations if any exist on the speed, capacity, and accuracy of processing. Measures of processing efficiency and accuracy provide insights about the operation of the individual stages. Examining the nature of processing errors is often also illuminating as well.

This chapter describes several sets of memory data from sequence repetition tasks involving immediate recall of auditory and visual patterns. Therefore, it may be useful here to describe how the classic memory task of repeating back a list of spoken words is conceptualized within the information-processing framework. Figure 9.1 presents a simplified diagram of the information-processing stages of primary interest with regard to this task.

Within this framework, transduction of external sound energy patterns into the neural codes utilized by the sensory receptors of the perceptual systems constitutes the initial processing stage of an auditory immediate memory span task. For a cochlear implant user, the electronics of the CI speech processor substitute for function of the outer ear, middle ear, and cochlea. While the sensitivity and characteristic frequencies of these early processing stages constrain and shape to some degree, the sensory representation of incoming sound information, the representation of sound at this early sensory stage of processing is largely veridical, that is, there is relatively little transformation of the incoming representation.

Following this early step of sensory registration, however, is a processing stage in which sound not only is detected and brought into consciousness, but also is processed and recoded in relation to the listener's past auditory experiences. In this encoding stage, incoming information focused in conscious attention is recoded in one or more ways based on information already stored in the system. This encoding stage organizes incoming sensory information according to the perceiver's prior experiences, and one or more classifications of the input are arrived at based on stored knowledge in long-term memory.

The distinction between sensory detection and encoding or pattern recognition can be illustrated by the example of a listener being presented

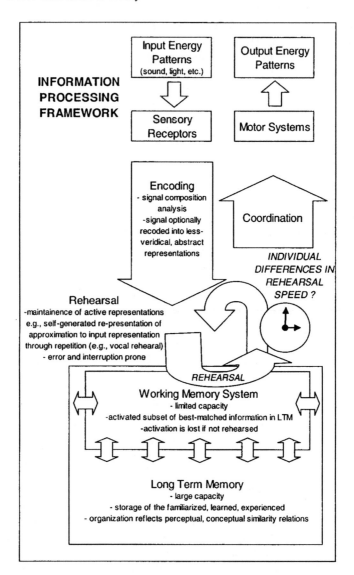

FIGURE 9.1. The foundational assumption of a cognitive psychologist's approach to understanding individual differences in speech and language outcome measures is to view the human nervous system as an information processor. An information processor is a system that encodes, stores, and manipulates various types of representations. Information can exist in several different forms at a number of levels of representation in the system, ranging from early sensory registration and the temporary encoding of information within working memory to the permanent storage of representations in long-term memory.

with a list of words spoken in a foreign language. In this case, the listener's auditory system is fully capable of detecting the acoustic signal, but without a system trained with similar inputs, the listener will experience difficulty identifying the individual speech sounds within each spoken word, will certainly be unable to re-represent the information using a semantic code, and will probably have substantial difficulty if immediately asked to imitate and reproduce these words in speech production.

Learned information, conceptualized as a long-term memory store, influences how early sensory information is encoded and held in active awareness. When the input is familiar, behavioral data suggest that explicit identification and awareness correspond to the maintained activation of a subset of stored representations in long-term memory. Novel and unfamiliar sensory inputs, however, require that new representations be created, even if these consist simply of new conjunctions of familiar inputs. Particularly for the processing of novel inputs, the role of working memory is believed to be vital as a system for the temporary storage and processing of information representations.

Working memory provides a storage mechanism for information to be held in conscious awareness and, optionally, maintained temporarily in this store. The amount of information that can be held in conscious awareness at any one time is quite limited, and the representations at this stage are vulnerable to rapid loss within 15 to 30 seconds both through passive activation decay and from the active interference of new incoming information that can "overwrite" prior information. The constraints on human information processing at the level of working memory and conscious awareness are quite striking and easily demonstrated. Presented with an unfamiliar sequence of digits, for example, most individuals encounter a great deal of difficulty accurately recalling a sequence longer than about eight or nine digits in length even after a delay of only a few seconds.

The limitations on retention of information in working memory, however, can be partially circumvented through the use of so-called control processes (Atkinson and Shiffrin 1968). The most basic of these control processes is "rehearsal." Rehearsal involves the ability of individuals to overcome, at least temporarily, the limitations on the working memory system by repeating to themselves an approximation of the original input containing the information of interest. When the task is to reproduce a word list, for example, rehearsal may take the form of simple verbal repetition of the input list.

Verbal rehearsal appears ubiquitously as a strategy used by individuals to maintain information in active awareness. The degree to which individuals require explicit instruction in order to use this strategy is a matter of debate; however, there is good evidence in the literature that verbal rehearsal is adopted as a memory maintainence strategy by normally developing children as young as 5 to 8 years of age (Flavell et al. 1966; McGilly and Siegler 1989). Interestingly, although individuals who are relatively unpracticed with verbal rehearsal typically rehearse aloud using overt speech, it has been found that when individuals rehearse "subvocally" by repeating the information internally but without actually moving the speech articulators, the same maintenance effects in memory are achieved (Landauer 1962; Standing and Curtis 1989). Moreover, it has been shown that individual differences in the rate and efficiency with which items are verbally rehearsed can predict much of the variability in working memory task performance within otherwise seemingly homogeneous groups of study participants (Baddeley et al. 1975).

Rote verbal rehearsal appears most suited for maintaining information that needs to be temporarily stored, but that may not be needed at a later point in time. Although for some types of information, rote rehearsal can help to build permanent long-term memory representations, in general other methods of strengthening the representation of new information in long-term memory are more effective, such as elaboration strategies for semantically incorporating new information with semantic information already stored in long-term memory. However, when relatively few opportunities for making semantic connections exist, as in the case of encoding the sequential order of randomly ordered items, verbal rehearsal can be an effective strategy for maintaining information for brief periods of time.

3.2 Role of Auditory Experience in Cognitive Development

An active area of research in cognitive psychology involves how information processing changes as a result of experience. Differences in processing between populations, such as children versus adults and experts versus novices, continue to be studied. Because what is stored in long-term memory helps to determine how new information is perceived and encoded, and because newly encoded information can then potentially also be incorporated into long-term memory itself, issues of memory and learning are closely related and are important to understanding differences between subjects, as well as changes in the performance of an individual subject over time.

By viewing human cognition and traditional areas of basic research such as sensation, perception, attention, memory, and learning as information processing within a larger integrated framework, cognitive scientists have obtained several important benefits. These include the development of experimental methodologies to study the processes that underlie these behaviors as well as the availability of a new theoretical framework that can be used to explain and predict variability in complex higher level behaviors such as speech and language in different clinical populations. The information-processing approach to human cognition has also provided the theoretical motivation for reformulating some long-standing problems as well as identifying new research questions that can be studied using this framework.

Although there are many good clinical reasons to study prelingually deaf children who have received cochlear implants and to try to understand the basis for the variability in outcome measures of speech and language, there are also several additional reasons for carrying out research with this unique population that touch on basic theoretical issues related to neural development and behavior. For a variety of moral and ethical reasons, researchers cannot carry out sensory deprivation experiments with young children. It is not possible to delay or withhold treatment for an illness or disability that has been identified and diagnosed. Thus, for studies of this kind, which are concerned with investigating the effects of early sensory experience on neural and behavioral development, it becomes necessary to rely on clinical populations that are receiving interventions of various kinds and hope that appropriate experimental designs can be developed that will yield new scientific knowledge.

Some of the broader theoretical questions about neural and cognitive development that this research addresses include: What effects does the absence of sound and auditory stimulation during the first few years of life have on the development of the basic elementary information processing mechanisms and skills used in speech and language? What effects does the introduction of sound and auditory stimulation by means of electrical hearing through a cochlear implant have on the subsequent development of speech and language processing after a period of auditory deprivation? What kind of a linguistic system (i.e., grammar) does a deaf child develop after receiving a cochlear implant? Is a deaf child's language delayed but otherwise typical of normal-hearing children or is it disordered or impaired in some fundamental way relative to normal-hearing, typically developing, age-matched peers because of neural reorganization? These are only a few of the theoretical questions this work focuses on.

4. Looking at the "Stars"—Analysis of the Exceptionally Skilled Implant Users

We began first by looking at the exceptionally good users of cochlear implants—the so-called stars (Pisoni et al. 1997; Pisoni et al. 2000). These are the children who did extraordinarily well with their cochlear implants after only 2 years of implantation. The stars appeared to acquire spoken language relatively quickly and easily and seemed to be on a developmental trajectory that closely paralleled normal-hearing children. At first glance, these children look like normal-hearing and normally developing children who simply have language delays (Svirsky et al. 2000a,b).

4.1 Traditional Audiological Outcome Measures

To learn more about the stars, we analyzed outcome data from the children who scored exceptionally well on the Phonetically Balanced Kindergarten (PBK) test (Haskins 1949) 2 years after receiving their implant. The PBK test is an open-set test of spoken word recognition (Meyer and Pisoni 1999). Among clinicians, the PBK test is considered to be very difficult for prelingually deaf children compared to other, closed-set perceptual tests that are routinely used in standard assessment batteries to assess outcome and benefit of cochlear implantation (Kirk et al. 1995; Zwolan et al. 1997). Open-set tests of speech perception like the PBK test assess spoken word recognition and lexical discrimination and require children to search and retrieve the phonological representations of the test words from their lexical long-term memory. These particular types of word recognition tests are extremely difficult for hearing-impaired children and adults with cochlear implants because the procedures and task demands require the listener to perceive and encode fine phonetic differences based entirely on the available sensory information present in the speech signal without the aid of any external context or retrieval cues. Basically, the listener is required to discriminate, select, and then identify a unique phonological pattern from a very large number of equivalence classes in long-term lexical memory (Luce and Pisoni 1998).

The PBK score was then used as the criterial variable to initially identify and select two distinct groups of children for subsequent analysis using an extreme-groups design. One group consisted of children who were exceptionally good cochlear implant users—the so-called stars. These were children who scored in the top 20% on the PBK test after 2 years of implant use. A second group of children was selected for comparison. The children in this group scored in the bottom 20% on the PBK test 2 years postimplant and were unable to recognize any of the test words when they were presented in isolation using an open-set format.

After these children were selected and sorted into two groups based on their PBK scores 2 years postimplantation, we examined their performance on a variety of other outcome measures that were available for each child as part of our large-scale longitudinal project. These outcome measures included tests of speech feature perception, language comprehension, word recognition, receptive vocabulary knowledge, receptive and expressive language development, and speech intelligibility. Descriptive analyses were carried out first to compare differences between the two groups on these different measures. Then correlations were computed among the various outcome measures to look at relationships and commonalities between the measures.

The results of our descriptive analyses after 1 year of implant use revealed several interesting findings about the exceptionally good users of cochlear implants. First, we found that although the stars showed better performance on some outcome measures such as speech feature perception, spoken language comprehension, spoken word recognition, and speech intelligibility than the comparison group, the two groups of children did not differ from each other on other measures of receptive vocabulary knowledge (tested in their preferred communication mode), nonverbal intelligence, visual-motor integration, or visual attention (Pisoni et al. 1997). We also found that some outcome measures of performance such as open-set spoken word recognition and speech intelligibility continued to improve over the course of 3 years, whereas other outcome measures such as the perception of the voicing and place speech feature contrasts obtained using simple two-alternative closed set tasks remained fairly stable after the first year.

These findings demonstrate that the stars differ in selective ways from the comparison group of subjects. Whatever differences are revealed by other descriptive measures, it is clear that the results are not due to some global difference in overall performance levels between the two groups. More importantly, we found that the stars also displayed exceptionally good performance on another open-set test of spoken word recognition, the Lexical Neighborhood Test (LNT) (Kirk et al. 1995). This result demonstrates that the superior lexical discrimination skills of these children are not due to the specific words on the PBK test or the particular methods used to administer the test. Instead, the differences appear to be related to a common set of basic information processing operations and procedures that are used by these children to carry out open-set word recognition tasks. Among the component elementary information processing skills needed for this task are encoding, storage, verbal rehearsal, imitation, and speech production. These are the same basic information-processing skills that are used in the other clinical outcome measures that assess performance in these children after implantation.

The results of our correlational analyses of the test scores for the stars 1 year after implantation revealed a consistent pattern of strong and highly significant intercorrelations among several of the dependent variables, particularly measures of spoken word recognition, language development, and speech intelligibility, suggesting a common underlying source of variance that is shared by these measures (Pisoni et al. 1997). The same patterns of intercorrelations were not observed for the comparison group. One common source of variance found in the correlational analyses of the stars appeared to be related to the processing of spoken words and to the encoding, storage, and verbal rehearsal of the phonological representations of words.

Of particular interest was the unexpected finding of strong correlations of several outcome measures with speech intelligibility scores obtained for these children, suggesting transfer of knowledge between speech perception and production and the use of a common shared representational system for receptive and expressive language functions (Shadmehr and Holcomb 1997). The results suggested that the exceptionally good performance of the stars may be due to their superior spoken language processing abilities, specifically their ability to perceive, encode, and retrieve phonological representations of spoken words from lexical memory and use these phonological representations in a variety of different language-processing tasks, especially tasks that depend on the rapid decomposition and reassembly of the sound patterns of spoken words such as lexical retrieval, verbal rehearsal, and speech production. Our working hypothesis was that this particular source of variance might reflect "modality-specific" elementary information processing operations that are involved in the phonological encoding of sensory inputs and the construction of phonological representations of spoken language.

While the results of our initial correlational analyses pointed to several new directions for future research on individual differences, the data available on these children were based on traditional audiological outcome measures that were collected as part of the annual assessments in a longitudinal study. All of the scores on these tests are "end-point" measures of performance and as such they reflect the final product of perceptual and linguistic analysis. Process measures of performance, that is, measures of what children do with the sensory information provided by their cochlear implant, were not part of the standard research protocol, so it was impossible to investigate differences in speed, fluency, or information-processing capacity—behavioral measures that are known to reveal significant individual differences among normally developing individuals (Kane and Engle 2002). It is very likely that fundamental differences in neural and cognitive information-processing skills may underlie the individual differences observed between the two groups of children in our initial study.

4.2 New Process Measures of Performance

The analyses we carried out on the speech feature perception, spoken word recognition, language comprehension, vocabulary knowledge, and language development scores revealed that a child who displayed exceptionally good performance on the PBK test also showed very good scores on a variety of other speech and language measures as well. We consider these new results to be theoretically important. The differences in the clinical outcome measures observed between the two groups of children suggested that it may be possible to determine more precisely how the stars differ from the comparison group of poorly performing cochlear implant users. Knowledge of the factors that are responsible for individual differences in performance among deaf children who receive cochlear implants, particularly the variables that underlie the extraordinarily good performance of the stars, may be useful in helping the children who are not doing as well with their implants at an earlier point in development following implantation. Moreover, research on the nature of the individual differences may have direct clinical relevance in terms of intervention in recommending specific changes in the child's language-learning environment and in modifying the nature of the interactions a child has with his parents, teachers, and speech therapists, who provide the primary language model for the child. Research on individual differences in these children may also help by providing clinicians and parents with a principled basis for generating realistic expectations about outcome measures, particularly measures of speech feature perception, spoken word recognition, language development, and speech intelligibility.

The key component that links the different stages of information processing and serves as the interface or "gateway" between the initial sensory input and stored knowledge in long-term memory is the construct of working memory. Investigations of the properties of working memory (Carpenter et al. 1994; Gupta and MacWhinney 1997; Baddeley et al. 1998; Kane and Engle 2002) may provide new insights into the nature and locus of the individual differences observed among users of cochlear implants. Unfortunately, at the time our analyses of the stars were carried out, we did not have any data on working memory available to test this hypothesis. Since that time, however, several new studies have been carried out with our collaborators at the Central Institute for the Deaf (CID) in St. Louis to study working memory in children with cochlear implants (Pisoni and Geers 2000; Cleary, Pisoni and Geers 2001; Pisoni and Cleary 2003). The results of these experiments are summarized below.

At a superficial level, it seems reasonable to conclude that the children who "hear" better through their cochlear implant simply learn language better and, subsequently, recognize words better. This generalization might be appropriate as an explanation of the perceptual data obtained from traditional audiological outcome measures of receptive function, but it is much more difficult to explain the observed differences in speech intelligibility and expressive language development simply on the basis of better hearing and language-processing skills without a more detailed description of the underlying linguistic skills and abilities that are used in speech production.

To account for the differences in speech intelligibility and expressive language development, it is necessary to assume that an individual child has acquired an underlying linguistic system or language model that mediates between speech perception and speech production. Without assuming a common linguistic system—a grammar—we would have no principled reason to expect a child's receptive and expressive language abilities to be as closely coupled as they typically are in normal-hearing, normally developing children. It is well known that reciprocal links exist among speech perception, speech production, and a whole range of language-related abilities; these interconnections reflect the child's linguistic knowledge of phonology, morphology, and syntax. Speech perception, spoken word recognition, and language comprehension are not isolated, autonomous perceptual abilities, independent of the child's developing linguistic system. Thus, explanations of the superior performance of the stars framed only in terms of hearing, audibility, or sensory discrimination abilities cannot provide a satisfactory theoretical account of all of the results reported above or an adequate description of how early auditory and linguistic experience affects speech perception and language development in these children. Some other set of nonsensory cognitive processes may be responsible for the commonalities in speech perception and production observed across these diverse outcome measures.

To understand and explain the individual differences in outcome measures in these children, several additional performance measures are needed to assess how deaf children with cochlear implants process, code, and represent the sensory, perceptual, and linguistic information they receive through their implants, and how they use this information in a variety of behavioral tasks. In addition, we need to study the cognitive processes of memory, learning, attention, and automaticity in this clinical population.

5. Measures of Information Processing Capacity

To obtain some initial measures of information-processing capacity, several different working memory tasks were included as part of a test battery used in a large-scale project designed to study the development of speech, language, and reading skills in children with cochlear implants (Geers and Brenner 2003; Geers et al. 2003a,b). The children studied in this project were all 8 or 9 years of age and had become deaf before the age of 3. Additionally, all of the children had received a cochlear implant by the age of 5 years and had used their implants for at least 3.75 years at the time of testing. Thus, many of the traditional demographic factors known to contribute to clinical outcome measures were relatively controlled within this sample. The first measure of working memory studied in this group was immediate memory span for spoken digits.

5.1 Immediate Memory Span for Spoken Digits

Using the test lists and procedures from the Wechsler Intelligence Scale for Children, 3rd edition (WISC-III) (Wechsler 1991), forward and backward auditory digit spans were obtained from four groups of 8- and 9-year-old deaf children with cochlear implants. A total of 176 children were tested at the CID during the summers of 1997, 1998, 1999, and 2000. Forward and backward digit spans were also collected from an additional group of 44 normal-hearing 8- and 9-year-old children. These typically developing children were tested in Bloomington, Indiana, and served as a comparison group.

The WISC-III memory span task requires the child to repeat back lists of digits that are spoken by an experimenter at a rate of approximately one digit per second (Wechsler 1991). In the digits-forward section of the task, the child is required to simply repeat back each list as heard. In the digits-

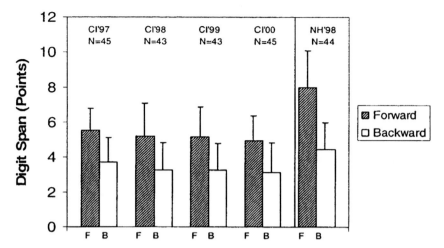

FIGURE 9.2. Wechsler Intelligence Scale for Children (WISC) digit spans scored by points for the four groups of 8- and 9-year-old children with cochlear implants (CI) and for a comparison group of 8- and 9-year-old normal-hearing (NH) children. Forward digit spans are shown by the shaded bars, backward digit spans by the open bars. Error bars indicate one standard deviation from the mean.

backward section of the task, the child is told to "say the list backward." In both parts of the WISC task, the lists begin with two items, and the increase in length after successful repetition until a child gets two lists incorrect at a given length, at which time testing stops. Points are awarded for each list correctly repeated. There is no partial credit. The task was administered using live-voice presentation with the face of the clinician visible to the child.

A summary of the digit span results obtained for all five groups of children is shown in Figure 9.2. Forward and backward digit spans are shown separately for each group. The mean scores from the children with cochlear implants are shown in the four sets of bars on the left, presented separately by year of testing; the mean scores from the normal-hearing children are shown on the right. Each child's digit span in points was calculated by summing the number of lists correctly recalled at each list length. The score for forward digit span can vary between zero and 16; the score for backward digit span can vary between zero and 14.

Inspection of the data shown in Figure 9.2 reveals an orderly and systematic pattern of the forward and backward digit spans for the deaf children with cochlear implants. All four groups are quite similar to each other; in each group, the forward digit span is longer than the backward digit span. The pattern is quite stable over the 4 years of testing despite the fact that these scores are based on independent groups of subjects. The difference in span length between forward and backward report was highly significant for the entire group of 176 deaf children and for each group taken separately ($p < .001$).

The digit spans from the normal-hearing children differ in several systematic ways from the data obtained from the children with cochlear implants. First, both digit spans are longer than the spans obtained from the children with cochlear implants. Second, the forward digit span for the normal-hearing children is much longer than the forward digit spans obtained from the children with cochlear implants. This latter finding is particularly important because it suggests atypical development of the deaf children's short-term memory capacity and points to several possible differences in the underlying processing mechanisms that are used to encode and maintain sequences of spoken digits in immediate memory.

Numerous studies over the years have suggested that forward digit spans can be used to index and assess initial coding strategies related to phonological processing and verbal rehearsal mechanisms used to maintain information in short-term memory for brief periods of time before retrieval and output response. In contrast, differences in backward digit spans are thought to reflect the contribution of controlled attention and operation of higher level "executive" processes that are used to recode, transform, and manipulate verbal information for later processing operations (Rudel and Denckla 1974; Rosen and Engle 1997).

The digit spans for the normal-hearing children shown in Figure 9.2 are age-appropriate and fall within the published norms for the WISC-III. However, the forward digit spans obtained from the deaf children with cochlear implants are atypical and suggest possible differences in encoding and/or verbal rehearsal processes used to maintain information in immediate memory. In particular, the forward digit spans reflect possible differences in processing capacity of immediate memory between the two groups of children. These differences may cascade and affect other information-processing tasks that make use of working memory and verbal rehearsal processes (Fry and Hale 1996). Because all of the behavioral tasks typically used to assess clinical speech and language outcomes following implantation make use of the component processes of working memory and verbal rehearsal, it seems reasonable to assume these tasks also reflect variation due to differences in working memory and processing capacity.

5.2 Correlations with Immediate Memory Span

To understand the basis for the observed differences in auditory digit span and the limitations in processing capacity in the children with cochlear implants, we examined the correlations between forward and backward digit spans and several speech and language outcome measures that were also obtained from the same children at CID. Of the various demographic measures available, the only one that correlated strongly and significantly with digit span was a measure called "communication mode," which quantifies the nature of the child's early sensory and linguistic experience after receiving a cochlear implant in terms of the degree of emphasis on oral language skills by teachers and therapists in the educational environment.

Each child's degree of exposure to oral-only communication methods was quantified by determining the type of communication environment experienced by the child in the year just prior to implantation, each of the first 3 years of CI use, and then in the year just prior to the current testing. A score was assigned to each year, with 1 corresponding to the use of "total communication" with a sign emphasis (that is, extensive use of manual signs in addition to spoken language), and 6 indicating an auditory-verbal environment with a strong emphasis on auditory communication without the aid of lipreading (see Geers and Brenner 2003 for details). Communication methods intermediate between these two extremes were assigned intermediate scores ranging from 2 to 5. These scores were then averaged over the five points in time. The mean communication mode score for the entire group over the five intervals was approximately 3.9 on this six-point scale. A wide range of communication mode backgrounds, however, was present within the sample (range of average communication mode scores, 1.0 to 6.0).

We found that forward digit span was positively correlated with communication mode ($r = .34, p < .001$); children in language learning environments that primarily emphasized oral skills displayed longer forward digit spans than children who were in TC environments. However, the correlation between digit span and communication mode was selective in nature because it was restricted only to the forward digit span scores; the backward digit spans were not significantly correlated with communication mode ($r = +.14, p > .05$).

To examine the effects of early experience on digit span in greater detail, a median split was carried out on the communication mode scores to create two subgroups: oral children and TC children. Figure 9.3 shows the digit spans plotted separately for the oral and TC children for each of the 4 years of testing at CID. Examination of the forward and backward digit spans for these two groups of children indicates that the oral group consistently displayed longer forward digit spans than the TC group. While the differences in forward digit span between the oral and TC groups were highly significant ($p < .001$), the differences in backward digit span were not significant. This pattern suggests that the effects of early sensory and linguistic experience on immediate memory is selective in nature and appears to be restricted to coding and verbal rehearsal processes that affect only the forward digit span task.

The difference in forward digit span between the oral and TC children present for each of the four groups suggests that forward digit spans are sensitive to the nature of the early sensory and linguistic experience that the child receives immediately after cochlear implantation. The differences

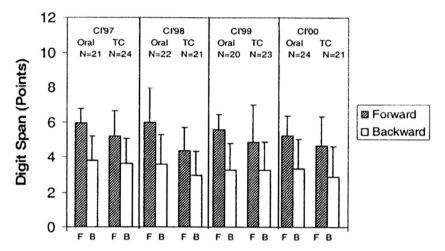

FIGURE 9.3. WISC digit spans scored by points for the four groups of 8- and 9-year-old children with cochlear implants, separated by communication mode. For each year, scores for the oral group are shown to the left of those for the total communication group. Forward digit spans are shown by the shaded bars, backward digit spans by the open bars. Error bars indicate one standard deviation from the mean.

observed in the forward digit spans could be due to several factors, such as more efficient encoding of the initial stimulus patterns into phonological representations in working memory, speed and efficiency of the verbal rehearsal processes that are used to maintain information in working memory, or possibly even speed of scanning and retrieval of phonological information in working memory after recognition has taken place. All three factors could influence measures of processing capacity, and any one of these could affect the number of digits correctly recalled from immediate memory in this task (Burkholder and Pisoni 2003).

Regardless of which factor or factors may be responsible for the differences observed above, these results clearly demonstrate that forward digit span is sensitive to the effects of early sensory and linguistic experience and suggest that several specific mechanisms in the information-processing system may be affected by the nature of the early experience the child receives after implantation. Although these results show that early experience in an environment that emphasizes oral language skills is associated with longer digit spans and increased information-processing capacities of working memory, without additional converging measures of performance, it is difficult to specify more precisely what elementary information-processing mechanisms are actually affected by early experience and which ones are responsible for the increases in forward digit spans observed in these particular children.

5.3 *Immediate Memory Span and Word Recognition*

The findings of our earlier study examining the exceptionally skilled cochlear implant users or stars suggested that the large individual differences that are consistently observed in a range of outcome measures of speech and language development in the pediatric CI population were related in some way to spoken word recognition skills and to tasks that make use of phonological representations of spoken words. While a number of demographic factors such as duration of deafness, length of device use, and age at implantation have been shown to be related to the individual variability in outcome, these variables are able to account for only a portion of the observed variance in performance. We reasoned that some other factor may be responsible for the wide range of performance observed in these deaf children.

Our current hypothesis is that a portion of the remaining unexplained variance can be accounted for in terms of individual differences in the elementary information-processing operations that are related to the speed and efficiency with which phonological representations of spoken words are maintained and retrieved from working memory after recognition and identification have taken place. Numerous studies of normal-hearing children over the past few years have demonstrated close links between phonological working memory and learning to recognize and understand new words (Gupta and MacWhinney 1997; Baddeley et al. 1998). Other research has found that vocabulary development and several important milestones in speech and language acquisition are also associated with differences in measures of phonological working memory (Gathercole et al. 1997; Adams and Gathercole 2000).

To determine if measures of immediate memory as indexed by the digit span task are related to spoken word recognition in deaf children following cochlear implantation, we examined the correlations between the WISC forward and backward digit span scores and three different measures of spoken word recognition that were also obtained from the children tested at CID in 1997 and 1998. A summary of the correlations between digit span and word recognition scores based on these 88 children is shown in Table 9.1 for the Word Intelligibility by Picture Identification (WIPI), LNT, and Bamford-Kowal-Bench (BKB) word recognition tests.

The WIPI is a closed-set test of spoken word recognition in which the child selects a word from among six alternative pictures (Ross and Lerman 1979). The LNT is an open-set test of word recognition and lexical discrimination that requires the child to imitate and reproduce an isolated word (Kirk et al. 1995). This test is similar to the well-known PBK test, although the vocabulary on the LNT was designed to control for familiarity. Lexical competition among the test items is also varied systematically to measure discrimination among phonetically similar words in the child's

TABLE 9.1. Correlations between digit span scores and three measures of word recognition ($n = 88$).

	Simple Bivariate Correlations		Partial correlations[a]	
	WISC forward digit span	WISC backward digit span	WISC forward digit span	WISC backward digit span
Closed-set word recognition (WIPI)	.50*	.33**	.31**	.13
Open-set word recognition (LNT-E)	.54*	.31**	.33**	.10
Open-set word recognition in sentences (BKB)	.56**	.37*	.35**	.12

*$p < .001$; **$p < .01$
[a]Statistically controlling for communication mode score, age at onset of deafness, duration of deafness, duration of cochlear implant (CI) use, number of active electrodes, VIDSPAC total segments correct (speech feature perception measure), age.
BKB: Bamford-Kowal-Bench test; LNT-E: Lexical Neighborhood Test-Easy; VIDSPAC: Video Game Test of Speech Contrast Perception; WIPI: Word Intelligibility by Picture Identification Test WISC: Wechsler Intelligence Scale for Children.

lexicon. The BKB is an open-set word recognition test in which key words are presented in spoken sentences (Bench et al. 1979).

Table 9.1 displays two sets of correlations. The first two columns show the simple bivariate correlations of the forward and backward digit spans with the three measures of word recognition. Examination of the correlations for both the forward and backward spans reveals that children who had longer WISC digit spans also displayed higher scores on all three word recognition tests. These correlations are all positive and reached statistical significance, although the correlations of forward digit span with the word recognition scores are consistently larger than the correlations found for the backward span.

The last two columns show the partial correlations among these same measures after statistically controlling for differences due to chronological age, communication mode, duration of deafness, duration of device use, age at onset of deafness, number of active electrodes, and speech feature discrimination. When these seven other contributing variables were statistically removed from the correlational analyses, the partial correlations between digit span and word recognition scores become smaller in magnitude overall. However, the correlations of the forward digit span with the three word recognition scores are still positive and statistically significant, while the correlations of the backward digit spans are now much weaker and no longer reach significance. These results demonstrate that children who have longer forward WISC digit spans also show higher word recognition scores, and this relationship is observed for all three word recognition tests even after the other sources of variance are removed.

Forward digit span accounted for approximately 11% of the currently unexplained variance in the word recognition scores, while the backward digit span accounted for only 1.4% of the variance in these scores. These results suggest the presence of a common underlying source of variance that is shared between forward digit span and measures of spoken word recognition and that is independent of other obvious mediating factors that have been found to contribute to the variation in these outcome measures.

5.4 Immediate Memory Span and Speaking Rate

While the correlations of the digit span scores with communication mode and spoken word recognition suggest fundamental differences in encoding and verbal rehearsal speed, these measures of immediate memory span and estimates of information processing capacity are not sufficient on their own to identify the underlying processing mechanism (or mechanisms) that are responsible for the individual differences. Additional converging measures are needed to pinpoint the locus of these processing differences more precisely. Fortunately, an additional set of behavioral measures was obtained from these same children for a completely different purpose and made available to us for several new analyses. These data consisted of a set of acoustic measurements of speech samples obtained from each child. These speech samples provided a unique opportunity for us to use converging measures to understand and explain the digit span results.

As part of the research project at CID, several speech production samples were obtained for each child to assess speech intelligibility and measure changes in articulation and phonological development following implantation (Tobey et al. 2000, 2003). The speech samples consisted of three sets of meaningful English sentences that were elicited using the stimulus materials and experimental procedures originally developed by McGarr (1983) to assess the speech intelligibility and articulation of deaf children. All of the utterances produced by the children were originally recorded and stored digitally for playback to groups of naive adult listeners who were asked to transcribe what they thought the children had said. In addition to the speech intelligibility scores that were obtained for each child using these playback procedures, we analyzed the duration measurements of the individual sentences in each set and used these measures as estimates of a child's articulation rate.

We knew from a large body of earlier research in the memory literature on normal-hearing typically-developing children that a child's articulation rate is closely related to speed of subvocal verbal rehearsal (Standing and Curtis 1989; Cowan et al. 1998). Numerous studies in the literature over the past 25 years have demonstrated strong relations between speaking rate and memory span for digits and words. The results of these studies have been replicated with several different populations and suggest that measures of an individual's speaking rate reflect articulation speed, which in turn can be used as an index of the rate of covert verbal rehearsal of phonological representations in working memory (Baddeley et al. 1975). Individuals who speak more quickly have been found to have longer immediate memory spans than individuals who speak more slowly.

Several different explanations of these findings have been proposed. One account assumes that more forgetting occurs from immediate memory at slower speaking rates because fewer words can be articulated and perceived within the same period of time. Another proposal assumes that the mechanism that controls speaking rate is the same one that regulates the speed of verbal rehearsal processes in immediate memory. Thus, more words can be maintained in immediate memory at faster verbal rehearsal speeds. Regardless of which view is correct, the relation observed between measures of speaking rate and immediate memory span is a reliable and robust finding reported in the literature on working memory that has been observed in several different populations of subjects.

The forward digit span scores for the 88 children tested in 1997 and 1998 are shown in Figure 9.4 along with estimates of their speaking rates obtained from measurements of the seven-syllable McGarr sentences. The digit spans are plotted on the ordinate; the average sentence durations are shown on the abscissa. The top panel shows mean sentence durations; the bottom panel shows the logarithmic transformations of the sentence durations. The pattern of results displayed in both figures is quite clear; children who produce sentences with longer durations speak more slowly and, in turn, display shorter forward digit spans. The correlations between forward digit span and both measures of sentence duration were strongly negative and highly significant ($r = -.63$ and $r = -.70; p < .001$, respectively). For backwards digit span, we observed somewhat smaller but still statistically significant correlations with sentence duration ($r = -.42$ and $r = -.42; p < .001$).

These results replicate with hearing-impaired children with cochlear implants findings previously reported for normally developing children and adults, namely, that individual differences in articulation rate predict a substantial amount of the individual variability observed in performance on a verbal digit span task. Children with cochlear implants who spoke more slowly tended to have shorter digit spans than implanted children who spoke more quickly.

5.5 Speaking Rate and Word Recognition

To determine if verbal rehearsal speed is also related to individual differences in word recognition performance, we examined the correlations between sentence durations and the three different measures of spoken word recognition described earlier. Table 9.2 shows the simple bivariate correlations between speaking rate and word recognition scores on the WIPI, LNT, and BKB. All of these correlations are strongly negative, suggesting once again that a common processing mechanism, which we assume is related to verbal rehearsal speed, may be the primary factor that underlies

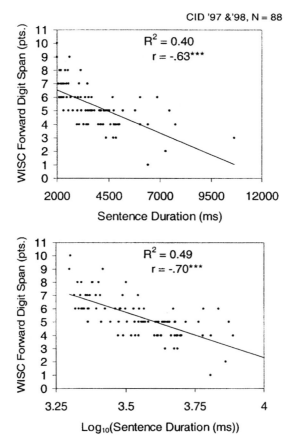

FIGURE 9.4. Scatterplots illustrating the relationship between average sentence duration for the seven-syllable McGarr sentences (abscissa) and WISC forward digit span scored by points (ordinate). Each data point represents an individual child. Nontransformed sentence durations are shown in the top panel, log-transformed sentence durations in the bottom panel. R^2 values indicate proportion of variance accounted for by the linear relation.

TABLE 9.2. Simple bivariate correlations between speaking rate and word recognition scores.

	Sentence duration	Log (sentence duration)
Closed-set word recognition (WIPI)	−.65*	−.69*
Open-set word recognition (LNT-E)	−.59*	−.66*
Open-set word recognition in sentences (BKB)	−.71*	−.78*

*$p < .001$

TABLE 9.3. Partial correlations between speaking rate and word recognition scores[a].

	Sentence duration	Log (sentence duration)
Closed-set word recognition (WIPI)	−.50*	−.55*
Open-set word recognition (LNT-E)	−.38*	−.47*
Open-set word recognition in sentences (BKB)	−.52*	−.64*

*$p < .001$.
[a]Statistically controlling for communication mode score, age of onset at deafness, duration of deafness, duration of CI use, number of active electrodes, VIDSPAC total segments correct (speech feature perception measure), age.

the variation and individual differences observed in all three word recognition tasks (see Pisoni and Cleary, 2003).

Table 9.3 shows a summary of the partial correlations computed between the two measures of speaking rate based on the McGarr sentence durations and the three measures of spoken word recognition performance. As in the earlier analyses, differences due to demographic factors and the contribution of other variables were statistically controlled for by computing partial correlations. In all cases, the correlations between speaking rate and spoken word recognition were still strongly negative and highly significant. Thus, slower speaking rates as measured by longer sentence durations are robustly associated with poorer word recognition scores on all three word recognition tests. Sentence duration accounted for approximately 25% of the currently unexplained residual variance in the word recognition scores after the other mediating variables were removed. These findings linking speaking rate and word recognition suggest that all three measures— digit span, speaking rate, and word recognition performance—are related because they share a common underlying source of variance.

To determine if digit span and sentence duration share a common process and the same underlying source of variance that relates them both to word recognition performance, we reanalyzed the intercorrelations between each pair of variables with the same set of the demographic and mediating variables statistically partialed out. When sentence duration was partialed out of the analysis, the correlations between digit span, either forward or backward, and each of the three measures of word recognition approached zero. This result indicates that the relationship between digit span and word recognition could be accounted for entirely in terms of individual differences in articulation rate and rehearsal speed. Furthermore, the negative correlations between sentence duration and word recognition were still present even after digit span was partialed out of the analysis, suggesting it was not working memory capacity per se that was the underlying factor, but rather individual differences in verbal rehearsal speed.

The pattern of results that emerges from these analyses suggests that the underlying process that is shared in common with sentence duration is related to the rate of information processing, specifically, to the speed of the verbal rehearsal process in working memory. This processing component of verbal rehearsal could reflect either the actual articulatory speed used to recycle and maintain phonological patterns in working memory or the time to retrieve and scan items already in working memory (Cowan et al. 1998). In either case, the common factor that links word recognition and speaking rate appears to be related to the speed of information-processing operations used to maintain phonological information in working memory. Thus, variation in performance in these two tasks can be traced to a common elementary cognitive process that is shared by both measures of performance.

These new findings demonstrating a relation between speaking rate and digit span permit us to identify for the first time a specific information-processing mechanism, the verbal rehearsal process, that appears to be responsible for the limitations on processing capacity. Processing limitations are present in a wide range of behavioral tasks that make use of verbal rehearsal and phonological processing skills to encode, store, maintain, and retrieve spoken words from working memory. We suggest that these fundamental information-processing operations are common components of all current outcome measures that are routinely used to assess both receptive and expressive language functions. The present findings suggest that the variability in performance on the traditional clinical outcome measures used to assess speech and language-processing skills in deaf children after cochlear implantation may reflect fundamental differences in the speed of information-processing operations such as verbal rehearsal and the rate of encoding phonological and lexical information in working memory.

5.6 Simon Sequence Reproduction: Immediate Memory Spans

The traditional methods for measuring working memory using digit spans require a subject to imitate and reproduce a sequence of test items using an overt verbal articulatory motor response. Because most deaf children with cochlear implants also have delays in speech development and display atypical articulation and speech motor control, it is possible that any differences observed in working memory using digit spans could be due to the nature of the response requirements during retrieval and output in addition to any possible differences in encoding, storage, and verbal rehearsal processes.

To eliminate the use of an overt articulatory-verbal response, we developed a new experimental methodology to measure immediate/working memory spans based on Milton Bradley's Simon, a popular memory game.

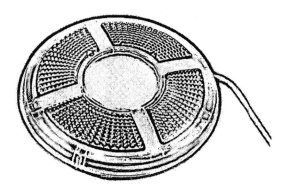

FIGURE 9.5. The memory game response box based on the popular Milton Bradley game Simon.

Figure 9.5 shows a display of the apparatus, which was interfaced to a computer. In this procedure, a subject is asked to simply "reproduce" a stimulus pattern by manually depressing a sequence of colored response panels on a four-alternative response box.

The Simon memory game procedure permitted us to manipulate the stimulus presentation conditions in several systematic ways while holding the response format constant. Specifically, three different stimulus presentation formats were employed. In the first condition, the target sequences to be reproduced consisted only of auditory color names (A). Under the second presentation condition, sequences of colored lights (L) were presented in the visual modality. In the remaining presentation condition, auditory color names were presented simultaneously with matching colored lights (A+L). The ability to vary the experimental procedure in this manner was quite useful because it provided a way to measure how several perceptual dimensions of the auditory and visual modalities are analyzed and processed, alone and in combination.

Forty-five hearing-impaired children who took part in the 2000 CID cochlear implant project were tested using the Simon memory game apparatus. Thirty-one of these children were able to complete all six conditions included in the testing session, and, furthermore, demonstrated the ability to identify the recorded color-name stimuli used in this task when these items were presented alone in isolation. Thirty-one normal-hearing children who were matched in terms of age and gender with the group of children with cochlear implants were also tested. Finally, 48 normal-hearing adults were recruited to serve as an additional comparison group (Karpicke and Pisoni 2000).

Of the six conditions tested in total, three of these measured the children's immediate memory skills. During the immediate memory task, the lights on the Simon were illuminated in temporal sequences that systematically increased in length as the subject progressed through successive trials in the experiment. Within each condition, the child started with a list length of one item. If two lists in a row at a given length were correctly reproduced, the next list presented was increased by one item in length. If on any trial the list was incorrectly reproduced, the next trial used a list that was one item shorter in length. This "adaptive tracking procedure" is similar to methods typically used in psychophysical testing (Levitt 1970). Sequences used for the Simon memory game task were generated pseudo-randomly by a computer program, with the stipulation that no single item would be repeated consecutively in a given list. We computed a weighted memory span score for each child by finding the proportion of lists correct at each list length and summing these proportions across all list lengths.

A summary of the results from the Simon sequence reproduction immediate memory task is shown in Figure 9.6 for the three groups of subjects. The normal-hearing adults are shown in the left panel, the normal-hearing age-matched children are shown in the middle panel, and the children with cochlear implants are shown in the right panel. Within each panel, the scores for auditory-only presentation (A) are shown on the left, scores for lights-only presentation (L) are shown in the middle, and scores for the combined auditory and lights presentation condition (A+L) are shown on the right.

Examination of weighted Simon memory span scores for the normal-hearing adults reveals several findings that can serve as a useful benchmark for evaluating performance differences within the other two groups of participants. First, we found a "modality effect" in presentation format. Auditory presentation of sequences of color names produced longer memory spans than visual presentation of sequences of colored lights ($p < .02$). Second, we found a "redundancy gain." When information from separate auditory and visual modalities was combined together and presented simultaneously, the memory span scores increased compared to presentation using only one sensory modality ($p < .02$ for auditory-only and $p < .001$ for visual-only).

The modality effect and the redundancy gains displayed here demonstrate that the Simon memory game procedure is a valid methodology for measuring immediate memory span in normal-hearing adults because it is able to measure subtle differences in the sensory modality used for presentation of the stimulus patterns. As in other studies of verbal short-term memory, longer Simon memory spans were found for auditory sequences compared to visual sequences, suggesting the use of phonological coding and verbal rehearsal strategies (Watkins et al. 1974; Penny 1989). In addition, the Simon memory spans were sensitive to cross-modality redundancies between stimulus dimensions when the same information about a stimulus pattern was presented simultaneously to more than one sensory modality. This latter finding demonstrates that adults are able not only to combine and integrate redundant sources of information across different sensory modalities but also to increase their working memory capacity

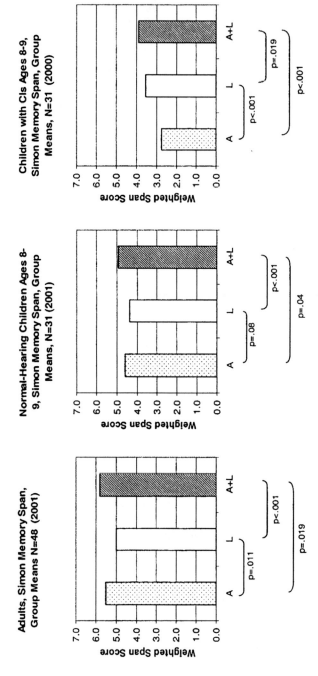

FIGURE 9.6. Mean immediate memory spans in each of the three conditions tested using the Simon response box. Scores for a group of normal-hearing adults are shown on the left, scores for normal-hearing 8- and 9-year-old children are shown in the center, and scores for a group of 8- and 9-year-old cochlear implant users are shown on the right. Speckled bars indicate mean spans in the auditory-only (A) condition, open bars indicate mean spans in the lights-only (L) condition, and shaded bars indicate mean spans in the auditory-plus-lights (A+L) condition.

when stimulus redundancies are present in both auditory and visual modalities simultaneously.

The middle panel of Figure 9.6 shows the results of the three presentation conditions for the group of normal-hearing 8- and 9-year-old children. Overall, the pattern of the Simon weighted span scores is quite similar to the findings obtained with the normal-hearing adults shown in the left panel, although there are several differences worth pointing out. First, the memory span scores for all three presentation conditions are uniformly lower than the scores obtained from the adults. Second, while the modality effect found with the adults is also present in these data, it is smaller in magnitude and only marginally significant ($p = .08$), suggesting possible developmental differences in the rate and efficiency of verbal rehearsal between adults and children in processing auditory and visual sequential patterns like those used in this task. The cross-modal redundancy gain observed with the adults was also found with the normal-hearing children, although it is also smaller in magnitude. Again, these differences may simply be due to age, maturation, and development.

The Simon memory span scores for the deaf children with cochlear implants are shown in the right panel of Figure 9.6 for the same three presentation conditions. Examination of the pattern of these memory span scores reveals several important differences from the span scores obtained for the normal-hearing children. First, the memory spans for all three presentation conditions were consistently lower overall than the spans from the corresponding conditions obtained for the normal-hearing children. Second, the modality effect observed in both the normal-hearing adults and normal-hearing children is reversed for the deaf children with cochlear implants; the memory spans for visual-only presentation were longer than for auditory-only presentation, and this difference was highly significant ($p < .001$). Third, although the cross-modal redundancy gain found for both the adults and normal-hearing children was also observed for the deaf children and was statistically significant for both conditions ($p < .001$ for auditory-only and $p < .02$ for visual-only), the size of the gain was much smaller. Moreover, the differences in the magnitude of the gain relative to performance in the auditory-only and visual-only conditions were also due to the reversal of the modality effect in the deaf children.

The results shown in Figure 9.6 for the visual-only presentation conditions are of special theoretical interest because the deaf children with cochlear implants displayed shorter memory spans than the normal-hearing children. This was an unexpected finding that adds support to the hypothesis that recoding and verbal rehearsal processes in working memory may play an important role in perception, learning, and memory in these children. Capacity limitations of working memory are closely tied to speed of processing information even for visually presented sequential patterns that are rapidly recoded and represented in memory in a phonological or articulatory code for certain kinds of sequential processing tasks. Verbal coding strategies may be automatic and mandatory in memory tasks that require immediate serial recall of temporal patterns that preserve item and order information (Gupta and MacWhinney 1997). Thus, although the visual patterns were presented using only sequences of colored lights, both groups of children may have attempted to recode the sequential patterns using verbal coding strategies to create stable phonological representations in working memory prior to response output in this task.

Although normal-hearing adults and normal-hearing children showed a similar pattern of memory span scores across the three presentation conditions, the deaf children displayed a fundamentally different pattern of results, suggesting that they may have used different encoding strategies and less efficient verbal rehearsal processes for maintaining temporal sequences in working memory. A period of auditory deprivation and the resulting absence of sound stimulation due to deafness during early stages of perceptual development may affect not only early sensory processing and perception but also subsequent encoding and verbal rehearsal processes in working memory (Burkholder and Pisoni 2003). The deaf children showed a reduced capacity to maintain temporal information in working memory even when that information was initially presented through the visual sensory modality. These findings on working memory spans for auditory and visual patterns obtained with the Simon memory game, which did not require overt verbal articulatory-motor responses, are consistent with the earlier memory span results obtained using the WISC digit spans, which showed systematic differences between the deaf children with cochlear implants and normal-hearing children (Pisoni and Cleary 2003).

To our knowledge, these are the first memory span data collected from deaf children with cochlear implants demonstrating specific effects on working memory capacity and verbal rehearsal processes without relying on an articulatory-based verbal response for output. Under all three presentation conditions, the children used the same manual response to reproduce the stimulus sequences. The deaf children also showed much smaller redundancy gains in the multimodal presentation condition, which suggests that in addition to differences in working memory capacity and rate of verbal rehearsal, their information-processing skills and abilities to perceive and encode multidimensional stimuli are atypical and compromised relative to age-matched normal-hearing children. The smaller redundancy gains observed in these deaf children may also be due to the reversal of the typical modality effect observed in studies of working memory that reflect verbal coding of the stimulus materials. The modality effect in short-term memory studies is generally thought to reflect phonological coding and verbal rehearsal strategies that actively maintain temporal order information of sequences of stimuli in immediate memory for short periods of time (Watkins et al. 1974).

5.7 Simon Sequence Reproduction: Learning Spans

The first version of our Simon memory game used novel sequences of color names or colored lights. All of the sequences were generated randomly on each trial to prevent any learning from occurring other than the routine adaptation that is normally observed in learning how to do a new task in a laboratory setting. Our primary goal was to obtain estimates of working memory capacity for temporal patterns that were not influenced by sequence repetition effects or idiosyncratic coding strategies that might increase memory capacity from trial to trial. Each test sequence was created on the fly by a random number generator so that the internal structure of a sequence of colors was always different and varied from trial to trial during the course of the experiment. Thus, there was no basis for any new learning to take place and the measures of Simon memory span can be used as estimates of capacity of immediate memory.

We have also used the same basic Simon memory game methodology to study sequence learning and to investigate the effects of long-term memory on coding and verbal rehearsal strategies in working memory. To assess the effects of learning on sequence reproduction, we examined performance under conditions in which the subject was given the opportunity to learn a long sequence of either auditory color-names (A), visual-only colored lights (L), or auditory color-names presented simultaneously with matching colored lights (A+L). In this new variant on the Simon procedure, the same stimulus pattern was repeated on each trial for an individual subject, but the pattern was increased in length by one item after each correct reproduction of the sequence, and decreased in length after each incorrect reproduction of the sequence. This methodology therefore provided an opportunity to study learning based on pattern repetition and to investigate how repetition is related to individual differences in immediate memory. In this procedure, as before, we computed a weighted learning span score for each participant by finding the proportion of lists correct at each list length and summing these proportions across all list lengths.

Figure 9.7 summarizes the results obtained in the Simon learning conditions for the same three presentation formats used in the earlier conditions, that is, auditory-only (A), lights-only (L), and auditory plus lights (A+L). The group means for the sequence learning conditions are shown on the right side of each panel in this figure; the corresponding set of memory span scores obtained earlier under the random presentation format for the same three presentation conditions are reproduced on the left side of each panel. The data for the normal-hearing adults are shown in the left panel, the data for the normal-hearing 8- and 9-year-old children are shown in the middle panel, and the data for the deaf children with cochlear implants are shown in the right panel.

Examination of the two sets of sequence reproduction scores shown within each panel reveals several consistent findings. First, just simply

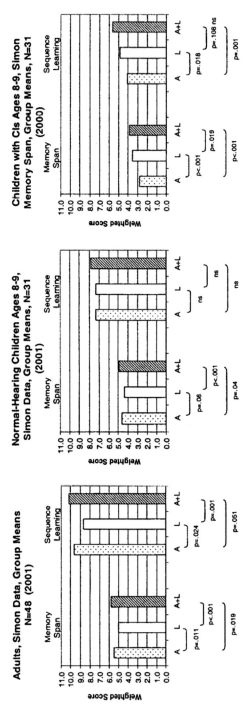

FIGURE 9.7. Mean immediate memory spans and mean sequence learning scores in each of the three conditions tested using the Simon response box. Scores for a group of normal-hearing adults are shown on the left, scores for normal-hearing 8- and 9-year-old children are shown in the center, and scores for a group of 8- and 9-year-old cochlear implant users are shown on the right. Speckled bars indicate mean scores in the auditory-only (A) condition, open bars indicate mean scores in the lights-only (L) condition, and shaded bars indicate mean scores in the auditory-plus-lights (A+L) condition. For each task, *p*-values for paired samples t-tests (uncorrected) between conditions are provided.

repeating the same stimulus sequence again produced robust learning effects for all three groups of subjects (Hebb 1961). This repetition effect can be seen clearly by comparing the three scores on the right side of each panel to the three scores on the left side. In every case, the learning span scores are higher than the memory span scores; repetition of a stimulus pattern improved sequence reproduction ability, although the magnitude of the learning effects differed across the three groups of subjects. The sequence reproduction scores for the adults in the learning condition are about twice the size of memory spans observed when the sequences were generated randomly from trial to trial. Although a repetition effect was also obtained with the deaf children who use implants, the size of their repetition effect was only about half the size of the repetition effect found for the normal-hearing children shown in the middle panel.

Second, the rank ordering of the three presentation conditions in the sequence learning conditions was similar to the rank ordering observed in the memory span conditions for all three groups of subjects. The repetition effect was largest for the A+L conditions for all three groups. For both the normal-hearing adults and children, we also observed the same modality effect in learning that was found for memory span; auditory presentation was better than visual presentation. And, as before, the deaf children showed a reversal of this modality effect for learning. For these children, visual presentation was better than auditory presentation. Although none of the pair-wise differences in the sequence learning conditions reached statistical significance for the normal-hearing children, the overall pattern of their learning spans was similar to their earlier memory span results and to the pattern of results observed with the adults.

To assess the magnitude of the repetition learning effects, we computed difference scores between the learning and memory conditions by subtracting the memory span scores from the learning span scores. The difference scores for the individual subjects in each group for the three presentation formats are displayed in Figure 9.8. Inspection of these distributions reveals a wide range of performance for all three groups of subjects. While most of the subjects in each group displayed some evidence of learning in terms of showing a positive repetition effect, there were a few subjects at the end of the distribution who either failed to show any learning at all or showed a small reversal of the predicted repetition effect. Although the number of these subjects was quite small in the adults and normal-hearing children, about one third of the deaf children with cochlear implants showed no repetition learning effect at all and obtained no benefit from having the same stimulus sequence repeated over again on each trial. These findings suggest that in addition to having a hearing impairment, some of these children may also have learning problems as well because they are unable to made use of simple pattern repetition to improve their performance for this task.

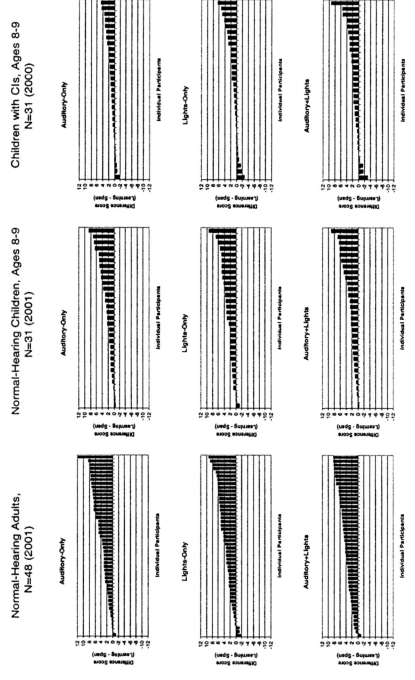

FIGURE 9.8. Difference scores for individual subjects showing sequence reproduction immediate memory spans minus sequence reproduction learning spans. Data for the auditory-only (A) condition is shown on the top, lights-only (L) condition in the middle, and auditory-plus-lights (A+L) condition, on the bottom. Data from normal-hearing adults are shown on the left, scores for normal-hearing 8- and 9-year-old children in the center, and scores for 8- and 9-year-old cochlear implant users on the right.

5.8 Correlations Among Measures of Immediate Memory Span

As we noted earlier in the section on redundancy gains under multi-modal presentation, the deaf children with cochlear implants did not use the informationally redundant auditory cues as well as the normal hearing children did to improve their immediate memory span capacity. This finding was observed even when the deaf children could identify all of the auditory color names presented in isolation. To understand the basis for these differences and to obtain some further insights into the nature of the coding strategies used by the deaf children in carrying out the Simon memory game task, we examined the intercorrelations between the multi-modal and the visual-only conditions of the Simon memory game for children with cochlear implants and for normal-hearing children (Cleary et al. 2001). We then also examined the correlations between the WISC digit spans and the memory span scores obtained from the Simon memory game.

Table 9.4 summarizes the intercorrelations among the two different sets of memory span scores, the forward and backward WISC digit span scores, and the Simon immediate memory span scores for the visual (V) and auditory plus visual (A+V) presentation conditions for both groups of children. The top rows display the correlations among these memory span measures for the deaf children with cochlear implants, and the bottom rows display the correlations for the normal-hearing children.

Examination of the intercorrelations of the Simon memory span scores for the deaf children reveals a strong positive correlation between the multi-modal and the visual-only conditions ($r = .71, p < .01$). If the deaf children were using the same visual-spatial coding strategy in both conditions of the Simon memory task, regardless of whether additional redundant auditory

TABLE 9.4. Intercorrelations between WISC forward digit span, sequence reproduction immediate memory span, and WISC backward digit span.

	WISC forward digit span	Immediate memory span, color-names-plus-lights	Immediate memory span, lights-only
Children with Cochlear Implants			
Immediate memory span, color-names-plus-lights	.09	—	
Immediate memory span, lights-only	−.02	.71**	—
WISC backward digit span	.52**	.46**	.48**
Normal-hearing children			
Immediate memory span, color-names-plus-lights	.58**	—	
Immediate memory span, lights-only	.01	.20	—
WISC backward digit span	.32*	.36*	.19

$*p < .05, **p < .01$.

information was available, one would predict that these two measures of memory span should be strongly correlated. This is exactly the pattern we found. In contrast, examination of the intercorrelations for the normal-hearing children reveals a different result. Here the correlation between the multimodal and visual-only conditions is much lower and nonsignificant ($r = .20, p > .05$). This pattern of results suggests that the normal-hearing children used two different coding strategies in the Simon memory game, a verbal-sequential strategy in the multimodal condition and a visual-spatial strategy in the visual-only condition, while the deaf children used the same visual-spatial coding strategy in both tasks.

Additional support for this explanation comes from an analysis of the correlations between the WISC digit span scores and the Simon memory game scores. As shown in Table 9.4, a different pattern of correlations was observed between these two sets of measures in each group of children, suggesting again that different coding and verbal rehearsal strategies were used to perform these two tasks. Although the WISC forward digit span scores were positively correlated with the memory game scores for the normal-hearing children, the correlation with forward span was observed in only the multimodal Simon condition ($r = +.58$) and not the visual-only Simon condition ($r = +.01$). This finding suggests that the normal-hearing children used the same verbal coding strategies in both the forward digit span task and the multimodal Simon memory span task. In contrast, the WISC forward digit spans and the memory game conditions were not correlated at all for the deaf children in either of these conditions.

The dissociations observed between these two memory tasks would be expected if the deaf children used one coding strategy in the multimodal condition of the Simon memory game and a different strategy in the WISC forward digit span task. The failure to find the same pattern of intercorrelations among the two different memory tasks in the deaf children with cochlear implants, taken together with the smaller redundancy gains observed in multimodal (A+L) Simon condition, suggests that the deaf children encoded and processed these temporal sequences in fundamentally different ways than the normal-hearing typically developing children. The deaf children apparently relied primarily on a visual-spatial coding strategy (Wilson and Emmorey 1998; Wilson 2001) to reproduce the sequences even when additional redundant auditory cues were available to help them improve their performance on this memory game task.

6. Theoretical and Clinical Significance

Taken together, the results of our research on working memory provide some new insights into the elementary information processing skills of deaf children with cochlear implants and the underlying cognitive factors that affect the development of their speech and language abilities. Our studies of immediate memory span using traditional digit span tests and the Simon memory game were specifically designed to obtain new process measures of performance that assessed specific subcomponents of working memory in order to understand the nature of the capacity limitations in encoding and processing sensory information. In this section, we briefly discuss the theoretical significance of our findings in light of the problems surrounding the enormous variability and individual differences on the clinical outcome measures of speech and language that have been consistently reported in the literature.

Detailed analyses of the links between three different sets of measures— the digit span scores, the sentence durations, and the word recognition scores—using partial correlations revealed that the common source of variance that was shared by all three of these tasks was processing speed, specifically, articulation speed and, by inference, speed of the verbal rehearsal process. We believe that this is a significant finding both theoretically and clinically because it provides converging evidence from several different behavioral measures obtained on the same children for the existence and operation of a common information-processing mechanism used for storage and maintenance of phonological information in working memory, and it suggests a principled explanation for the individual differences observed in a wide range of speech and language-processing tasks. A task analysis of the traditional test battery used for clinical assessment reveals that verbal rehearsal is a common subcomponent in each of the outcome measures used to assess speech perception, spoken word recognition, vocabulary, comprehension, and speech intelligibility.

We also found effects of early deafness and auditory deprivation on memory and learning of visual sequential patterns, a result that was initially unexpected when we began this research. The visual-only spans for the deaf children were shorter than the visual-only spans obtained from the age-matched normal-hearing children. This difference was found in both the Simon memory span experiment that used random sequences as well as the Simon learning experiment that used repeated sequences.

The results of the visual-only conditions are of special theoretical interest to us because they demonstrate that differences observed in working memory are not necessarily restricted only to temporal patterns perceived through the auditory sensory modality; differences in memory span were also found in tasks when both item and order information in sequential visual patterns must be preserved for a short period of time in immediate memory before response execution.

More detailed analyses of these results suggest that the normal-hearing children appeared to use two different coding strategies to carry out the Simon sequence memory task: a visual-spatial coding strategy for the visual patterns and a verbal-phonological coding strategy for the auditory patterns. In contrast, the hearing-impaired children may have used the same spatial coding and rehearsal strategy for both the visual and auditory patterns. Although for a normal-hearing child, verbal rehearsal is a processing strategy that emerges, develops, and becomes highly automatized over time and is routinely applied in an automatic and mandatory fashion for sequential patterns containing familiar linguistic stimuli like digits or color names, verbal rehearsal may be a relatively unpracticed and less automatized control strategy for a hearing-impaired child who has experienced a period of auditory deprivation at an early point in development.

7. Summary and Future Directions

The new findings presented in this chapter suggest that additional process measures of performance should be developed to study other aspects of cognition such as attention, categorization, learning, and memory—central cognitive processes that make use of the initial sensory input provided by a cochlear implant. One can imagine the construction of an entirely new battery of behavioral tests based on process measures of performance that could be used to assess the benefits of cochlear implantation and study the time course of development of these basic information-processing skills. Some of these new measures could be used to assess how well a listener is able to use the limited and impoverished sensory information conveyed through the cochlear implant. Other measures could be used to assess differences in processing speed, efficiency, and processing capacity. Additional measures of working memory span, verbal and visual-spatial coding and rehearsal strategies, controlled and automatic attention, and the development of automaticity may provide fundamental new knowledge about individual differences in the elementary cognitive processes that underlie the traditional end-point measures of outcome performance. At some point in the future, it may also be possible to develop a set of preimplant measures of performance based on measures of visual attention and memory for temporal sequences that could be useful in predicting outcome and benefit after implantation (see Bergeson et al. 2003). Given the new tools and experimental methodologies that are currently available from research in cognitive science, we believe these goals can be achieved in the next few years as the age of implantation becomes lower and more profoundly deaf children become candidates for cochlear implantation.

Acknowledgments. This work was supported by National Institutes of Health – NIH-NIDCD Training Grant T32DC00012 and NIH-NIDCD Research Grants R01DC00111 and R01DC00064 to Indiana University and NIH-NIDCD Research Grant R01DC03100 to the Central Institute for the Deaf. We would like to thank Ann Geers, Chris Brenner, Emily Tobey, and

Rosalie Uchanski for their help and assistance with this project. We also gratefully acknowledge the important contributions of Jeff Karpicke and Rose Burkholder in several aspects of data collection and thank Tonya Bergeson, Steven Chin, and Karen Iler Kirk for their comments on an earlier version of this chapter.

References

Adams AM, Gathercole SE (2000) Limitations in working memory: implications for language development. Int J Lang Commun Disord 35:95–116.

Ashcraft MH (1998) Fundamentals of Cognition. New York: Longman.

Ashcraft MH (2002) Cognition, 3rd ed. Upper Saddle River, NJ: Prentice Hall.

Atkinson RC, Shiffrin RM (1968) The control of short-term memory. Sci Am 225:82–90.

Baddeley AD, Thomson N, Buchanan M (1975) Word length and the structure of short-term memory. J Verb Learn Verb Behav 14:575–589.

Baddeley A, Gathercole SE, Papagno C (1998) The phonological loop as a language learning device. Psychol Rev 105:158–173.

Ball GF, Hulse SH (1998) Birdsong. Am Psychol 53:37–58.

Bench J, Kowal A, Bamford J (1979) The BKB (Bamford-Kowal-Bench) sentence lists for partially-hearing children. Br J Audiol 13:108–112.

Bergeson TR, Pisoni DB (in press) Audiovisual speech perception in deaf adults and children following cochlear implantation. In: Calvert G, Spence C, Stein BE, eds. Handbook of Multisensory Integration. Cambridge, MA: MIT Press.

Bergeson TR, Pisoni DB, Davis RAO (2003) A longitudinal study of audiovisual speech perception by children with hearing loss who have cochlear implants. Volta Rev 103:347–370.

Biederman I, Shiffar MM (1987) Sexing day-old chicks: a case study and expert systems analysis of a difficult perceptual-learning task. J Exp Psychol Learn 13:640–645.

Blamey PJ, Sarant JZ, Paatsch LE, Barry JG (2001) Relationships among speech perception, production, language, hearing loss, and age in children with impaired hearing. J Speech Lang Hear Res 44:264–285.

Burkholder R, Pisoni DB (2003) Speech timing and working memory in profoundly deaf children after cochlear implantation. J Exp Child Psychol 85:63–68.

Carpenter PA, Miyake A, Just MA (1994) Working memory constraints in comprehension: evidence from individual differences, aphasia, and aging. In: Gernsbacher MA, ed. Handbook of Psycholinguistics. San Diego, CA: Academic Press, pp. 1075–1122.

Cleary M, Pisoni DB, Geers AE (2001) Some measures of verbal and spatial working memory in eight- and nine-year-old hearing-impaired children with cochlear implants. Ear Hear 22:395–411.

Cowan N, Wood NL, Wood PK, Keller TA, Nugent LD, Keller CV (1998) Two separate verbal processing rates contributing to short-term memory span. J Exp Psychol Gen 127:141–160.

Cullington H, Hodges AV, Butts SL, Dolan-Ash S, Balkany TJ (2000) Comparison of language ability in children with cochlear implants placed in oral and total communication educational settings. Ann Otol Rhinol Laryngol Suppl 185: 121–123.

Ericsson KA, Pennington N (1993) The structure of memory performance in experts: implications for memory in everyday life. In: Davies GM, Logie RH, eds. Memory in Everyday Life. Amsterdam: Elsevier, pp. 241–282.

Flavell JH, Beach DH, Chinsky JM (1966) Spontaneous verbal rehearsal in a memory task as a function of age. Child Dev 37:283–299.

Fry AF, Hale S (1996) Processing speed, working memory, and fluid intelligence: evidence for a developmental cascade. Psychol Sci 7:237–241.

Fryauf-Bertschy H, Tyler R, Kelsay D, Gantz B, Woodworth G (1997) Cochlear implant use by prelingually deafened children: the influences of age at implantation and length of device use. J Speech Lang Hear Res 40:183–199.

Gantz BJ, Rubinstein JT, Tyler RS, Teagle HF, et al. (2000) Long-term results of cochlear implants in children with residual hearing. Ann Otol Rhinol Laryngol Suppl 185:33–36.

Gathercole SE, Hitch GJ, Service E, Martin AJ (1997) Phonological short-term memory and new word learning in children. Dev Psychol 33:966–979.

Gazzaniga MS (2000) The new cognitive neurosciences, 2nd ed. Cambridge, MA: MIT Press.

Geers A, Brenner C (2003) Background and educational characteristics of prelingually deaf children implanted by five years of age. Ear Hear 24(suppl):2S–14S.

Geers A, Brenner C, Davidson L (2003a) Factors associated with development of speech perception skills in children implanted by age five. Ear Hear 24(suppl):24S–35S.

Geers AE, Nicholas JG, Sedey AL (2003b) Language skills of children with early cochlear implantation. Ear Hear 24(suppl):46S–58S.

Giraud AL, Price CJ, Graham JM, Frackowiak RS (2001a) Functional plasticity of language-related brain areas after cochlear implantation. Brain 124:1307–1316.

Giraud AL, Price CJ, Graham JM, Truy E, Frackowiak RS (2001b) Cross-modal plasticity underpins language recovery after cochlear implantation. Neuron 30:657–663.

Gupta P, MacWhinney B (1997) Vocabulary acquisition and verbal short-term memory: computational and neural bases. Brain Lang 59:267–333.

Haber RN (1969) Information-Processing Approaches to Visual Perception. New York: Holt, Rinehart and Winston.

Haskins H (1949) A phonetically balanced test of speech discrimination for children. Unpublished master's thesis, Northwestern University, Evanston IL.

Hebb DO (1961) Distinctive features of learning in the higher animal. In: Delafresnaye JF, ed. Brain Mechanisms and Learning. New York: Oxford University Press.

Hodges AV, Dolan-Ash M, Balkany TJ, Schloffman JJ, Butts SL (1999) Speech perception results in children with cochlear implants: contributing factors. Otolaryngol Head Neck Surg 12:31–34.

Kail R (1984) The Development of Memory in Children, 2nd ed. New York: WH Freeman.

Kane MJ, Engle RW (2002) The role of prefrontal cortex in working-memory capacity, executive attention, and general fluid intelligence: an individual-differences perspective. Psychon Bull Rev 9:637–671.

Karpicke J, Pisoni DB (2000) Memory span and sequence learning using multimodal stimulus patterns: preliminary findings in normal-hearing adults. Research on Spoken Language Processing progress report No. 24. Bloomington, IN: Speech Research Laboratory, pp. 393–406.

Kirk KI, Pisoni DB, Osberger MJ (1995) Lexical effects on spoken word recognition by pediatric cochlear implant users. Ear Hear 16:470–481.

Kirk KI, Diefendorf AO, Pisoni DB, Robbins AM (1997) Assessing speech perception in children. In: Mendel LL, Danhauer JL, eds. Speech Perception Assessment. San Diego: Singular Press, pp. 101–132.

Kirk KI, Pisoni DB, Miyamoto RT (2000) Lexical discrimination by children with cochlear implants: effects of age at implantation and communication mode. In: Waltzman SB, Cohen NL, eds. Cochlear Implants. New York: Thieme, pp. 252–254.

Kirk KI, Miyamoto RT, Lento CL, Ying E, O'Neill T, Fears B (2002) Effects of age at implantation in young children. Ann Otol Rhinol Laryngol Suppl 189: 69–73.

Konishi M (1985) Birdsong: from behavior to neuron. Annu Rev Neurosci 8:125–170.

Konishi M, Nottebohm R (1969) Experimental studies in the ontogeny of avian vocalizations. In: Hinde RA, ed. Bird Vocalizations. New York: Cambridge University Press, pp. 29–48.

Lachman R, Lachman JL, Butterfield EC (1979) Cognitive Psychology and Information Processing: An Introduction. Hillsdale, NJ: Lawrence Erlbaum Associates.

Landauer TK (1962) Rate of implicit speech. Percept Motor Skill 15:646.

Lee DS, Lee JS, Oh SH, Kim SK, et al. (2001) Cross-modal plasticity and cochlear implants. Nature 409:149–150.

Levelt, WJM (1989) Speaking: From Intention to Articulation. ACT-MIT Press Series in Natural-Language Processing. Cambridge, MA: MIT Press.

Levitt H (1970) Transformed up-down methods in psychoacoustics. J Acoust Soc Am 49:467–477.

Lindsay PH, Norman DA (1977) Human Information Processing: An Introduction to Psychology, 2nd ed. New York: Academic Press.

Ling D (1993) Auditory-verbal options for children with hearing-impairment: helping pioneer an applied science. Volta Rev 95:187–196.

Luce PA, Pisoni DB (1998) Recognizing spoken words: the neighborhood activation model. Ear Hear 19:1–36.

Marler P, Peters S (1988) Sensitive periods for song acquisition from tape recordings and live tutors in the swamp sparrow, Melospiza-Georgiana. Ethology 77:76–84.

McGarr NS (1983) The intelligibility of deaf speech to experienced and inexperienced listeners. J Speech Hear Res 26:451–458.

McGilly K, Siegler RS (1989) How children choose among serial recall strategies. Child Dev 60:172–182.

Meyer TA, Pisoni DB (1999) Some computational analyses of the PBK test: effects of frequency and lexical density on spoken word recognition. Ear Hear 20:363–371.

Miyamoto RT, Svirsky MA, Robbins AM (1997) Enhancement of expressive language in prelingually deaf children with cochlear implants. Acta Otolaryngol 117:154–157.

Neisser U (1967) Cognitive Psychology. New York: Appleton-Century-Crofts.

NIH Consensus Conference (1995) Cochlear implants in adults and children. JAMA 274:1955–1961.

Nittrouer S (2002) From ear to cortex: a perspective on what clinicians need to understand about speech perception and language processing. Lang Speech Hear Ser 33:237–252.

Penny CG (1989) Modality effects and the structure of short-term verbal memory. Mem Cognit 17:398–422.

Pisoni DB (2000) Cognitive factors and cochlear implants: some thoughts on perception, learning, and memory in speech perception. Ear Hear 21:70–78.

Pisoni DB, Cleary M (2003) Measures of working memory span and verbal rehearsal speed in deaf children after cochlear implantation. Ear Hear 24(suppl): 106S–120S.

Pisoni DB, Cleary M, Geers AE, Tobey EA (2000) Individual differences in effectiveness of cochlear implants in children who are prelingually deaf: new process measures of performance. Volta Rev 101:111–164.

Pisoni DB, Geers A (2000) Working memory in deaf children with cochlear implants: correlations between digit span and measures of spoken language processing. Ann Otol Rhinol Laryngol Suppl 109:92–93.

Pisoni DB, Svirsky MA, Kirk KI, Miyamoto RT (1997) Looking at the "Stars": a first report on the intercorrelations among measures of speech perception, intelligibility, and language development in pediatric cochlear implant users. Research on Spoken Language Processing progress report No. 21 (1996–1997). Bloomington, IN: Speech Research Laboratory, pp. 51–91.

Ponton CW, Eggermont JJ (2001) Of kittens and kids: altered cortical maturation following profound deafness and cochlear implant use. Audiol Neurootol 6:363–380.

Ponton CW, Don M, Eggermont JJ, Waring MD, Kwong B, Masuda A (1996a) Auditory system plasticity in children after long periods of complete deafness. Neuroreport 8:61–65.

Ponton CW, Don M, Eggermont JJ, Waring MD, Masuda A (1996b) Maturation of human cortical auditory function: differences between normal-hearing children and children with cochlear implants. Ear Hear 17:430–437.

Posner MI (1989) Foundations of Cognitive Science. Cambridge MA: MIT Press.

Reitman WR (1965) Cognition and Thought: An Information-Processing Approach. New York: Wiley.

Robbins A, Kirk KI (1996) Speech perception assessment and performance in pediatric cochlear implant users. Semin Hear 17:353–369.

Robbins AM, Svirsky M, Osberger MJ, Pisoni DB (1998) Beyond the audiogram: the role of functional assessments. In: Bess F, ed. Children with Hearing Impairment. Nashville, TN: Bill Wilkerson Center Press, pp. 105–124.

Rosen VM, Engle RW (1997) Forward and backward serial recall. Intelligence 25:37–47.

Ross M, Lerman J (1979) A picture identification test for hearing-impaired children. J Speech Hear Res 13:44–53.

Rudel RG, Denckla MB (1974) Relation of forward and backward digit repetition to neurological impairment in children with learning disability. Neuropsychologia 12:109–118.

Sarant JZ, Blamey PJ, Dowell RC, Clark GM, Gibson WPR (2001) Variation in speech perception scores among children with cochlear implants. Ear Hear 22:18–28.

Shadmehr R, Holcomb HH (1997) Neural correlates of motor memory and consolidation. Science 277:821–825.

Sharma A, Dorman MF, Spahr AJ (2002a) A sensitive period for the development of the central auditory system in children with cochlear implants: implications for age of implantation. Ear Hear 23:532–539.

Sharma A, Dorman MF, Spahr AJ (2002b) Rapid development of cortical auditory evoked potentials after early cochlear implantation. Neuroreport 13:1365–1368.

Siegler RS (1998) Children's Thinking, 3rd ed. Upper Saddle River, NJ: Prentice Hall.

Skinner MW, Ketten DR, Holden LK, Harding GW, et al. (2002) CT-derived estimation of cochlear morphology and electrode array position in relation to word recognition in Nucleus-22 recipients. J Assoc Res Otolaryngol 3:332–350.

Standing L, Curtis L (1989) Subvocalization rate versus other predictors of the memory span. Psychol Rep 65:487–495.

Sur M, Leamey CA (2001) Development and plasticity of cortical areas and networks. Nat Rev Neurosci 2:251–262.

Sur M, Pallas SL, Roe AW (1990) Cross-modal plasticity in cortical development: differentiation and specification of sensory neocortex. Trends Neurosci 13:227–233.

Svirsky MA, Robbins AM, Kirk KI, Pisoni DB, Miyamoto RT (2000a) Language development in profoundly deaf children with cochlear implants. Psychol Sci 11:153–158.

Svirsky MA, Sloan RB, Caldwell M, Miyamoto RT (2000b) Speech intelligibility of prelingually deaf children with multichannel cochlear implants. Ann Otol Rhinol Laryngol Suppl 185:123–125.

Tait M, Lutman ME, Robinson K (2000) Preimplant measures of preverbal communicative behavior as predictors of cochlear implant outcomes in children. Ear Hear 21:18–24.

Tobey EA, Geers AE, Morchower B, Perrin J, et al. (2000) Factors associated with speech intelligibility in children with cochlear implants. Ann Otol Rhinol Laryngol Suppl 185:28–30.

Tobey EA, Geers AE, Brenner C, Altuna D, Gabbert G (2003) Factors associated with development of speech production skills in children implanted by age five. Ear Hear 24(suppl):36S–45S.

Watkins MJ, Watkins OC, Crowder RG (1974) The modality effect in free and serial recall as a function of phonological similarity. J Verb Learn Verb Behav 13:430–447.

Wechsler D (1991) Wechsler Intelligence Scale for Children, 3rd ed. (WISC-III). San Antonio, TX: The Psychological Corporation.

Wilson M (2001) The case for sensorimotor coding in working memory. Psychon Bull Rev 8:44–57.

Wilson M, Emmorey K (1998) A "word length effect" for sign language: further evidence for the role of language in structuring working memory. Mem Cognition 26:584–590.

Yoshinaga-Itano C, Apuzzo ML (1998) Identification of hearing loss after age 18 months is not early enough. Am Ann Deaf 143:380–387.

Yoshinaga-Itano C, Sedey AL, Coulter DK, Mehl AL (1998) Language of early- and later-identified children with hearing loss. Pediatrics 102:1161–1171.

Zwolan TA, Zimmerman-Phillips S, Asbaugh CJ, Hieber SJ, Kileny PR, Telian SA (1997) Cochlear implantation of children with minimal open-set speech recognition skills. Ear Hear 18:240–251.

Index

Printed in the United States
53570LVS00001B/94-96